New Waves in Philosophy

Series Editors: **Vincent F. Hendricks** and **Duncan Pritchard**

Titles include

Otávio Bueno and Øystein Linnebo (*editors*)
NEW WAVES IN PHILOSOPHY OF MATHEMATICS

Boudewijn DeBruin and Christopher F. Zurn (*editors*)
NEW WAVES IN POLITICAL PHILOSOPHY

Vincent F. Hendricks and Duncan Pritchard (*editors*)
NEW WAVES IN EPISTEMOLOGY

Yujin Nagasawa and Erik J. Wielenberg (*editors*)
NEW WAVES IN PHILOSOPHY OF RELIGION

Jan Kyrre Berg Olsen, Evan Selinger and Søren Riis (*editors*)
NEW WAVES IN PHILOSOPHY OF TECHNOLOGY

Thomas S. Petersen, Jesper Ryberg and Clark Wolf (*editors*)
NEW WAVES IN APPLIED ETHICS

Kathleen Stock and Katherine Thomson-Jones (*editors*)
NEW WAVES IN AESTHETICS

Forthcoming:

Jesús Aguilar, Andrei A. Buckareff and Keith Frankish (*editors*)
NEW WAVES IN PHILOSOPHY OF ACTION

Thom Brooks (*editor*)
NEW WAVES IN ETHICS

Allan Hazlett (*editor*)
NEW WAVES IN METAPHYSICS

P.D. Magnus and Jacob Busch (*editors*)
NEW WAVES IN PHILOSOPHY OF SCIENCE

Nikolaj Pedersen and Cory Wright (*editors*)
NEW WAVES IN TRUTH

Sarah Sawyer (*editor*)
NEW WAVES IN PHILOSOPHY OF LANGUAGE

Future Volumes

New Waves in Philosophy of Mind

New Waves in Meta-Ethics

New Waves in Formal Philosophy

New Waves in Philosophy of Law

New Waves in Philosophy
Series Standing Order ISBN 978–0–230–53797–2 (hardcover)
Series Standing Order ISBN 978–0–230–53798–9 (paperback)
(*outside North America only*)

You can receive future titles in this series as they are published by placing a standing order. Please contact your bookseller or, in case of difficulty, write to us at the address below with your name and address, the title of the series and the ISBN quoted above.

Customer Services Department, Macmillan Distribution Ltd, Houndmills, Basingstoke, Hampshire RG21 6XS, England

New Waves in Philosophy of Mathematics

Edited by

Otávio Bueno
University of Miami, USA

Øystein Linnebo
University of Bristol, UK

First published 2009 by
PALGRAVE MACMILLAN

Palgrave Macmillan in the UK is an imprint of Macmillan Publishers Limited,
registered in England, company number 785998, of Houndmills, Basingstoke,
Hampshire RG21 6XS.

Palgrave Macmillan in the US is a division of St Martin's Press LLC,
175 Fifth Avenue, New York, NY 10010.

Palgrave Macmillan is the global academic imprint of the above companies
and has companies and representatives throughout the world.

Palgrave® and Macmillan® are registered trademarks in the United States,
the United Kingdom, Europe and other countries.

ISBN-13: 978–0–230–21942–7 hardback
ISBN-13: 978–0–230–21943–4 paperback

This book is printed on paper suitable for recycling and made from fully
managed and sustained forest sources. Logging, pulping and manufacturing
processes are expected to conform to the environmental regulations of the
country of origin.

A catalogue record for this book is available from the British Library.

A catalog record for this book is available from the Library of Congress.

10   9   8   7   6   5   4   3   2   1
18 17 16 15 14 13 12 11 10 09

Printed and bound in Great Britain by
CPI Antony Rowe, Chippenham and Eastbourne

Contents

Series Editors' Foreword

New Waves in Philosophy Series

The aim of this series is to gather the young and up-and-coming scholars in philosophy to give their view on the subject now and in the years to come, and to serve a documentary purpose, that is, "this is what they said then, and this is what happened". It will also provide a snapshot of cutting-edge research that will be of vital interest to researchers and students working in all subject areas of philosophy.

The goal of the series is to have a New Waves volume in every one of the main areas of philosophy. We would like to thank Palgrave Macmillan for taking on this project in particular, and the entire *New Waves in Philosophy* series in general.

Vincent F. Hendricks and Duncan Pritchard
Editors

Acknowledgements

Many people and institutions have helped make this volume possible, and we would like to take this opportunity to thank them. On April 24–26, 2008 we organized a conference on "New Waves in Philosophy of Mathematics" at the University of Miami (UM) at which earlier versions of the papers collected here were presented. The conference was made possible by a grant from the British Academy, as well as generous support from the University of Miami's College of Arts and Sciences, the UM Philosophy Department, and the UM Graduate School. We thank them all for their support.

Institutions are what they are in virtue of the people that make them all they can be. Without the help and support of the following people, the conference couldn't have taken place, and this collection couldn't have the shape it currently has. So it is a pleasure to thank Dean Michael Halleran and Associate Deans Daniel Pauls and Perry Roberts from the UM College of Arts and Sciences; Harvey Siegel, the Chair of the UM Philosophy Department, and Dean Terri Scandura from the UM Graduate School. They have all made the difference. Thanks also go to Alex Puente, Lisa Israel, and Ben Burgis for making sure that everything in the conference ran smoothly. We really appreciate their help. We also thank Ben Burgis for preparing the index, and Associate Dean Perry Roberts for supporting his work on the project.

The general editors of the series *New Waves in Philosophy*, Vincent Hendricks and Duncan Pritchard, have also been extremely valuable sources of help and feedback throughout the process.

Finally, we would like to thank all the contributors for their work, excitement, and the insight they all brought to this project. It has been really a pleasure to work with them.

Otávio Bueno and Øystein Linnebo

List of Contributors

Alan Baker, Swarthmore College, USA

Otávio Bueno, University of Miami, USA

Mark Colyvan, University of Sydney, Australia

Roy Cook, University of Minnesota, USA

Thomas Hofweber, University of North Carolina, Chapel Hill, USA

Peter Koellner, Harvard University, USA

Hannes Leitgeb, University of Bristol, UK

Mary Leng, University of Liverpool, UK

Øystein Linnebo, University of Bristol, UK

Alexander Paseau, Wadham College, University of Oxford, UK

Christopher Pincock, Purdue University, USA

Agustín Rayo, Massachusetts Institute of Technology, USA

Gabriel Uzquiano, Pembroke College, University of Oxford, UK

Introduction

Philosophy of Mathematics: Old and New

Otávio Bueno and Øystein Linnebo

This volume contains essays by 13 up-and-coming researchers in the philosophy of mathematics, who have been invited to write on what they take to be the right philosophical account of mathematics, examining along the way where they think the philosophy of mathematics is and ought to be going. As one might expect, a rich and diverse picture emerges. Some broader tendencies can nevertheless be detected: there is increasing attention to the practice, language, and psychology of mathematics, a move to reassess the orthodoxy, as well as inspiration from philosophical logic.

In order to describe what is new in a field, we need some understanding of what is old. We will accordingly begin with a quick overview of earlier work in the philosophy of mathematics before attempting to characterize the tendencies that emerge in this volume and describing how the various essays fit into this pattern.

The beginning of modern philosophy of mathematics

Modern philosophy of mathematics can plausibly be taken to have begun with the pioneering work of Gottlob Frege in the late nineteenth century. Having first developed modern formal logic—essentially what we now know as *higher-order logic*—Frege was able to articulate and pursue his philosophical goals with unprecedented precision and rigor. His primary goal was to show that arithmetic and analysis are just "highly developed logic", that is, that all the basic notions of these branches of mathematics can be defined in purely logical terms and that, *modulo* these definitions, mathematical theorems are nothing but logical truths. This would show that large parts of mathematics are analytic rather than synthetic *a priori*, as Kant so influentially had argued. Although later philosophers of mathematics have often disagreed with Frege's substantive views, the precision and rigor which Frege introduced into the debate have remained a widely accepted ideal.

Another great influence on modern philosophy of mathematics came from the logical paradoxes that were discovered in the late nineteenth and early twentieth century. An early casualty was the foundation for mathematics that Frege himself had proposed, which Russell's paradox revealed to be inconsistent. More generally, the paradoxes prompted a broad examination of foundational and philosophical questions concerning mathematics to which both philosophers and mathematicians contributed.

This examination led to the articulation of what we now regard as the three classical philosophies of mathematics. *Logicists* agree with Frege that much or all of mathematics is reducible to pure logic. But because of the paradoxes they tie this claim to a carefully circumscribed system of logic. The classical development of this view is found in Russell and Whitehead's famous *Principia Mathematica*. *Intuitionists* disagree with the logicists and argue that the only legitimate foundation for mathematics is provided by pure intuition and mental constructions. This philosophical conception of mathematics is given a sophisticated development by L.E.J. Brouwer and Arendt Heyting, who show that it leads to a revision not just of classical mathematics but even of the classical logic developed by Frege. Finally, *formalism* was developed and defended by the great mathematician David Hilbert. On the standard interpretation of this view, only finitary mathematics has content—which it obtains from pure intuition—whereas infinitary mathematics is a formal game with uninterpreted symbols. Hilbert's goal was to prove the consistency of infinitary mathematics on the basis of finitary mathematics alone. This goal is shown to be unattainable by Gödel's second incompleteness theorem, which says roughly that any consistent formal system of a certain mathematical strength cannot prove its own consistency, let alone that of stronger systems.

The recent past

Until around 1960, the discussion of the three classical philosophies of mathematics and their descendents dominated much of the debate (although Wittgenstein's views too received much attention). In the three decades that followed, much of the agenda was set by W.V. Quine and Paul Benacerraf.

According to Quine, mathematics isn't essentially different from theoretical physics. Both go beyond what can be observed by means of our unaided senses. And both are justified by their contribution to the prediction and explanation of states of affairs that can be thus observed. Sets are therefore said to be epistemologically on a par with electrons. In this way Quine uses his holistic form of empiricism to support a platonistic interpretation of mathematics, according to which there really are abstract mathematical objects. This approach relies on what has become known as the "indispensability argument", which was explicitly articulated and developed by Hilary Putnam. According to this argument, we have reason to believe in the existence of sets, just as we have reason to believe in the existence of electrons, because both sorts of objects are indispensable to our best scientific theories of the world.

Benacerraf's influence is due primarily to two seminal articles. In "What Numbers Could Not Be" (1965), Benacerraf challenges all attempts to identify the natural numbers with sets and argues (following Dedekind and others) that all that matters to mathematics is *the structure* of the natural number system, not any "intrinsic properties" of these numbers. Considerations of this sort have given rise to various forms of *mathematical structuralism*. In particular, Geoffrey Hellman has developed an eliminative form of structuralism according to which mathematics is concerned with the structures that are and could be instantiated by concrete objects,[1] whereas Charles Parsons, Michael Resnik, and Stewart Shapiro defend non-eliminative forms of structuralism which take mathematical objects to be real yet to be nothing more than positions in structures.

Platonistic views have met with epistemological challenges for as long as they have been around. Benacerraf's second paper, "Mathematical Truth" (1973), articulates a striking version of an epistemological challenge to platonism about mathematics. If mathematics is about abstract objects, how is mathematical knowledge possible? Knowledge of some class of objects typically requires some form of contact or interaction with these objects. But if mathematical objects are abstract, no such contact or interaction seems possible. Many philosophers (most famously Hartry Field) have regarded Benacerraf's challenge as an argument for *nominalism* (that is, the view that there are no abstract objects). The quarter century following the publication of "Mathematical Truth" saw a great variety of nominalistic accounts of mathematics, associated with Field, Hellman, Charles Chihara, and others. Field's nominalism was also meant to respond to the indispensability argument by showing how to reformulate scientific theories without quantification over mathematical objects.

The contemporary debate

The views defended by Quine and the challenges so powerfully formulated by Benacerraf are still with us. The indispensability argument is still being debated and has recently been defended by Mark Colyvan and sympathetically discussed by Alan Baker. Although mathematical structuralism has received more criticism than support over the past decade, it has no doubt secured a position as one of the canonical philosophies of mathematics. And nominalism continues to exert a strong influence on many philosophers.

The past 10–15 years have seen various innovations and new orientations as well. New forms of fictionalism and nominalism have been developed by Jody Azzouni, Stephen Yablo, and others, aimed at doing better justice to actual mathematical language and thought than traditional forms of error theory or fictionalism. There has also been a revival of interest in logicism, where a neo-Fregean view has been defended by Bob Hale and Crispin Wright and received careful philosophical scrutiny and technical analysis by George Boolos and Richard Heck.

A particularly important new orientation goes by the name of "naturalism". The starting point is Quinean naturalism, which rejects any form of "first philosophy" in favor of a conception of philosophy as informed by and continuous with the natural sciences. Naturalistic approaches to mathematics share Quine's disdain for any form of philosophy which aspires to be prior to or more fundamental than some successful scientific practice. John Burgess, Gideon Rosen, and especially Penelope Maddy have argued that mathematical questions allow of no deeper form of justification than what is operative in actual mathematical practice. This has led to increased attention to the practice and methodology of mathematics, as exemplified for instance by the work of Maddy and Paolo Mancosu.

All of the concerns described in this section are represented in the present volume. Another theme, which emerges in this volume is an increasing engagement with neighboring disciplines such as linguistics, psychology, and logic. In what follows, we will attempt to place the contributions to this volume in a larger intellectual landscape. Although the categories with which we operate are not sharply defined, and although many of the essays discuss questions belonging to several of these categories, it is our hope that this will provide readers with a useful overview. These categories have also been used to structure the collection into five corresponding parts.

Part I. Reassessing the orthodoxy in the philosophy of mathematics

The two chapters that make up Part I of the volume critically examine the orthodoxy in the philosophy of mathematics.

Roy Cook's essay explores and defends Frege's combination of logicism and mathematical platonism. He also argues that the neo-Fregean view defended by Hale and Wright shares more features of the original Fregean view than is generally realized; in particular, that both Frege and the neo-Fregeans articulate an epistemic notion of analyticity (unlike Kant's and Quine's semantic notions) and seek to show that much of mathematics is analytic in this sense. For Fregeans old and new, to be analytic is to be justifiable on the basis of logic and definitions alone and thus to have absolutely general validity. The chief difference is that, for Frege, "logic" includes not just second-order logic but also the notorious Basic Law V, whereas for the neo-Fregeans, "definitions" include not just explicit definitions but also implicit ones such as abstraction principles.

According to conventional wisdom, Benacerraf established that numbers cannot be sets. Alexander Paseau challenges this view, exploring and defending the opposing Quinean view that the numbers can be reduced to sets. This view enjoys ontological, ideological, and axiomatic economy. But it also faces various objections, which Paseau discusses and rejects. Perhaps the most serious objection is that a reductionist interpretation is incompatible with what speakers mean with their own arithmetical vocabulary. Paseau offers two responses. Firstly, he denies that speakers have transparent knowledge of the referents of their words, in arithmetic or elsewhere. Secondly, he argues that any "hermeneutic" concern with getting actual meanings right must be secondary to the overall theoretical virtues of a proposed account of arithmetic.

Part II. The question of realism in mathematics

The question of realism and anti-realism in mathematics has been with us for a long time and is unlikely to go away any time soon. The contributions to Part II reassess the question and its significance, while developing new approaches to the problems that are raised.

Otávio Bueno's contribution is a defense of mathematical fictionalism. He identifies five desiderata that an account of mathematics must meet in order to make sense of mathematical practice. After arguing that current versions of platonism and nominalism fail to satisfy the desiderata, he outlines two versions of mathematical fictionalism, which meet them. One version is based on an empiricist view of science and has the additional benefit of providing a unified account of both mathematics and science. The other version is based on the metaphysics of fiction and articulates what can be considered a truly fictionalist account of mathematics. Bueno argues that both versions of fictionalism satisfy the desiderata and he thinks that they are best developed together. He concludes that mathematical fictionalism is alive and well.

Peter Koellner discusses the question of pluralism in mathematics. A radical pluralist position is defended by Carnap, who regards it as merely a matter of practical convenience, rather than objective truth, whether we should accept certain quite elementary arithmetical statements (so-called Π_1^0-sentences). Koellner critiques Carnap's position and argues that his radical form of pluralism is untenable. To address the question whether a more reasonable pluralism might be defensible, he develops a new approach to the question of pluralism inspired by a comparison with physics. In both cases, we have a certain pool of "data": respectively observational generalizations and Π_1^0-sentences. Two theories are "observationally equivalent" if they have the same entailments concerning "the data". We then face the "problem of selection": which of the many incompatible theories that are "observationally adequate" should we select as true? Koellner describes some constraints on the problem of selection which arise from set theory itself and argues that these constraints rule out the common view that the set-theoretic independence results alone suffice to secure pluralism. The chapter ends with a discussion of a "bifurcation scenario", that is, a scenario about possible developments in set theory which would arguably support a form of set-theoretic pluralism. This scenario is based on a recent result with Hugh Woodin and has the virtue that whether it pans out is sensitive to actual developments in mathematics.

Mary Leng's contribution is a critical discussion of "algebraic" approaches to mathematics. This class of approaches is explained in terms of the classic Frege–Hilbert debate, where Frege argued that mathematical axioms are assertions about some mathematical reality, and Hilbert had the more algebraic view that axioms rather define or circumscribe their subject matter. Some of the major philosophical approaches to mathematics belong to the algebraic camp. Leng argues that these algebraic approaches are able to address Benacerraf's two challenges to mathematical platonism but that the algebraic approaches face two new challenges, namely explaining the modal notions on which they crucially rely, and accounting for mixed mathematical–empirical claims.

Part III. Mathematical practice and the methodology of mathematics

Making sense of mathematical practice and the methodology of mathematics are crucial components of a philosophical account of mathematics. Intriguing issues emerge when these components are taken seriously: from the role of mathematical explanation in mathematical practice, through the issue of the status of inconsistent mathematical theories, to the complexities of understanding applied mathematics. These issues are discussed in Part III.

Alan Baker's chapter opens by calling attention to the increased interest among philosophers of mathematics in the concept of mathematical explanation, which has long been ignored. A central thesis of the chapter is that the notion of mathematical explanation can be illuminated by a study of the concept of an accidental mathematical fact. Since mathematical facts are assumed to be necessary, mathematical accidents cannot be explained in terms of the notion of contingency. Baker first explains the notion of "a mathematical coincidence" in terms of the equality of two quantities lacking a unified explanation. A mathematical truth is then said to be accidental to the extent that it lacks a unified, non-disjunctive proof.

Mark Colyvan discusses some problems posed by inconsistent mathematical theories. For instance, naïve set theory is inconsistent, and arguably, so is eighteenth century calculus. How can reasoning in such theories be represented? Clearly, we don't proceed to derive every sentence whatsoever, as is permitted in classical logic. Paraconsistent logics provide a way of avoiding this trivialization. Such logics enable Colyvan to ask some philo-sophical questions. Firstly, what happens when the Quinean indispensability argument is applied to inconsistent but nevertheless useful mathemati-cal theories? Colyvan suggests that theorists may then have had reason to believe in the existence of inconsistent mathematical objects. He also outlines an account of applied mathematics intended to be capable of accounting for the applicability of inconsistent mathematics.

Christopher Pincock exemplifies the "naturalistic turn" in the philoso-phy of mathematics, which is characterized by a greater attention to actual mathematical practice. He argues that applied mathematics differs from pure mathematics in its priorities, methods, and standards of adequacy. These claims are illustrated by means of a case study, namely so-called "boundary layer theory". He shows how in applied mathematics wholly mathemati-cal justifications are not always available, as there is also a need to rely on experiment and observation. There may also be semantic and meta-physical differences between pure and applied mathematics based on these differences.

Part IV. Mathematical language and the psychology of mathematics

Many philosophers argue that a philosophical interpretation of mathemat-ics must be sensitive to the way mathematical language is used and to the psychology of mathematical thought. This concern plays a prominent role in the contribution to Part IV of the volume.

Hofweber argues that progress in the philosophy of mathematics is ham-pered by excessive and misguided use of formal tools. He claims that the role for formal tools in the philosophy of mathematics is limited, at least as concerns the philosophical project of answering questions about truth, knowledge, and fact in mathematics. Hofweber points out that many ques-tions in the philosophy of mathematics are empirical, for example, whether fictionalism is the correct interpretation of actual mathematical language and practice. He also warns that, although the language of Peano Arithmetic has very desirable mathematical properties (such simplicity and inferen-tial adequacy), it does not provide an adequate linguistic analysis of actual arithmetical language. However, formal tools would have an important role to play in the philosophy of mathematics, Hofweber argues, if the axioms of certain parts of mathematics function constitutively rather than descriptively, while the axioms of arithmetic are descriptive rather than constitutive.

Øystein Linnebo discusses two competing accounts of how the natural numbers are individuated. His preferred notion of individuation is a seman-tic one which is concerned with our most fundamental ways of singling out objects for reference in thought or in language. According to the *cardinal account* (defended by Fregeans and other logicists), a natural number is singled out by means of a concept or plurality which instantiates the number in question. According to the *ordinal account* (defended by structuralists and constructivists), a natural number is singled out by means of a numeral which occupies the corresponding position in a sequence of numerals. Linnebo criticizes the cardinal account for being linguistically and psychologically implausible. He then develops a version of the ordinal account which countenances numbers as objects but only in a very lightweight sense.

Agustín Rayo's article defends the view that mathematical sentences have trivial truth-conditions; that is, that every mathematical truth is true in every intelligible scenario. This view would ensure that mathematics is epistemologically tractable. The notion of an intelligible scenario is explained in terms of two notions of identity: ordinary identity between individuals and an identity-like relation between properties (such as: to be water *just is* to be H_2O). A sentence has trivial truth-conditions just in case it holds in every scenario compatible with all true identities. If trivialism is right and mathematical sentences have trivial truth-conditions, why is mathematical knowledge so valuable? Rayo answers that its value lies in an improved ability to distinguish intelligible from unintelligible scenarios.

Part V. From philosophical logic to the philosophy of mathematics

Finally, some contributors have approached the philosophy of mathematics using tools and methods from philosophical logic. Their work forms Part V of the collection.

Hannes Leitgeb studies the notion of informal provability, which he distinguishes from formal provability. The former notion is legitimate, Leitgeb argues, because it plays an important role in ordinary mathematical practice. Next he provides an interpretation and defense of some of Gödel's cryptic remarks about the role of intuition in the epistemology of mathematics. This is based on a broadly naturalistic account of cognitive representations of mathematical structures. Leitgeb then turns to the logic of informal provability, agreeing with Gödel that this logic is at least S4 and making some suggestions about how one may come to extend this logic even further. The chapter closes with some tentative suggestions about how we may come to establish that there are true but informally unprovable mathematical statements.

Gabriel Uzquiano discusses the threat which the phenomenon of "indefinite extensibility" poses to the existence of an all-comprehensive domain of quantification. The claim that there is an all-comprehensive domain of quantification is first distinguished from a strong form of metaphysical realism with which it is often conflated. A more neutral formulation of the claim, Uzquiano suggests, is that there are some objects such that every object is one of them. He then explores a challenge to this version of the claim. Assume any plurality of objects form a set. Then there cannot be an all-comprehensive domain; for if there were, there would be a universal set, which by Russell's paradox would lead to a contradiction. Can the assumption that every plurality forms a set be incorporated into set theory? Uzquiano argues that this is possible only at the cost of denying some important features of contemporary set theory.

Concluding remarks

As this brief outline indicates, the contributions to this volume offer "new waves" to the philosophy of mathematics in two senses. Several contributions revisit established philosophical issues about mathematics but offer new ways of framing, conceptualizing, and approaching these issues. Other contributions raise new philosophical issues about mathematics and its practice that haven't been addressed as forcefully as they should have. And several contributions do both. Since the chapters that follow speak for themselves, we leave it as an exercise to the reader to identify which contributions do what. We hope this may be an interesting way of enjoying the journey, while the waves will take you to new, exciting, and often rather unexpected places.

Note

1. Hellman's modal-structural interpretation of mathematics was also inspired by an earlier view advanced by Putnam in his defense of "mathematics without foundations".

Part I

Reassessing the Orthodoxy in the Philosophy of Mathematics

1

New Waves on an Old Beach: Fregean Philosophy of Mathematics Today

Roy T. Cook

1 Motivations philosophical and personal

We shall assume throughout this essay that platonism[1] regarding the subject matter of mathematics is correct. This is not to say that arguments for platonism – that is, arguments for the existence of abstract objects such as numbers, sets, Hilbert spaces, and so on that comprise the subject matter of mathematics – are either uninteresting or unneeded. Nevertheless, I am not personally interested in such arguments. The reason for this is simple: I have never[2] doubted that there are abstract objects, and I have never doubted that mathematics is about (some of) them. While arguments for the existence of abstract objects are philosophically important, such debates hold little interest for me when my intuitions already fall so strongly on one side.

What I do find to be both important and interesting is addressing certain problems that plague platonist accounts of mathematics. The central, and most important such problem, of course, is the epistemological one: How can we interact with abstract objects in such a way (or any way!) as to come to have knowledge of them?

There are a number of more precise epistemological questions that arise once one adopts some form of platonism regarding the subject matter of mathematics. These include the following:

[a] How can we have the appropriate sort of connection with abstract objects necessary for knowledge, when we cannot have any causal interaction with these objects? (In other words, what sort of "interaction" do we have with abstract objects?)

[b] How do we explain the special epistemological status that such knowledge seems to enjoy? (In other words, how do we explain the a priority, certainty, etc. that mathematical knowledge seems to have, at least in ideal cases?)

[c] How does our explanation of the epistemology of abstract objects in general, and mathematical objects in particular, "mesh" with the rest of our

13

epistemology? (In other words, how is platonism compatible with the obvious utility of mathematics in empirical applications?)[3]

Adequately answering these questions is obviously important for the platonist since platonism without a decently worked out epistemology is little more than mysticism. In addition, solving these puzzles will defuse many of the most common arguments against platonism since more often than not they hinge on the epistemological troubles that the platonist faces.[4]

Thus, a defense of platonism depends upon a satisfactory account of the epistemology of abstract objects. The obvious next question to ask is as follows: Where should we look to find such an account? The answer, which I shall spend the remainder of this essay developing and defending, is that we should look to the work of Gottlob Frege.

Frege's logicism is, in the wake of Russell's paradox and similar set-theoretic pathologies, typically dismissed as a failure. While there is no doubt that Frege's formal system is inconsistent, I would like to suggest that rejecting the philosophical underpinnings of his logicist project is a clear case of throwing out the baby with the bathwater. Instead, I will argue that Frege was, for the most part, correct about both the ontology and the epistemology of arithmetic. The technical details must be finessed, of course, since the methods Frege uses in the *Grundlagen* and *Grundgesetze* are inadequate to the task. Recent work by neo-Fregeans has gone quite a ways towards determining exactly what form this finessing should take, however.

Thus, our main focus here will be to determine exactly what was right about Frege's logicist project, and to sketch how *these* aspects of nineteenth century logicism have been retained in twenty-first century neo-Fregeanism. Two additional issues motivate the discussion to follow, however.

The first is that both explications and critiques of neo-Fregeanism tend to draw the connection between Frege's project and its more recent reformulation by emphasizing the similarities in the technical details of the projects – namely, the use of abstraction principles to derive the axioms of particular mathematical theories. While these technical parallels exist, I think that this is the wrong place to look for the important links between the two projects. Instead, the truly important connections between Frege's logicism and neo-Fregeanism lie in the philosophical frameworks underlying these projects – in particular, the idea that the analyticity of mathematics can be explained in terms of a reduction of mathematics to logic plus definitions, and the accompanying idea that the epistemological problems facing platonism can be solved by attending to such a reduction.

Second, in giving talks and writing papers on neo-Fregean accounts of mathematics, I often find myself defending the very approach, instead of focusing on the details of the particular issues I am attempting to address. The reason is relatively simple: My audience hears the terms "neo-Fregeanism" or "neo-logicism" and take this terminology seriously, thinking that I am discussing some variant of Frege's original view. But of course, they will interject, Frege's logicism, as we noted a few paragraphs earlier, was obliterated by Russell's paradox. Surely, they ask, I am not trying to revive this doomed project? In the face of such objections, I often find myself disingenuously explaining that neo-Fregeanism (or neo-logicism) is not a new version of logicism at all, but instead is a different project altogether, with only superficial similarities to Frege's original undertaking (in other words, I find myself drawing the connections between the two views solely in terms of the technical similarities). It is time to uncover this dishonesty:[5] Contrary to what is often claimed, neo-Fregeanism *is* a version of logicism. Thus, in this chapter I will strive to tease out exactly what aspects of neo-Fregeanism make it so.

Finally, I should note two limitations to the discussion. First, length constraints prevent me from addressing in great detail many of the interesting technical and philosophical issues that arise with regard to logicist and neo-Fregean accounts of mathematics.[6] In this essay, I am instead focusing on general issues regarding the epistemological framework provided by Fregean ideas regarding the nature of mathematics. Second, I am, again as a result of space considerations, restricting my attention to formal and philosophical developments of arithmetic.[7]

2   Fregean philosophy of mathematics

Before examining how we can and should adopt Fregean ideas regarding the epistemology of mathematics, a brief survey of the relevant bits of Frege's logicism is in order. As is well known, one of Frege's primary goals was to defend arithmetic (and analysis, etc., but not geometry) from Kant's charge of syntheticity. Equally well known is Frege's strategy: Arithmetic can be reduced to logic. Logic is analytic. So arithmetic inherits its analyticity in virtue of the reduction.

Frege's attempt to reduce mathematics to logic proceeds in the following two steps: First, Frege showed that arithmetical concepts such as cardinal number, successor, and so on are definable in terms of (purportedly) purely logical notions. Second, Frege demonstrated how, given these definitions, the truths of arithmetic turn out to be logical truths.

As noted above, both Frege and his commentators and critics typically characterize all of this as a response to Kant's accusations against arithmetic. See MacFarlane (2002) for a useful examination of the ins and outs of this aspect of Frege's project. Interestingly, however, Kant himself might have thought that Frege's project is a bit of a cheat, since Kant was quite explicit regarding his view that existential propositions could not be analytic:

But if, on the other hand, we admit, as every reasonable person must, that all existential propositions are synthetic, how can we profess to maintain that the predicate of existence cannot be rejected without contradiction? This is a feature which is found only in analytic propositions, and is indeed precisely what constitutes their analytic character. (1965, p. 504).[8]

Of course, Kant's notion of what counted as analytic was likely tied to his ideas regarding logic – in particular, to the limitations imposed on his thought by the acceptance of Aristotelian categorical logic (or perhaps term logic more generally) as the whole of logic. Frege, having effectively invented modern quantificational logic in his *Begriffschrift* and elsewhere, had a much broader notion of what could count as a logical principle, and as a result could more easily countenance existential claims as logical truths, and thus analytic.[9]

Frege's reconstruction of arithmetic as logic begins with his notorious *Basic Law V*:

$$BLV: (\forall X)(\forall Y)[\S X = \S Y \leftrightarrow (\forall z)(Xz \leftrightarrow Yz)]^{10}$$

which assigns a unique object – an extension, or course-of-values (what we might call a set) to each concept (§ is, in the above formulation, a function – an abstraction operator – mapping concepts to their extensions). Frege then defines the cardinal numbers as a certain subspecies of extensions:

My definition is therefore as follows:
The number which belongs to the concept F is the extension of the concept "equal to the concept F". (1980, §68, pp. 79–80)[11]

In symbols

$$\#P = \S((\exists Y)(x = \S(Y) \wedge Y \approx P))$$

(where "F $\approx$ G" abbreviates the second-order statement asserting that there is a one-to-one onto mapping from the F's to the G's). Given this definition, it is relatively straightforward to derive *Hume's Principle*:

$$(HP): (\forall X)(\forall Y)[\#(X) = \#(Y) \leftrightarrow X \approx Y]$$

As is well-known, the Peano axioms, including the second-order induction axiom, follow from (HP) (plus natural definitions of zero, successor, etc.) – this result is known as *Frege's Theorem*. Thus, full second-order *Peano Arithmetic* follows from second-order logic supplemented with BLV! Of course, Frege's hopes for showing that arithmetic is analytic were smashed by Bertrand Russell's (1902) demonstration that the logic of the *Grundgesetze* is inconsistent.

This is, in essence, the story that is typically told regarding the rise and fall of Fregean logicism. Since the technical problem resides in one of Frege's first principles – BLV – anything that is built upon this principle – that is, all of Frege's subsequent definitions and deductions – is suspect at best.[12] Thus, the reader might be forgiven at this stage for wondering what was *right* about Frege's project.

In order to see what was correct about Frege's approach, we need to look more closely at the philosophical framework underlying the technical apparatus developed in the *Begriffschrift* and the *Grundgesetze*. As already noted, Frege's goal was to show that arithmetic was analytic, and his attempted reduction of arithmetic to logic was merely a means to this end. Thus, the natural question to ask is as follows: Why did Frege think that the analyticity of arithmetic was so important? To answer this question, we first need to note that Frege understood analyticity as an epistemic notion:

> Now these distinctions between a priori and a posteriori, synthetic and analytic, concern, as I see it, not the content of the judgement but *the justification for making the judgement*. Where there is no such justification, the possibility of drawing the distinctions vanishes. An a priori error is thus as complete a nonsense as, say, a blue concept. When a proposition is called a posteriori or analytic in my sense, this is not a judgement about the conditions, psychological, physiological and physical, which have made it possible to form the content of the proposition in our consciousness; nor is it a judgement about the way in which some other man has come, perhaps erroneously, to believe it true; rather, *it is a judgement about the ultimate ground upon which rests the justification for holding it to be true*. (Frege (1980), §3, p. 3, emphasis added)

In other words, on Frege's view whether or not a judgement is analytic depends on the type of justification which that judgement possesses. A bit later in the same section of the *Grundlagen* Frege provides an explicit definition of analyticity:

> The problem becomes, in fact, that of finding the proof of the proposition, and of following it up right back to the primitive truths. *If, in carrying out this process, we come only on general logical laws and on definitions, then the truth is an analytic one*, bearing in mind that we must take account also of all propositions upon which the admissibility of any of the definitions depends. *If, however, it is impossible to give the proof without making use of truths which are not of a general logical nature, but belong to some special science, then the proposition is a synthetic one.* For a truth to be a posteriori, it must be impossible to construct a proof of it without including an appeal to facts, i.e., to truths which cannot be proved and are not general, since they contain assertions about particular objects. But

if, on the contrary, its proof can be derived exclusively from general laws, which themselves neither need not admit of proof, then the truth is a priori. (Frege (1980), §3, p. 4, emphasis added)

Thus, for Frege, a judgement Φ is analytic if and only if Φ has a proof that depends solely upon logical laws and definitions, and a judgement Φ is *a priori* if and only if Φ has a proof that depends only upon self-evident, general truths. All logical laws and definitions are self-evident general truths, but not vice versa: the difference between *a priori* judgements and analytic judgements hinges on the fact that Frege believed that there were self-evident general truths that were neither definitions nor logical laws, and thus that there were *a priori* truths that were not analytic (the laws of geometry are the paradigm example since they are, for Frege, self-evident general truths which depend on geometrical intuition).[13]

A partial answer to our question is now apparent. Frege thought the analyticity of arithmetic was important since if arithmetic were synthetic, then it would depend on the truths of some special science. But why did Frege think it was important to show that the truths of arithmetic do not depend on principles from a special science (especially if these truths might be of a general nature, and thus *a priori*)? The answer lies in Frege's formulation of, and acceptance of, what has come to be called *Frege's Constraint*:

FC: The correct analysis of a mathematical concept proceeds from consideration of applications of that concept.[14]

Frege fleshes out the constraint in some detail in the *Grundgesetze*:

We retain the conception of real number as a relation of quantities... but dissociate it from geometrical or any other specific kinds of quantities and thereby approach more recent efforts. At the same time, on the other hand, we avoid the drawback showing up in the latter approaches, namely that any relation to measurement is either completely ignored or patched on solely from the outside without any internal connection grounded in the nature of the number itself... our hope is thus neither to lose our grip on the applicability of [analysis] in specific areas of knowledge, nor to contaminate it with the objects, concepts, and relation taken from those areas and so to threaten its peculiar nature and independence. The display of such possibilities of application is something one should have the right to expect from [analysis] notwithstanding that that application is not itself its subject matter. (1903, §159).[15]

Although this passage is taken from a later portion of the *Grundgesetze* where Frege is working out the details of his account of real analysis, it is clear that the point is general: Both arithmetic and analysis are completely general in their applications; analysis applies to any collection of quantities, regard-less of what those quantities are quantities of, and arithmetic applies to any plurality of objects, regardless of the nature of those objects.[16]

After noting that we can "for the purposes of conceptual thought always assume the contrary of one or the other of the geometrical axioms" (1980, §14, p. 20), Frege notes that we do not seem to be able to do the same with arithmetic, and concludes that the reason for this difference lies in the complete generality and everywhere applicability of arithmetic:

> Can the same be said of the fundamental propositions of the science of number? Here, we have only to try denying any one of them, and com-plete confusion ensues. Even to think at all seems no longer possible. The basis of arithmetic lies deeper, it seems, than that of any of the empirical sciences, and even that of geometry. The truths of arithmetic govern all that is numerable. This is the widest domain of all, for to it belongs not only the actual, not only the intuitable, but everything thinkable. Should not the laws of number, then, be connected very intimately with the laws of thought? (1980, §14, p. 21)

If our account of arithmetic (and, eventually, analysis) depends solely on logic and definitions (which on Frege's view possess the requisite generality, being the "laws of the laws of nature" (1980, §87, p. 99)), then this gen-erality is unsurprising and easily accounted for. If, on the other hand, our reconstruction of arithmetic depends on the principles of some special sci-ence, then we are left in the dark regarding why arithmetic should have applications outside of the subject matter of that special science.

Thus, the analyticity of arithmetic is not just a matter of status for Frege – that is the analytic truths are somehow "better", or more secure, than syn-thetic ones. Instead, in Frege's understanding, one of the central features of arithmetic – its complete generality and everywhere applicability – cannot be explained other than in terms of the analyticity of the laws of arithmetic.

Frege goes on to provide a detailed positive account of the nature of cardi-nal numbers – one which both allows for their analyticity (in Frege's sense of "analytic") and answers the epistemological questions with which we began this essay. He first observes that applications of cardinal numbers are typ-ically intimately tied up with a certain sort of second-level concept – that is, a concept that holds of first-level concepts (where first-level concepts are concepts that hold of objects):

> ...the content of a statement of number is an assertion about a concept. (Frege 1980, §46, p. 59)

In other words, there is an intimate connection between the object that is the cardinal number three, and the second-level concept that holds of any first-level concept that holds of exactly three objects – let us call this second-level concept "*three-ity*". The relation of *three-ity* to arbitrary triples of objects can be represented graphically as the following:

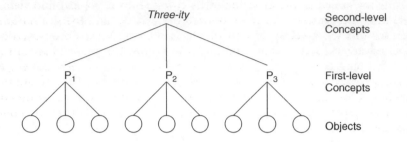

The problem then becomes determining exactly what the relationship is between *three-ity* and the cardinal number three, and Frege's solution depends critically on BLV. In short, BLV allows us to replace first-level concepts with objects – their extensions – that can serve as proxies for them. We can replace all of the first-level concepts in the above diagram with their extensions, and replace the second-level concept *three-ity* with the first-level concept Q such that Q holds of an object if and only if it is the extension of a concept that *three-ity* holds of. We obtain the following simplified picture:

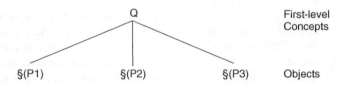

In order to obtain the cardinal number three, we just pull the "extensions-as-proxies-for-first-level-concepts" trick again, taking the extension of Q. Note that this object is exactly the extension that Frege's definition of number, provided above, identifies as the number three.

This reduction of second-level concepts to objects accomplishes two things for Frege. The first is that it provides a non-arbitrary identification of the cardinal numbers.[17] Second, and more importantly, perhaps, it provides for answers to the epistemological questions that began this essay. The crucial insight is that Frege's reduction goes both ways: Talk about certain second-level concepts can be reconstrued as talk about the cardinal numbers, but equally straightforward – given BLV, at least – is the reconstrual of talk about cardinal numbers in terms of talk about second-level concepts.

As a result, knowledge of the abstract objects that are the subject matter of arithmetic – that is, the cardinal numbers – does not, on Frege's picture, require us to have some sort of quasi-mystical rapport with causally inefficacious abstract objects. Instead, such knowledge is, in the final analysis, merely knowledge about the characteristics of (in fact, the logical characteristics of) second-level concepts.[18] On Frege's view, arithmetical knowledge obtains its special status in virtue of its equivalence to logical knowledge regarding second-level concepts – in other words, arithmetical knowledge inherits whatever special epistemological advantages are enjoyed by logic. Finally, for Frege, the epistemology of arithmetic is just a part of the larger epistemology of logic and, since logic is completely general and everywhere applicable – that is, logic consists, in Frege's words, of "the laws of the laws of nature", arithmetic is – and in particular, applications of arithmetic are – completely general and everywhere applicable as well. This last point is summed up nicely by Frege himself:

> The laws of number will not... need to stand up to practical test if they are to be applicable to the external world; for in the external world, in the whole of space and all that therein is, there are no concepts, no properties of concepts, no numbers. The laws of numbers, therefore, are not really applicable to external things; they are not laws of nature. They are, however, applicable to judgements holding good of things in the external world: they are laws of the laws of nature. They assert not connexions between phenomena, but connexions between judgements; and among judgements are included the laws of nature. (Frege 1980, 87)

This is all well and good, of course, as reconstruction of what Frege was up to. But as noted, the project depends crucially on BLV. As our discussion of Frege's analysis of the nature of cardinal numbers makes clear, BLV is not needed merely for its proof-theoretic strength (which is considerable indeed!). In addition, the reduction of first-level concepts to their object-level proxies provided by BLV is central to Frege's identification of the cardinal numbers. Thus, the central question facing us, which I shall attempt to answer in the next section, is how the neo-Fregean can retain the promising logicist epistemological story while simultaneously rejecting BLV.

3 The Neo-Fregean picture

Before beginning, an important caveat is in order: The discussion in this section is not meant to be report of the official Hale and Wright view, as presented in Hale and Wright (2001),[19] for example. Instead, I am merely presenting *my* interpretation of what the neo-Fregean view ought to look like. There will no doubt be details where either Hale or Wright or both will disagree. I am confident, however, that these points of disagreement are in most cases fairly minor, and I will flag such possible points of contention in the footnotes.

In my understanding of the neo-Fregean project, the primary goal is to explain the apriority of mathematics – and sort out other puzzling epistemological issues – via a defense of the analyticity of arithmetic. As Crispin Wright puts it

> *Frege's Theorem* will ensure... that the fundamental laws of arithmetic can be derived within a system of second order logic augmented by a principle whose role is to explain, if not exactly to define, the general notion of identity of cardinal number... If such an explanatory principle... can be regarded as analytic, then that should suffice... to demonstrate the analyticity of arithmetic. Even if that term is found troubling, as for instance by George Boolos, it will remain that *Hume's Principle* – like any principle serving to implicitly define a certain concept – will be available *without significant epistemological presupposition*... Such an epistemological route would be an outcome still worth describing as logicism. (1997, pp. 210–211, emphasis added)

The strategy should sound familiar since it is exactly the strategy that Frege adopted in order to solve these problems. In addition, Wright notes that if one is uncomfortable with the notion of analyticity – perhaps due to Quinean qualms, about which more later – then we can nevertheless view the neo-Fregean account of arithmetic as providing a means to understanding the concepts of arithmetic "without significant epistemological presupposition". Thus, Wright, like Frege, seems to understand the notion of analyticity (or whatever analogous notion is at work here) epistemologically. The differences between Frege's logicism and modern neo-Fregean "logicism" stem from the latter's rejection of BLV, and the resulting additions that are required in order to make up for its loss.

From a technical perspective, the neo-Fregean account is straightforward and amounts to nothing more than an adoption of part, but not all, of the formal work developed in Frege's *Grundgesetze*. Instead of deriving (HP) from other, more fundamental theses, the neo-Fregean accepts its truth primitively – (HP) is an implicit definition of the concept cardinal number. Once (HP) is laid down, however, the project proceeds exactly as Frege's, at least from a technical perspective. Given suitable definitions of zero, successor, and the like, *Frege's Theorem* ensures that the second-order Peano axioms follow from (HP).

The real work comes in recasting the philosophical framework that this technical achievement is meant to serve. The most important thing to notice is that the neo-Fregean, like Frege, understands analyticity to mean logic plus definitions. The neo-Fregean just means something different by both "logic" and "definitions".

The logic adopted by the neo-Fregean – standard second-order logic, or perhaps some free-logic variant of it – is substantially weaker than the system developed by Frege in the *Begriffschrift* and *Grundgesetze*. As a result, the neo-Fregean must make up for this loss with a corresponding gain somewhere else, if he is to be successful in carrying out the sort of reconstruction of arithmetic sketched in the previous section. Thus, the second difference between the Fregean and neo-Fregean understandings of analyticity stems from a liberalization of allowable definitions. Frege is quite clear about the fact that the only definitions allowed in the *Grundgesetze* (with the possible exception of BLV itself, if we view it as a definition of extensions) are explicit definitions such as his definition of cardinal number in terms of extensions. The neo-Fregean, on the other hand, allows implicit definitions such as (HP) and other abstraction principles to count as legitimate definitions. We can view the move from Frege's logicism to neo-logicism as consisting primarily of a shift from what we might call Frege-Analytic:[20]

> FA: A statement is Frege-Analytic if and only if it can be proven from (*Grundgesetze*) logic and (explicit) definitions.

to what we might call neo-Frege Analytic:

> NFA: A statement is Neo-Frege-Analytic if and only if it can be proven from (second-order) logic and (implicit) definitions.[21]

This difference is nowhere more evident than in the different attitudes taken by Frege and the neo-Fregeans towards (HP). On the one hand, Frege explicitly rejects (HP) as a definition of the concept of cardinal number in §66 of the *Grundlagen*. Neo-Fregeans, on the other hand, hold up (HP) as a paradigm instance of what a good implicit definition of a mathematical concept should look like.

Before moving on, it is worth noting that our brief definition of NFA is rather inadequate: The neo-Fregean does not think that any statement that can be proven from logic plus *any* implicit definition is analytic in the relevant sense. Instead, the implicit definition must be an abstraction principle – that is, it must be of the same general logical form as (HP) and BLV:

$$(\forall \alpha)(\forall \beta)[@(\alpha) = \S(\beta) \leftrightarrow E(\alpha, \beta)]$$

where α and β range over entities, or sequences of entities, of an appropriate sort (typically objects or first-level concepts or relations) and E is an equivalence relation on the sort of entity ranged over by α and β.[22] Furthermore, being an abstraction principle is still not enough since BLV is of the requisite logical form yet it is not a legitimate definition upon which analytic truths can depend. We shall return to this issue a bit later in the chapter.

The notion of analyticity at work in neo-Fregeanism is thus an epistemic notion. As such, it should be clearly distinguished from semantic notions such as the one attacked by Quine in "Two Dogmas of Empiricism" (1960) and elsewhere. Quine's understanding of analyticity, which we can call Quine-Analytic, is something along the lines of

> Quine-Analytic: A statement is Quine-Analytic if and only if it is true in virtue of the meanings of its constituent expressions.

Setting aside metaphysical worries regarding how meanings could make any statement true (other than, perhaps, statements about meanings), this understanding of analyticity is the one most commonly discussed and argued about today. Analyticity in this sense, however, is a semantic issue, quite distinct from the notion that Frege had (and the neo-Fregeans have) in mind. Quine is quite explicit about this semantic reading of analyticity.

> A statement is analytic when it is true by virtue of meanings and independently of fact. (1960, p. 21)

Although the notion that Quine is interested in is distinct from the notion mobilized by Frege and the neo-Fregeans, it nevertheless has a distinguished pedigree, since Quine-Analyticity seems to be much closer than Frege's notion to what Kant had in mind:

> In all judgements in which the relation of a subject to the predicate is thought... this relation is possible in two different ways. Either the predicate B belongs to the subject A, as something which is (covertly) contained in this concept A; or B lies outside the concept A, although it does indeed stand in connection with it. In the one case I entitle the judgement analytic, in the other synthetic. (1965, p. 48)

It might be the case that Quine's arguments show that the analytic/synthetic distinction is untenable on his, and Kant's, semantic reading. For the sake of argument, and for the sake of moving on, let us grant that they do. Even so, there are no reasons for thinking that the same problems plague the Fregean or neo-Fregean notions of analyticity – after all, if the neo-Fregean can sharply delineate exactly what counts as logic and what counts as an acceptable implicit definition, then the resulting class of analytic truths will be sharply delineated as well.

This is not to say that the neo-Fregean has provided a completely adequate account of the divide between the analytic truths and everything else. Although the relevant notion of logic is relatively non-problematic (standard second-order consequence as detailed in, e.g., Shapiro (1991)), the neo-Fregean still owes an account specifying which abstraction principles are acceptable. The observation that, after a tremendous amount of ink has been spilled, such a criterion has not been forthcoming is known in the literature as the Bad Company Objection.

In a nutshell, the Bad Company Objection is merely the observation that some abstraction principles, such as BLV, are not candidates for acceptability since they are incompatible with presumably acceptable principles such as (HP). Stated this way, a partial solution to the problem is already evident: Acceptable implicit definitions must be consistent, and consistent with (HP). Unfortunately, this is not enough since there are pairs of abstraction principles that are (individually) consistent with (HP), but not consistent with each other. Stricter criteria are necessary.

Note that our current lack of a sharp border between good and bad abstraction principles is not due to the sort of Quinean worries mentioned earlier regarding the coherence of there being any such sharp border. Instead, the problem is (so the neo-Fregean hopes!) merely an epistemological one: The neo-Fregean remains convinced (rightly, in my opinion) that there is such a sharp border, but has failed, as of yet, to locate it. Thus, the neo-Fregan's primary goal is to show that arithmetic is analytic, yet the neo-Fregean has yet to provide a precise definition of (one half of) analyticity.

The present essay is not the place to provide a full summary of all of the twists and turns that have been taken in attempting to address this problem. Nevertheless, a closer look at two proposed criteria will be helpful. The debate has, for the most part, taken the form of suggested criteria for marking out the good from the bad, followed by clever counterexamples to the suggested criteria, followed by amended criteria, and so on. For example, Crispin Wright develops the idea that acceptable abstraction principles must be conservative as follows:

> A legitimate abstraction, in short, ought to do no more than introduce a concept by fixing truth-conditions for statements concerning instances of that concept ... How many sometime, someplace zebras there are is a matter between that concept and the world. No principle which merely assigns truth-conditions to statements concerning objects of a quite unrelated, abstract kind – and no legitimate second-order abstraction can do any more than that – can possibly have any bearing on the matter ... What is at stake ... is, in effect, conservativeness in (something close to) the sense of that notion deployed in Hartry Field's exposition of his nominalism. (1997, p. 296)

Abstraction principles ought to be conservative, according to Wright, since they should have no bearing on, for example, the number of zebras there might be in the world. This fits well with the idea that acceptable abstraction principles must be completely general and everywhere applicable, as discussed above. Unfortunately, it turns out that there are pairwise inconsistent, conservative abstraction principles. See Weir (2004) for examples.

Unfortunately, as more and more criteria have been suggested, and have turned out to be inadequate, the literature on the subject has taken a decidedly technical turn.[23] As a result, motivations for proposed criteria seem to have lost their connection with the epistemology underlying the neo-Fregean project.

For example, the passage first introducing the idea that abstraction principles must be irenic – that is, that any acceptable abstraction principle must be consistent with all conservative abstraction principles, comes from an appendix to Wright's "Is Hume's Principle Analytic":

> Why not just say that pairwise incompatible but individually conservative abstractions are ruled out – however the incompatibility is demonstrated – and have done with it? (1999, p. 328)

It is not completely clear that Wright meant this offhand query to be a serious suggestion for how to draw the acceptable/non-acceptable line, but the notion of irenicity and the closely connected notion of stability[24] have been taken up in the literature as the default account(s) of what counts as a good abstraction. See, for example, Weir (2004). The problem, of course, is that irenicity does not seem to have the same sort of epistemological motivation as conservativeness. On the contrary, the inference underlying the restriction to irenic abstraction principles seems to be little more than a fallacy: faced with pairwise incompatible, conservative abstraction principles, logic dictates that at least one must be unacceptable, but as of yet we have no data determining which of the two is unacceptable. Thus, we conclude that both are.[25]

The problem, I think, is that when faced with this extremely difficult problem – drawing a sharp distinction between good and bad abstraction principles, attention has strayed from the philosophical question regarding what epistemological characteristics might make an abstraction principle a good one, to what formal characteristics might supply us with a powerful yet consistent collection of abstraction principles. If the sketch of neo-Freganism given above is correct, and the primary motivations for neo-Fregean accounts of arithmetic (and other mathematical theories) are epistemological, then this is a mistake.

Instead, we should be thinking about the epistemology of implicit definitions in general, and abstraction principles in particular. After all, what is important about neo-Fregeanism is the epistemological story it allows us to tell about arithmetic. Thus, whatever formal criterion for acceptability we eventually decide on, the success of neo-Fregeanism will hinge on defending the claim that all abstraction principles that satisfy that criterion also have the requisite epistemological properties. The most promising way to

obtain such a criterion is to require our search to proceed on epistemological, and not merely technical, grounds. In other words, we should first
determine what epistemological characteristics abstraction principles must
have in order to play the foundational role sketched for them in the previous
sections, and then formulate an account of acceptable abstraction principles
in terms of these epistemological characteristics. The hope is that if we sort
out the epistemology, then Bad Company will take care of itself.

The present essay is not the place to attempt a full solution to this problem, or even to give a detailed sketch of how one might go. But we can
formulate a suggestion: If we take Frege's constraint and its consequences
seriously, then one of the primary motivations for reducing various mathematical theories to logic plus definitions – that is, one of the primary
motivations for being a neo-Fregean – is that such a methodology promises
to shed light on the complete generality and everywhere applicability of
mathematical theories such as arithmetic. Thus, acceptable abstraction principles will be those that, among other things perhaps, are completely general
and everywhere applicable[26] – that is, very loosely put, they provide us with
no information whatsoever other than providing us with definitions of the
mathematical concepts involved.[27] Looked at in this way, Wright's suggestion that abstraction principles must be conservative seems on the right
track, but it is clearly not enough (that is, it is necessary but not sufficient).

Setting aside these worries about Bad Company, however (and noting that
there is little disagreement regarding which side of the divide (HP) will
eventually fall on) it is clear that, with this notion of analyticity to hand,
the neo-Fregean can provide a reconstruction of arithmetic that retains
the significant epistemological benefits of Frege's original account. The first
step is acceptance of Frege's Constraint.[28] In particular, the neo-Fregean
accepts Frege's thought that the correct analysis of a mathematical concept proceeds from consideration of its applications. Thus, the neo-Fregean
agrees with Frege that at least some applications of cardinal numbers (such
as adjectival applications involving second-level concepts) are not possible merely because of some "coincidental" structure-preserving mapping
between sequences of objects and the cardinal numbers themselves, but
instead arise as part and parcel of what the cardinal numbers are.

More specifically, the neo-Fregean agrees with the fundamental Fregean
observation that applications[29] of arithmetic intimately involve second-level
concepts (such as *three-ity*, the second-level concept that holds of all first-
level concepts that hold of exactly three objects), and that in addition
such applications are completely general and everywhere applicable (i.e.
the second-level concept *three-ity* holds of *every* concept with exactly three
instances).

Thus, the neo-Fregean, like Frege, owes us an account of how such
second-level concepts connect to the cardinal numbers. In particular, the

neo-Fregean must provide some account of how the second-level concept *three-ity* (i.e.) is connected to the cardinal number three. So, as before, we begin with the following picture:

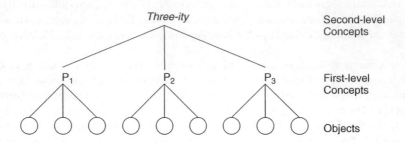

Unlike Frege, however, the neo-Fregean does not have access to the extensions provided by BLV, and so cannot carry out the "reduction" of second-level talk to object-level talk that is so central to Frege's technical project.

Fortunately, the neo-Fregean has not just abandoned BLV, but instead has traded it for a more liberal view of acceptable definitions. As a result, the neo-Fregean simply lays down (HP), stipulating the existence of a function mapping concepts to objects. The (HP) is, on the neo-Fregean picture, an implicit definition of the cardinal numbers, and, as such, it tells us that each first-level concept that is an instance of *three-ity* gets mapped onto the same object: the cardinal number three. We thus obtain the following structure:

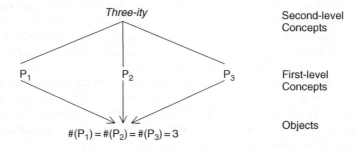

The details are different, but the end result is the same. The neo-Fregean explanation of the relation between the cardinal number three and the relevant second-level concept is provided by the fact that three is the cardinal number of exactly the first-level concepts that are instances of *three-ity*, and the existence of this cardinal number is guaranteed by (HP)'s status as an acceptable implicit definition. While this is, from a technical perspective, a different analysis of cardinal number than the one provided by

Frege, it provides the same end result: an analysis of the nature of mathematical objects, and of the connection between mathematical objects and second-level concepts such as *three-ity*.

In addition to providing an account of the cardinal numbers, the neo-Fregean construction provides answers to the epistemological questions that began this essay. Unsurprisingly, the answers the neo-Fregean provides are quite similar to, although not quite the same as, the answers provided by Frege.

First, on the neo-Fregean picture, as on the Fregean one, knowledge of the abstract objects that are the subject matter of arithmetic – that is, the cardinal numbers – does not require us to have some sort of quasi-mystical rapport with causally inefficacious abstract objects. Instead, such knowledge is, in the final analysis, merely knowledge about the first- and second-level concepts since (HP) allows us to reconstrue claims about cardinal numbers as claims about the concepts whose cardinal numbers they are. In a certain sense, as in Frege's original construction, this is the whole point of the project: to deal with the epistemological issues plaguing platonism by drawing fundamental connections between knowledge about abstract objects and knowledge about concepts.

Of course, on the neo-Fregean picture, the sort of knowledge required in order to have the corresponding knowledge about cardinal numbers is not merely knowledge of logical truths (plus explicit definitions), as it was on Frege's picture. Nevertheless, knowledge of arithmetic is still, according to the neo-Fregean, merely knowledge of logic plus (acceptable) implicit definitions.

Second, on the neo-Fregean view, arithmetical knowledge has the special status that it does in virtue of the fact that it follows from logical knowledge plus implicit definitions: in particular, that it follows from standard second-order logic plus (HP). Thus, arithmetical knowledge inherits whatever special epistemological advantages are enjoyed by standard second order logic and by (HP).[30]

Finally, for the neo-Fregean, the epistemology of arithmetic is just a part of the larger epistemology of logic plus definitions. Since both standard second-order logic and the implicit definitions accepted by the neo-Fregean are completely general and everywhere applicable, applications of arithmetic are topic neutral and everywhere applicable as well.

It is worth fleshing out these last aspects of the epistemology of the neo-Fregean view in a bit more detail. Recall that, for Frege, the reduction of arithmetic to logic was merely a means to an end – Frege's important target was not showing that arithmetic was reducible to logic, but was instead showing that arithmetic was *a priori* and absolutely general. The reduction to logic was merely a means to this end. The neo-Fregean can follow Frege's lead here, since the goal is to address the pressing epistemological problems that platonist accounts of mathematics face. As a result, although we have, throughout this essay, characterized the neo-Fregean account as one where arithmetic can be reduced to logic plus (implicit) definitions, where the term "logic" was understood to refer to standard second-order logic, the neo-Fregean does not need these second-order formalisms to turn out to be "logic" in the strict sense. All the neo-Fregean needs, in the end, is that the "logical consequences" and "logical truths" coded up in standard second-order systems be *a priori* knowable, completely general, and everywhere applicable. It is at least conceivable that second-order logic could be all of these things without being logic. Thus, it is not clear that the neo-Fregean needs to get bogged down by well-known arguments over the logical status of second-order logic such as those discussed in Shapiro (1991).

4   Conclusion

The important similarities between Frege's logicism and contemporary neo-Freganism are not the obvious technical parallels, but instead concern the shared epistemological project: the task of dealing adequately with the epistemological puzzles that arise on a face-value, platonist reading of arithmetic. In addition, once we understand the neo-Fregean's critical move – the abandonment of BLV and the corresponding liberalization of the notion of acceptable definition – as one that allows him to retain the central Fregean idea that arithmetic can be reduced to logic plus definitions, we see that neo-Fregeanism is a species of logicism after all.

Notes

1. I follow what has come to be semi-standard practice of using "platonism" with a lower-case "p" to refer to the view that mathematical theories are about collections of abstract objects, or structurally complex abstract objects, and reserve "Platonism" for discussions of Plato's actual views, which presumably have something much like platonism as a part.
2. Okay, almost never. I did flirt with nominalism for a few months as an undergrad after reading too much Quine.
3. This last question was famously raised by Paul Benacerraf in "Mathematical Truth" (1973).
4. For someone like myself whose intuitions point so strongly towards platonism as the correct account of mathematical ontology, plausible arguments against platonism are exactly as disturbing as arguments for platonism are uninteresting.
5. I have been as dishonest as anyone else about this, even in print: In the past, I have railed against the misapprehension that neo-Fregeanism was a species of logicism, even going so far as to suggest that the term "neo-Fregeanism", and especially the alternative appellation "neo-logicism" should be abandoned. See, for example, page xvi of the introduction to *The Arché Papers on the Mathematics of Abstraction*, Cook (2007).
6. The reader interested in pursuing these topics in more detail should consult Hale and Wright (2001), MacBride (2003), and the papers collected in Cook (2007).

7. Many of the papers collected in Cook (2007) concern extending the methods of neo-logicism to other mathematical theories – particularly set theory and analysis.
8. This passage occurs in a discussion of the ontological proof – hence the reference to the existence predicate. Nevertheless, there is nothing here that suggests that Kant's belief that existential claims are synthetic is not completely general.

Interestingly, one of George Boolos' most well-known and influential criticisms of neo-Fregeanism is merely a reworking of this Kantian thought in modern terminology:

> ... it was a central tenet of logical positivism that the truths of mathematics were analytic. Positivism was dead by 1960 and the more traditional view, that analytic truths cannot entail the existence either of particular objects, or of too many objects, has held sway since. (1997, p. 305)

9. Of course, the connections between Kant's notions of logic and analyticity and Frege's understanding of these same notions are much more complicated than this brief summary suggests. For a more thorough discussion of these issues, the reader is urged to consult MacFarlane (2002).
10. Here and below I translate Frege's formulas into modern notation.
11. The definition given in the *Grundlagen* is ambiguous, as it can be read as taking a cardinal number to be the extension of a first-level concept holding of extensions (as we interpret it here), or it can be interpreted as taking a cardinal number to be the extension of a second-level concept holding of first-level concepts. Frege's formal definition, provided in the *Grundgesetze*, is equivalent to the interpretation adopted here.
12. The reader is urged to consult Heck (1993), and Boolos and Heck (1998) for careful examinations of how much of Frege's formal work can be reconstructed within consistent subsystems of Frege's logic.
13. It is worth noting that late in his life, after the discovery of Russell's paradox, Frege clung to the view that there are three sources of knowledge: empirical, logical, and geometrical/temporal. See Frege (1979a).
14. It should be noted that there is some disagreement regarding exactly what Frege's constraint amounts to (see e.g. Hale (2000), Wright (2000) for two options). The present reading is a relatively weak interpretation of Frege's intentions.
15. The translation of this passage is taken from Wright (2000).
16. It is important to understand the appropriate force of Frege's Constraint: The claim is that we *can* apply arithmetic (or analysis) to any collection of objects, not that we *should*. In other words, applying arithmetic to a particular plurality is always possible, although it might not be the most fruitful manner in which to investigate those objects.
17. Of course, this assumes that there is no problem in identifying extensions themselves. Frege, of course, famously believed that extensions are immune to the Caesar Problem, but it is unclear why. For a detailed discussion of this issue, see Heck (2005).
18. It should be noted that the level of satisfaction one receives from this explanation of the epistemology of arithmetic will depend on one's views about concepts, and in particular, on whether or not one views concepts as something shared and accessible in a manner in which abstract objects such as cardinal numbers are not.

19. Of course, in reality there is no such thing as the "official" Hale and Wright view anyway, since these prominent proponents of neo-Fregeanism seem to disagree on a number of important points, including the role of Frege's Constraint.

20. The parallel with Boghossian's (1997) discussion of the difference between Frege-analytic versus Carnap-analytic is intentional.

21. Here lies one of the first possible points of disagreement between the present formulation of neo-Fregeanism and the account(s) given by Hale and Wright in (2001) and elsewhere: It is not clear that Hale and Wright believe that all analytic truths are obtained solely from logic plus definitions (although they would agree, presumably, that anything obtainable from logic and definitions is analytic). Nothing substantial hinges on this terminological point, however.

22. Of course, a much more detailed story needs to be told regarding why (some) abstraction principles can play the special epistemological role that the neo-logicist would have them play. MacBride (2003) contains a thorough discussion of the current state of play. I will not attempt to address this issue directly here, although much of the material at the end of the chapter on Bad Company has a clear, albeit indirect, bearing on the issue.

23. From a methodological point of view this technical turn is understandable. Often, when faced with a problem that involves both settling on a technical solution to some problem and providing a philosophical account of that solution as correct, the easiest place to start is with the technical side of the problem, since often the formal work is both a good deal easier than the philosophical work and, once carried out, can serve as a useful guide for how the philosophical account should proceed. This does not seem to be the case in the present instance, however.

24. An abstraction principle AP is stable if and only if there is some cardinal κ such that AP has models of size γ, for all $\gamma \geq \kappa$. Stewart Shapiro has recently observed that the proof that irenicity and stability are equivalent, given in Weir (2004), is faulty, and no straightforward repair is forthcoming.

25. This is not to say that stability (or perhaps irenicity) is not the correct criteria. The point, rather, is that we seem to have no good philosophical reasons for thinking that stability is the correct criteria (since mathematical elegance falls short of being a philosophical reason). In fact, I firmly believe that stability is, in fact, the correct criteria, and I am, as a result, busy attempting to formulate a philosophical, epistemological argument to that effect.

26. Some ways of characterizing the logical/non-logical divide, such as those that draw the line in terms of permutation invariance, are clearly motivated by some-thing like the "completely general and everywhere applicable" thought. Thus, the problem becomes one of adapting such criteria for logicality to the case at hand – acceptable abstraction principles. Doing so is non-trivial, although Fine (2002) contains much promising work, both technical and philosophical, along these lines.

27. Of course, it is important to note that on the neo-Fregean picture, the definitions in question typically will entail various existential claims regarding the objects being defined.

28. Here lies a point where Crispin Wright, but not Bob Hale, might object to my characterization of neo-Fregeanism since *Frege's Constraint* seems to play a much less central role in his development of the view. Compare the treatment of the issue in Hale (2000) and Wright (2000).

29. Of course, the claim here need not be that all applications of arithmetic involve these second-order concepts in a fundamental manner since we can apply

arithmetic in all sorts of "weird" ways. The point is, rather, that straightforward, "natural" applications of arithmetic fundamentally involve such second-order concepts.

30. Of course, in the end, some story will need to be told about exactly what sort of special epistemological status is enjoyed by (HP), whether that status is the same from the special epistemological status of logic, and, if not, what sort of epistemological status is enjoyed by claims that depend both on logic and (HP).

References

Benacerraf, P. (1973), "Mathematical Truth", *Journal of Philosophy* 70: 661–680.
Boghossian, P. (1997), "Analyticity", in Hale and Wright [1997]: 331–367.
Boolos, G. (1997), "Is Hume's Principle Analytic?", in Boolos [1998]: 301–314, reprinted in Cook [2007]: 3–15.
Boolos, G. (1998), *Logic, Logic, and Logic*, Cambridge, MA, Harvard University Press.
Boolos, G. and Heck, R. (1998). "*Die Grundlagen der Arithmetik* §82 – 83", in Boolos [1998]: 315–338.
Cook, R. (ed.) (2007), *The Arché Papers on the Mathematics of Abstraction*, Dordrecht, Springer.
Field, H. (1984), "Is Mathematical Knowledge Just Logical Knowledge?" *The Philosophical Review XCIII*: 509–552.
Fine, K. (2002), *The Limits of Abstraction*, Oxford, Oxford University Press.
Frege, G. (1893, 1903), *Grundgezetze der Arithmetik I & II*, Hildesheim, Olms.
Frege, G. (1979a), "Sources of Knowledge of Mathematics and the Mathematical Natural Sciences", in Frege [1979b]: 269–270.
Frege, G. (1979b), *Posthumous Writings*, Chicago, University of Chicago Press.
Frege, G. (1980), *Die Grundlagen der Arithmetic*, J. Austin (trans.), Evanston, IL, Northwestern University Press.
Frege, G. (1997), *The Frege Reader*, M. Beaney (ed.), Oxford, Blackwell.
Hale, R. (2000), "Reals by Abstraction", *Philosophia Mathematica* 8: 100–123, reprinted in Cook [2007]: 175–196.
Hale, R. and Wright, C. (1997), *A Companion to the Philosophy of Language*, Oxford, Blackwell.
Hale, R. and Wright, C. (2001), *The Reason's Proper Study*, Oxford, Oxford University Press.
Heck, R. (ed.) (1997), *Language, Thought, and Logic: Essays in Honour of Michael Dummett*. Oxford, New York: Oxford University Press.
Heck, R. (1993), "The Development of Arithmetic in Frege's *Grundgesetze der Arithmetik*", *Journal of Symbolic Logic* 10: 153–174.
Heck, R. (2005), "Julius Caesar and Basic Law V", *Dialectica* 59: 161–178.
Kant, I. (1965), *The Critique of Pure Reason*, Norman Smith (trans.), New York, St Martins Press.
MacBride, F. (2003), "Speaking with Shadows: A Study of Neo-Fregeanism", *British Journal of the Philosophy of Science* 54: 103–163.
MacFarlane, J. (2002), "Frege, Kant, and the Logic in Logicism", *The Philosophical Review* 111: 25–65.
Quine, W. (1960), "Two Dogmas of Empiricism", *The Philosophical Review* 60: 20–43.
Russell, B. (1902), "Letter to Frege", in van Heijenoort [1967]: 124–125.
Shapiro, S. (1991), *Foundations Without Foundationalism: The Case for Second-order Logic*, Oxford, Oxford University Press.

van Heijenoort, J. (1967), *From Frege to Gödel: A Sourcebook in Mathematical Logic*, Cambridge MA, Harvard University Press.

Weir, A. (2004), "Neo-Fregeanism: An Embarassment of Riches", *Notre Dame Journal of Formal Logic* 44: 13–48, reprinted in Cook [2007]: 383–420.

Wright, C. (1997), "On the Philosophical Significance of Frege's Theorem", in Heck [1997]: 201–244, reprinted in Hale and Wright [2001]: 272–306.

Wright, C. (1999), "Is Hume's Principle Analytic?", *Notre Dame Journal of Formal Logic* 40: 6–30, reprinted in Hale and Wright [2001]: 307–333 and in Cook [2007]: 17–43.

Wright, C. (2000), "Neo-Fregean Foundations for Real Analysis: Some Reflections on Frege's Constraint", *Notre Dame Journal of Formal Logic* 41: 317–334, reprinted in Cook [2007]: 253–272.

2
Reducing Arithmetic to Set Theory

Alexander Paseau

The revival of the philosophy of mathematics in the 60s following its post-1931 slump left us with two conflicting positions on arithmetic's ontological relationship to set theory. W.V. Quine's view, presented in *Word and Object* (1960), was that numbers are sets. The opposing view was advanced in another milestone of twentieth-century philosophy of mathematics, Paul Benacerraf's "What Numbers Could Not Be" (1965): one of the things numbers could not be, it explained, was sets; the other thing numbers could not be, even more dramatically, was objects. Curiously, although Benacerraf's article appeared in the heyday of Quine's influence, it declined to engage the Quinean position squarely, even seemed to think it was not its business to do so. Despite that, in my experience, most philosophers believe that Benacerraf's article put paid to the reductionist view that numbers are sets (though perhaps not the view that numbers are objects). My chapter will attempt to overturn this orthodoxy.

1 Reductionism

Benacerraf observes that there are many potential reductions of arithmetic to set theory. The most familiar are the "Ernie" (von Neumann) interpretation, which takes 0 as the empty set and the successor of x as $x \cup \{x\}$, and the "Johnny" (Zermelo) interpretation, which also takes 0 as the empty set but the successor of x as $\{x\}$—with New World irreverence, the set theorists' first names have been switched round and Americanised. No two such accounts can be correct (since e.g. 2 cannot be equal to both $\{\emptyset, \{\emptyset\}\}$ and $\{\{\emptyset\}\}$), and since there is no principled way to choose between them, "any feature of an account that identifies a number with a set is a superfluous feature of the account (i.e. not one that is grounded in our concept of number)" (Benacerraf 1998, p. 52). Hence, numbers are not sets.

We shall defend reductionism from Benacerraf's and other arguments. Our main defence will be that speakers might not have transparent knowledge of the referents of their number terms, and that any concern with getting actual meanings right must be subordinate to the overall theoretical virtues of a proposed account of arithmetic. We first outline some of reductionism's varieties: two objectual versions, a structural version and a partial-denotation version. These reductionisms all share a commitment to an ontology that includes sets but not sui generis numbers, differing only on the semantics for arithmetic.

1.1 Objectual reductionisms

Objectual reductionism is the doctrine's classic form, and the one Benacerraf had in mind when presenting his argument. Objectual reductions associate with each number a set and interpret the arithmetical claim "$\phi_A(N_1, N_2, \ldots)$" as the corresponding set-theoretic claim "$\phi_S(S_1, S_2, \ldots)$". For example, "$N_1 > 0$" and "$N_1 + 1 \neq N_1$" are interpreted as "$\emptyset \in S_1$" and "$S_1 \cup \{S_1\} \neq S_1$" on the von Neumann interpretation. A successful reduction thus shows that numbers are sets, and that claims about arithmetic are nothing other than claims about (some) sets. Objectual reductions come in two flavours: context-independent and context-dependent. The former identifies numbers with some sets once and for all, independently of context. The latter identifies numbers with different sets in different contexts, for example, von Neumann finite ordinals in the context of one proof, Zermelo ordinals in the context of another.[1] On some people's terminology, only objectual reductionisms truly deserve the label "reductionism"—see below.[2]

1.2 Structural reductionism

Rather than picking a particular reduction (in all contexts or context-dependently), structuralists quantify over all set-theoretic omega-sequences. A semi-formal interpretation of "$0 \neq 1$", for example, would be: "for any set X, for any (set-theoretic) function S_X on X and member 0_X of X, if the Peano axioms hold of the tuple $(X, S_X, 0_X)$, then $0_X \neq S_X(0_X)$". This type of reductionism is sometimes known as set-theoretic structuralism.[3,4]

1.3 Partial-denotation reductionism

A partial-denotation reduction attempts to combine objectual semantics with the idea that arithmetical terms can stand for the entities they would stand for in the von Neumann objectual reduction *and* the Zermelo one *and* ...—but only partially. A set-theoretic interpretation partially accords with arithmetical language if each term partially denotes the entity the interpretation assigns to it; and there are many such interpretations. Thus, the ω-sequence of numerals partially denotes several ω-sequences of sets.[5] Field (1974) is the original source for partial-denotation semantics, for mathematics and more generally. He argues there that "gavagai" might partially denote *rabbit* and *undetached rabbit part*; pre-relativistic tokenings of "mass" might partially denote relativistic mass as well as rest mass, and so on.

What is the motivation for reductionism? It tends to be economy, of various stripes. First, ontological economy: the reductionist believes in fewer things and, depending on her view, fewer types or fewer higher-level types of things than the non-reductionist, since she believes in sets rather than sets and sui generis numbers.[6] Second, ideological economy: reductionism reduces the number of primitive predicates our total theory employs. Third, axiomatic economy: the list of basic principles is reduced or simplified.[7] Reductionists also sometimes cite other motivations, for example that their view avoids or answers vexing inter-categorial identity questions such as whether the natural number 2 is identical to the rational number 2 or the real number 2 or the complex number $2 + 0i$.[8] Our interest is in whether reductionism can rebut Benacerraf's arguments, so we shall not evaluate these motivations but take them as given.

Reductionism as here understood is inspired by Quine but is free from Quinean idiosyncrasies. Quine developed his set theory NF as a rival to iterative set theories such as Zermelo–Fraenkel set theory with Choice (ZFC) and von Neumann–Bernays–Gödel class theory (NBG). Indeed, he believed that the only intrinsically plausible conception of set is the naïve one and that, following the discovery of the set paradoxes associated with that conception, the choice of set theory is purely pragmatic. Reductionism is not committed to this view nor to any particular reducing set theory, so long as it is stronger than arithmetic.[9] Reductionists also need not follow Quine in his scientific naturalism, which takes science to be authoritative and accordingly sees mathematics as true only to the extent that it is indispensably applied. Nor need reductionism be committed to ontological relativity, indeterminacy of translation, semantic behaviourism and holism, the thesis that first-order logic is the only true logic, and Quine's sometimes deflationary conception of ontology. The only Quinean doctrine, now widely shared, reductionists must unarguably accept is that their stated motivations are epistemic and can guide the choice of total theory.[10]

2 Reductionism's response

How do our four reductionisms handle Benacerraf's argument? The argument is aimed at reductions that aspire to reveal what number-talk meant all along, namely, meaning analyses. There is nothing in our current concept of number to favour the von Neumann over the Zermelo reduction, and since both cannot be true, neither is. An explication, in contrast, is immune to this problem. Given that set theory can do the job of arithmetic, it dispenses with a commitment to sui generis numbers. In particular, that there is nothing in our current concept of number to favour one reduction over another does not detract from such a reduction's desirability, since it is not constrained to respect our current concept. What remains is the minor issue of arbitrariness in the semantics. Let us see how this plays out for each reductionism.

2.1   Context-independent objectual reductionism

The decision to choose any reduction over any other is arbitrary. But that is a theoretical deficit so minor, perhaps even null, that it is overridden by the greater benefit of avoiding commitment to sui generis numbers, reducing the stock of one's ideological primitives, and so on. Compare the story of Buridan's ass (or rather Aristotle's man—see *De Caelo* 295b). This unfortunate creature died of inanition as a result of not breaking the tie between a bucket of water to its left and a stack of hay an equal distance to its right, at a time when it was equally hungry and thirsty. The two choices were perfectly symmetric, and there was no reason to prefer one to the other. *But* there was reason to choose one of them. Rationality counsels Buridan's ass to break the tie arbitrarily: in this situation, it is more rational to make an arbitrary choice than no choice. The story of Buridan's ass is, I hope, imaginary, but that does not matter, and if you like your examples a touch more realistic, observe that the choice faced by Buridan's ass is approximated on a daily basis by shoppers in a supermarket choosing between multiple versions of the same good; though in their case, quite reasonably, the paralysis of choice tends to be shorter-lived and less fatal.

Similarly, as the reductionist sees it, it is more rational to make an arbitrary selection between two equally good set-theoretic objectual semantics than to refuse to reduce arithmetic to set theory. The benefits of adopting some reduction or other outweigh the arbitrariness of choosing any particular one. Multiple reducibility is therefore not a significant obstacle for this kind of reductionist, just as the multiplicity of choice should not have been for Buridan's ass.

For a semantic analogy, suppose a mother of identical twin girls is delivered of them at exactly the same moment.[11] She had long ago settled on "Olympia" for her first-born girl's name and "Theodora" for the second-born; but as they were born simultaneously, she is in a pickle. What to do? An arbitrary choice is called for: say, name the baby on the left "Olympia" and the one on the right "Theodora". Given that each must have a name—what a pity to go through life nameless—it is rational for her to make an arbitrary decision. This decision is purely semantic and no "ontological" choice is involved; the mother is simply choosing one name assignment over another for some given entities, persons in this case. As above, rationality recommends arbitrarily breaking the tie with a terminological decision. Similarly, objectual reductionists should arbitrarily break the tie and fasten on to a particular assignment of numerals to sets.

One might object that the analogy between the number case and the twins case is imperfect, in particular that number terms have an established antecedent usage in a body of claims taken to be truths whereas in our example "Olympia" and "Theodora" do not. The analogy is of course imperfect—it does not hold in several respects. (Yet, notice that the twins story could be told in such a way that there is an established body of truths containing "Olympia" and "Theodora" prior to their referents' birth.) In Section 3, we shall consider a version of this objection based on the fact that number terms have a well-established antecedent use.

It goes without saying that reductionists (objectual or otherwise) need not advocate any surface change to normal mathematical practice. Their claim is not that mathematicians must have the set-theoretic reduction explicitly in mind when doing mathematics, but that they should be committed to it. It is a philosophical claim about mathematics, not a prescription about quotidian practice—though that is compatible with its also being "hermeneutic" (meaning-preserving), as we shall see.[12]

2.2 Context-dependent objectual reductionism

Here the choice of which reduction to use is seen as turning on pragmatic aspects of the context. For example, if finite arithmetic is considered as a fragment of transfinite arithmetic, von Neumann semantics is preferable to Zermelo's, since the former but not the latter extends neatly into the transfinite. For most contexts, some arbitrariness still remains in picking a particular semantics (e.g. extendability into the transfinite does not uniquely privilege the von Neumann semantics). As with the context-independent version, proponents of this view maintain that this arbitrariness is null or negligible compared to the gain in reducing arithmetic to set theory—a gain in ontological, ideological and axiomatic economy, the avoidance of inter-categorial identity problems, and whatever else motivates them.

2.3 Structural reductionism

On the surface, it seems that a structural reduction, unlike objectual ones, avoids arbitrariness. We need no longer arbitrarily choose set-theoretic representatives for the numbers; we can instead quantify over all omega-sequences. However, any structural reduction must reduce the ordered pair relation (more generally, ordered tuple relations) to set theory in some arbitrary fashion. For consider the successor relation S's set-theoretic representation: the reducing set must contain all and only ordered pairs whose first element is a member of S's domain and whose second element is the first's image under S. But there is no unique way of cashing out $\langle a, b \rangle$ in purely set-theoretic terms. One can stipulate that $\langle a, b \rangle$ is $\{\{a\}, \{a, b\}\}$, or $\{\{\{a\}\}, \{a, b\}\}$, or .... This issue recurs with the question of how to explicate higher tuples in terms of pairs. For example, is the triple $\langle a, b, c \rangle$ to be equated with $\langle a, \langle b, c \rangle \rangle$ or $\langle \langle a, b \rangle, c \rangle$?[13,14]

The structural reductionist, therefore, also accepts some arbitrariness, assuming she prefers to reduce mathematics to set theory rather than set theory plus a primitive theory of n-tuples (and, given her reductionist tendencies, she presumably does). She relocates rather than avoids the arbitrariness. However, as before, she sees any loss incurred by terminological arbitrariness as outweighed by the reduction's substantive gains.

2.4   Partial-denotation reductionism

A partial-denotation reduction is in the same boat as a structural reduction. Though it avoids choosing one particular ω-sequence of sets as the denotation of the numerals, it avails itself of ordered n-tuples, indeed of infinite-tuples, since it claims that $\langle "0", "1", "2", \ldots \rangle$ partially denotes $\langle \varnothing, \{\varnothing\}, \{\varnothing, \{\varnothing\}\}, \ldots \rangle$ and partially denotes $\langle \varnothing, \{\varnothing\}, \{\{\varnothing\}\}, \ldots \rangle$ and .... It therefore has to settle on some arbitrary reduction of the various ordered tuple relations to set theory. Once more, the resulting benefits are deemed to outweigh the arbitrariness of choice.

3   Objections

The question of which of the four reductionisms is to be preferred is not one we shall investigate. It is secondary to the ontological question of whether to be a reductionist in the first place. In the assessment of reductionism as a generic position, the debate between the various semantic proposals is internecine. Of course, if it turns out that each species is unattractive, so is the genus. However, that is not the case. As the reductionist sees it, the deficiencies associated with the most attractive reductionism are relatively minor and are outweighed by reductionism's general advantages.

Arguably, only the two objectual reductions count as reductions proper, because only they respect arithmetic's surface semantics. Even so, philosophies that reject sui generis numbers and take arithmetical truths to be about sets clearly form some sort of philosophical natural kind, which is why we have assimilated them. Having said that, we shall hereon concentrate on reductionism's first version, its classic variant, and defend it against objections.

3.1   Semantic objections

Hartry Field sketches two arguments against objectual reductions in the article in which he sets out partial-denotation semantics. The first does not apply to our case of interest,[15] so we focus on his second argument, which does explicitly address it:

> The reductivist might try to escape this difficulty [Benacerraf's multiplicity problem] by saying that it is not important to his purposes to hold that number-theoretic terms referred to these correlated objects all along;

it is sufficient (he might say) that we be able to *replace* number-theoretic talk by talk of the correlated objects. (Quine has suggested this line of escape in 1960, sections 53 and 54.) But this will not do, for it suggests that earlier number theorists such as Euler and Gauss were not referring to anything when they used numerals, and (barring a novel account of mathematical truth) this would imply the unpalatable conclusion that virtually nothing that they said in number theory was true. (1974, p. 214)

To spell this out, suppose upon reading *Word and Object* I decide that from this day forth the only mathematical entities I shall commit myself to are sets. In particular, I arbitrarily decide that from now on "0" in my mouth denotes Ø, "1" the set {Ø}, "2" the set {Ø, {Ø}}, and so on. Thus, I am an objectual reductionist, of the context-independent stripe, who arbitrarily chooses the von Neumann reduction to avoid commitment to sui generis numbers. As explained, the multiple reducibility objection does not undermine my position. But now I am faced with a problem of how to interpret *others*: other mathematicians, past and present, and past selves (myself before reading *Word and Object*). My semantic decision cannot be binding on them—the meaning of my words now is to some degree up to me, but the meaning of their words is not—and there is nothing in their concept of number to prefer one set-theoretic interpretation over another. Thus, it seems that their arithmetical talk cannot be interpreted as referring to sets. It should instead be interpreted as intending to refer to sui generis numbers, and so by my lights it is founded on a mistaken presupposition. Depending on my theory of presupposition, then, I must construe their arithmetical statements as false or truth-valueless; either way, not true. But surely it is absurd to suppose that others' utterances of "$1 + 1 = 2$" are not true.

Now, this consequence does not seem to me fatal for a philosophy of mathematics. For example, an error theory that declares that our mathematical discourse presupposes platonism yet that platonism is false, with the result that "$1 + 1 = 2$" as uttered by most people though mathematically correct is literally false, does not seem to me to be ruled out of court. Indeed, such an error theory seems to be Field's own position in the later *Science Without Numbers* (1980). Moreover, if Field (1974) is right, a partial-denotation reduction overcomes the difficulty.

However, a stronger response is available. Suppose that, following my decision to adopt a certain set-theoretic semantics for arithmetical language, I interpret others—Euler, Gauss and so on—in the very same way. On this semantics, Gauss's statement of the quadratic reciprocity theorem comes out true. What are the objections to this? One could accuse me of putting words in Gauss's mouth. His proof was about prime numbers, not sets. But if I think that prime numbers just are sets, I am right to think that his proof was about sets after all. Speakers do not know all the properties of their subjects of discourse. When the ancient Greeks spoke about the sun, they spoke, unknowingly, about a hydrogen–helium star that generates its energy by nuclear fusion. The ancient Egyptians and Mayans were not only ignorant of the sun's properties, which they worshipped; they were also wrong about them. In general, it is not attractive to suppose that a speaker cannot refer to something if she lacks knowledge of some of its properties or is wrong about a few of them. For familiar reasons not worth rehearsing, this contention represents a wrong turn in semantics.

The objection thus trades on the fact that "about" in the locution "*S* is talking about *X*" has a direct and indirect sense. In the direct sense, the statement is true only when *X* is presented in a way that *S* herself is familiar with or indeed *S* herself has used (in the latter case, the indirect discourse represents the direct discourse *verbatim*). For example, this is the sense in which, if *S* utters "Prince Albert was married to Victoria" but is foreign to London, the report "*S* is talking about Prince Albert" would be true though "*S* is talking about the man to whom the Hyde Park memorial is dedicated" would be false. In the indirect sense, in contrast, we can truly say that *S* is speaking of *X*, even if $\ulcorner X \urcorner$ is a way of picking out *X* that *S* neither used nor was familiar with, as in the Hyde Park memorial example. It is in this indirect sense that Gauss was talking about sets.[16]

Field's argument was that even if there is no first-person multiplicity problem—I can break the tie arbitrarily—there remains a third-person multiplicity problem. Yet, once my set-theoretic reduction is on the table, there is a genuine reason for preferring it to others, namely that it accords with my interlocutor's wish to mean by her number words what others, myself included, mean by them, and her wish for her arithmetic discourse to be interpreted as true.[17] Following the arbitrary choice of an explication for myself, a non-arbitrary reduction for others presents itself. Breaking the first-person symmetry also breaks the third-person symmetry.

Now, I don't deny that there are mathematicians who do not intend, explicitly or implicitly, to mean by their number talk whatever others mean by it. Nor do I deny that there are a few self-conscious arithmetical sui generists who cling to their interpretation so tenaciously that their arithmetical discourse cannot be interpreted according to whatever is the reductionist's understanding. These stridently non-deferential arithmeticians cannot be understood in this way. From the reductionist perspective, only an error theory is right for them, coupled with a pragmatic story distinguishing their mathematically correct utterances from the incorrect ones. But it should be obvious that very few of the billions of arithmeticians, amateur and professional, fall into these categories. For all but the most philosophically *parti pris* mathematical users, the intention to say something true (when uttering something mathematically correct) overrides any intention that their arithmetical discourse be interpreted along particular metaphysical lines. As long as the speaker has the intention to be interpreted truthfully, and this intention overrides any intentions to put a certain metaphysical spin on her arithmetical discourse, the reductionist is right to construe her as referring to sets. And as long as the intention to be interpreted as one's peers is similarly present and prevailing, the reductionist is right to construe the speaker as referring to the sets the reductionist's utterances are about. That accounts for all *normal* users of arithmetic, that is, philosophically casual speakers who are willing to submit themselves to the arbitration of the linguistic community for the meanings of their words. And Field's argument was clearly based on the attractions of the thought that it is wrong to interpret a normal arithmetician as consistently uttering untruths.

This is not to say that no other interpretative constraints are in play. Euler's intentions are best described by saying that they are to express true statements and to mean what others mean by their number words when used in a sufficiently similar way to his own. But there are limits to what we can take him to mean. The eccentric interpretation that takes, say, a carrot to be the referent of "2" in Euler's mouth and a parsnip that of his "3" (and makes corresponding changes for the predicates, quantifiers and function symbols) is categorically ruled out, given others' usage. An eccentric reductionist of this kind has no choice but to be an error theorist about Euler's claims involving "2" and "3". But less eccentric reductionists need not interpret Euler's arithmetical claims error-theoretically and may respect his intended truth-values.

What if several speakers in the community make different choices? What if I interpret numerals in the von Neumann way but you, as an equally convinced reductionist who decided to break the tie in another arbitrary way, interpret them as Zermelo ordinals? Since there is no longer a privileged interpretation, doesn't the third-person symmetry objection re-arise?

It is not clear that there can be an exact tie from a given reductionist's perspective. The kind of situation envisaged is one in which A is a von Neumann reductionist and B a Zermelo reductionist, there are no other reductionists around, and A and B are identical in all other respects (none is more of an expert or more linguistically deferred to by the community, etc.). From a God's eye view (on the running reductionist assumption that there are no sui generis numbers), the situation is symmetric and there is no reason to gloss normal arithmeticians' utterances as the A-interpretation or B-interpretation. But now look at it from A's point of view: A wants to preserve as much agreement between her utterances and those of normal arithmeticians—in particular this is more important to her than preserving as much agreement between B's utterances and those of normal arithmeticians; she has chosen the von Neumann interpretation for her arithmetical utterances; if she chooses that same interpretation for others' utterances, she gets to respect the agreement constraint; and no other symmetry-breaking constraints are in play. It follows from all this that she should choose the von Neumann interpretation for others' arithmetical utterances. The key is that A, not being God, always has a reason to prefer her own interpretation in cases that are symmetric from God's point of view. The reason is her prevailing wish to mean by her number terms what most (normal) speakers mean by theirs.[18]

Of course, it's possible that A's recognition of the God's-eye-view symmetry of her interpretation with the B-interpretation could loosen her resolve to interpret normal speakers in her chosen way. She may well decide to adopt B's interpretation instead. Or, she may decide to consult with B so as to adopt a shared semantics. In general, if pressures of uniformity are sufficiently felt, reductionists might collectively opt for a particular reduction (perhaps for mathematical reasons). This could happen by fiat, as a coordinated response to this problem. But as explained, there is no inconsistency in A and B sticking to their guns in the given scenario, A maintaining the A-interpretation and B the B-interpretation.

Two points underlying our discussion bear emphasis. The first is that the proper semantics for someone's discourse may not fit all her dispositions. In interpreting a speaker's words to mean one thing rather than another, we look for the best fit, judged by certain criteria; this best fit may not be a perfect fit. Thus, it is a captious objection to a suggested interpretation of a speaker's words that it is not a perfect fit despite its respecting her prevailing semantic intentions. This point should be familiar not only from the philosophy of language but also from everyday experience of translation and interpretation. The second point is that most mathematicians, past and present, tend not to have marked opinions about whether their utterances are objectual or structural, or if objectual, whether they are about *sui generis* entities or not. Within certain constraints,[19] mathematicians are more or less happy to write a blank cheque as far as the interpretation of their mathematical utterances is concerned. In any case, what is beyond doubt is that semantically vehement mathematicians are few and far between. This gives an interpreter greater licence in construing others' claims about numbers than in construing their claims about, say, tables and chairs.

3.2   Slippery slope objection

In a later phase, generally regarded as degenerate, Quine (1976) pushed his set-reductionism to apparent breaking point by insisting that everything—physical objects as well as numbers and other mathematical entities—should be reducible to set theory. For example, chairs are sets involved in the set-theoretic relation of being sat on by persons, who are also a kind of set. This omnivorous reductionism seems to be a *reductio* of set-reductionism—a parody of philosophy, even. Surely, if there's anything we believe, it is that we are not mathematical objects. Thus, the slippery slope objection—set-reductionism about number theory slides into an absurd set reductionism about everything.

A proper assessment of this objection would involve a case by case evaluation of each reductionism's merits. We cannot undertake that here, so we content ourselves with two observations. First, no axiomatic economy is achieved by a reduction of the physical to set theory. Unlike the case of arithmetic and the rest of mathematics, the corresponding physical principles have to be imported into set theory rather than derived. And whether the reduction of the physical to set theory achieves an ideological economy is unclear since it depends on one's theory of properties. This illustrates the point that the assessment of reductionism has to be piecemeal. The economy achieved depends on the particulars of the case.

Second, omni-reductionism does not obey a general constraint on reduction. There is an entrenched semantic constraint that the interpretation of mathematical kind terms such as "set" and "number" cannot overlap with that of concrete kind terms such as "person" and "chair". To put it non-linguistically, if philosophy is the best systematisation of our beliefs, there are constraints on how that systematisation proceeds. The systematisation that takes persons to be sets strays too far from *our* beliefs to pass as their systematisation. The contrast is with the case of "set" and "number", where no such semantic constraint is in place, or perhaps only a very attenuated one. There are no similarly entrenched principles.

In general, then, there are domain-specific reasons for or against reductionism, and one cannot naively leap from number-theoretic reductionism to omni-reductionism. Whether a given reductionism is acceptable or not depends on the balance of forces. That the proper balance points to reductionism about numbers and other mathematical objects but not reductionism about persons is prima facie consistent, as the case for person to set reductionism has yet to be made. In sum, yes, reductionist pressures, once given in to, do point towards omni-reductionism. But no, they might not take us to that unpalatable end point, because they might well be weaker than the pressures behind number-to-set reductionism, and because there are countervailing reasons against taking reductionism that far.[20]

3.3 Epistemological objection

The objection is simply that arithmetic cannot be reduced to set theory because the epistemology of arithmetic is more secure than that of set theory.

The response to this is straightforward: on the reductionist view, arithmetic is a branch of but not the whole of set theory.[21] The epistemology of arithmetic is accordingly the epistemology of a branch of set theory, not that of set theory as a whole.

On a traditional conception of a foundation for mathematics (held by Euclid and indeed by almost everyone else until the twentieth century), the foundation upon which mathematics is built is self-evidently acceptable to all rational beings. Reductionism need not go hand in hand with this kind of foundationalism: one can accept the former and reject the latter. Different reductionists have different stories to tell about what justifies set theory; yet none of them need buy into the view that set theory is self-evident or indeed more credible than the branches of mathematics it reduces. In fact, foundationalism may be turned on its head. Since all the branches of mathematics can be modelled in set theory, set theory inherits all their uncertainties. That ZFC entails the interpretation of branch B in set theory, B^s, means that any reasons to doubt B's consistency are inherited by ZFC. Hence, no logically omniscient subject can assign greater credibility to ZFC's consistency than to B's (assuming that $\text{Con}(B)$ and $\text{Con}(B^s)$ have the same degree of rational credibility and that the proof of ZFC's entailment of B^s is transparent, as it typically is). Conversely, good reason to believe in the consistency of set theory constitutes good reason to believe in the consistency of other mathematical theories.

In sum, set theory is the least likely branch of mathematics to be consistent. It does not claim to be more credible than the other branches of mathematics and it fails to meet the foundationalist ideal of self-evidence. That is compatible with its serving as the single theory into which all of mathematics is reduced. Reductionism is not foundationalism.

4 Meaning analysis versus explication

For each of the four reductionisms, the multiplicity problem, fatal for meaning analyses, is converted into a problem of terminological arbitrariness. I have explained that from the reductionist point of view this problem is trifling: given some labels and some objects, it is a question of how to assign the labels to the objects. It is rational for reductionists to choose some assignment, arbitrarily. This, I take it, was Quine's view of the matter, though to my knowledge he never responded to Benacerraf's article.[22] We then sketched why other objections to context-independent objectual reductionism, reductionism's classic form, do not succeed.

It was not lost on Benacerraf that his arguments appear not to threaten the Quinean stance. In a retrospective article on "What Numbers Could Not Be", he wrote

> Even if the realistically driven reductionist is undermined by *l'embarras du choix*, not so with the holistic Occamite, who is not beholden to any notion of "getting it right" that transcends the best theory that survives ontic paring. That is what keeps Quine, Field, and their friends in business. (1998, p. 56)

However, as I understand Benacerraf, although he recognises that his arguments may not have undermined the Quinean stance, he nevertheless objects that that stance is not worth taking.[23] Quine is interested in what he calls an "explication" of arithmetic to set theory rather than a meaning analysis (1960, pp. 258–259). He tells us not what our arithmetic is about but what it should be: his philosophy of mathematics is prescriptive rather than hermeneutic. As long as set theory can do the job of arithmetic, one can dispense with commitment to sui generis numbers; what matters is serviceability. On Benacerraf's view, however, the philosophy of mathematics should be hermeneutic: it should tell us what *our* theories, with their current meanings, are about, and how we know them. Philosophy of mathematics is either about our mathematics—arithmetic, group theory, analysis, and so on, as uttered in mathematicians' mouths—or it is about nothing. Hence, a meaning analysis, "appears to be the only line of inquiry that seems at all sensitive to arithmetical practice—the context of use of the arithmetical expressions" (1998, p. 57); the analysis of number was "constrained as an exercise that aimed, following Frege, at identifying 'which objects the numbers really were'" (1998, p. 48); and, decisively, "our primary philosophical task is hermeneutic" (1998, p. 49 fn. 18). This explains why a key premise of the argument is the statement quoted in Section 1, now with italics marking the appropriate emphasis, "any feature of an account that identifies a number with a set is a superfluous feature of the account (i.e. not one that is grounded in *our concept* of number)" (1998, p. 52). In short, Benacerraf attempts to wrong-foot reductionism by urging that though an explication may overcome the multiplicity problem, the philosophy of mathematics should not primarily concern itself with explications.

There is a quick response to this metaphilosophical objection: *our proposed reductions are in fact hermeneutic*. What goes for Euler and Gauss goes for other arithmeticians too: as explained, the objectual reductionist, say of the von Neumann stripe, may take other normal speakers of arithmetic to refer to the von Neumann finite ordinals. By definition, a normal speaker of arithmetic is anyone in the reductionist's linguistic community whose intentions to refer to whatever others refer to by their number terms in this community and to speak truthfully when uttering mathematically correct claims override his or her metaphysical views about the numbers (and nothing else undermines these intentions). As a matter of empirical fact, almost all speakers of arithmetic are normal. Thus, (objectual) reductionism as articulated here does *not* preach a think-with-the-learned-but-speak-with-the-vulgar sermon, recommending surface verbal agreement combined with semantic disagreement. It is compatible with the claim that reductionists mean the very same thing by their number discourse as normal arithmeticians mean by theirs.

Should explicative reduction turn out not to be hermeneutic, however, a further line of defence is in place. Even if an explication does not honour our pre-theoretic beliefs about numbers, it might still be our best theory of numbers, and we should not be any more beholden to pre-theoretic beliefs in arithmetic than elsewhere (e.g. natural science). Benacerraf's objection is misconceived precisely because it imposes such a constraint.

To spell out this second line of defence, call our current total theory T^{actual}, and suppose for argument's sake that T^{actual} is different from the reductionist theory T^{red}, perhaps because our actual arithmetical discourse makes an indefeasible commitment to *sui generis* numbers. (These assumptions will apply throughout the rest of this section.) The hermeneutic constraint is then that a philosophy of mathematics that issues in a theory T is correct if and only if $T = T^{actual}$. Since $T^{red} \neq T^{actual}$, it follows that T^{red} is not an acceptable philosophy of mathematics.

The objection to this constraint is straightforward. If T^{red} is agreed to be theoretically superior to T^{actual}, surely it is irrational to refuse to replace the latter with the former. And as the reductionist sees it, T^{red} *is* in fact superior to T^{actual}. We may well be interested in the question of what various statements of our actual mathematics mean—that is, in determining what T^{actual} is—just as we might be interested in the question of what the various statements in phlogiston theory mean. But that does not imply that we should reject reductionism. If it is demonstrated to phlogiston theorists that oxidation theory is scientifically superior to phlogiston theory, they should accept the theory of oxidation.[24] Hence, it seems that the only way to engage with the reductionist is to argue that her theory is not the best one.

A direct challenge to T^{red}'s theoretical superiority might for instance challenge (global) ontological economy's claim to be a principle of theory choice or that of other reductionist motivations. Now, since my aim here is not to assess the motivations for reductionism but rather to explicate it and to answer the multiplicity objection against it, I shall set such questions to one side. Our interest is in investigating whether *if* one accepts the reductionist principles one should be worried by Benacerraf's multiplicity objection. The interest of this investigation naturally turns on there being something to be said for the reductionist principles, otherwise the present chapter would be an exercise in strengthening a corner of logical space no one should be interested in occupying. As a matter of fact, the majority of philosophers do find a criterion of ideological economy attractive, and would prefer, for example, a theory with a single primitive predicate to one with 513, even if the two theories are spatiotemporally equivalent. Mathematicians also tend to prize theories with fewer primitives, which points to T^{red} being mathematically superior to $T^{non-red}$ and arguably scientists do so too. Ditto for axiomatic economy, and perhaps ontological economy as well (though this is more controversial).[25] Moreover, it is a good question to ask those who dismiss economy principles whether they think there are *any* norms of inquiry other than empirical adequacy. If you take this line, it is hard to see why, for example, you should prefer the theory of relativity to the "theory" which simply lists the theory of relativity's empirical consequences. Though I have not attempted to appease those who see the reductionist motivations as muddled, I hope these necessarily brief remarks point to some of the difficulties associated with that perspective.

One way in which a fan of meaning-analysis-only might try to defeat the reduction-as-explication view is to say that any reductionist theory must be inferior to our actual non-reductionist theory because it lacks the theoretical virtue of expressing the statements of our actual arithmetic. The importance of this factor in theory choice is questionable—can conservatism have that much weight?—but in any case the objection misconstrues the present dialectical context. The dialectical stage at which our discussion takes place is when it has been provisionally accepted that T^{red} (the theory that sets but not sui generis numbers exist) is theoretically superior to $T^{non-red}$ (the theory that both exist) and the multiplicity problem is then pressed. However, to argue that T^{red} cannot be judged superior to $T^{non-red}$ because $T^{non-red} = T^{actual}$ is to put into question the reductionist motivations by denying that T^{red} is superior to $T^{non-red}$ *modulo* the multiplicity problem. This denial is no part of Benacerraf's understanding of his own argument, or else "What Numbers Could Not Be" would have contained sustained arguments against principles of economy and other reductionist motivations, or arguments in favour of strong conservatism as a methodological principle.

A similar reply can be made to the objection that how good a theory is as a theory of numbers depends on its being a theory about *the numbers* rather than something else, and that T^{red} is a theory about sets but not *sui generis* numbers. This objection also misconstrues the present dialectical context. It has already been granted that a reductionist theory is superior to any non-reductionist theory modulo the multiplicity problem. After all, this problem is supposed to hit you if you think T^{red_1} superior to $T^{non-red}$, T^{red_2} superior to $T^{non-red}$, and so on; you then notice that no two of these reductionist theories can be true and, seeing no substantive reason to prefer any particular one, conclude that none is true. The multiplicity problem was never intended to apply to you if you did not think that T^{red_1}, T^{red_2}, and so on were better than $T^{non-red}$ in the first place. If you are already in that position, Benacerraf's argument is superfluous.

I have heard it said (compare one of the quotations from Benacerraf) that reductionism does not answer the question, "but what are numbers *really*?". Not so. Reductionism as here developed does in fact answer that question: it says that numbers really are sets. Or, to put it metalinguistically, that the numerals really denote sets. To many, there is a whiff of conventionalism about reductionism. Yet far from being conventionalist, reductionism as articulated here is a form of realism. Its conventional element lies only in the decision of which labels to use for various parts of our ontology.[26] Once we have settled on some principles of theory choice (say empirical adequacy, principles of economy, etc.), the ontology we should rationally accept is no longer up to us but is determined by these principles.[27] Compare an analogous question about the earlier twins example: "who is Olympia *really*?". The fact that her mother made an arbitrary choice about the denotation of

"Olympia" at birth does not imply some sort of anti-realism about Olympia; the arbitrarily chosen baby *really* is Olympia.[28]

Finally, I turn to two considerations Benacerraf himself offers in passing against Quine (1998, p. 56). He first rhetorically asks what the standpoint is from which we judge our total theory. The answer is surely from the standpoint of our current theory (where else?). We make judgments of the relative superiority of two theories T_1 and T_2 (one of which may be our current theory) using our current theory. For example, if we accept a qualitative version of ontological economy, which enjoins us to minimise kinds, then if according to our current theory T_1 is committed to 159 kinds of things but T_2 to only 23, it follows that T_2 is preferable to T_1 in this respect. Likewise with other theoretical desiderata. We use our current theory to arbitrate between the potential total theories under our purview, and pick whichever it sanctions as best. How to weight and aggregate various respects to return a final verdict is of course a tricky business, but theory choice is not a simple algorithmic process on anyone's view. That we are dealing with global rather than local theories may make the assessment more difficult, but I have not come across any cogent arguments to the effect that global theory choice is impossible, and Benacerraf himself does not appeal to any such argument.

Benacerraf's second consideration is that two total theories could pass the tests equally well. What if theory T is just as theoretically virtuous as T'? In that case, the multiplicity problem seems to re-arise, since there is no reason to prefer T over T' or vice-versa.

A first response is that it must be rational to adopt one of the following strategies. Arbitrarily pick one of the two total theories T and T'; alternate between T and T' (depending on context); give each of T and T' 50% credence.[29] Arguably, the last response is correct. But there is no need to privilege any of these here: our claim is only that at least one of them is rational. For what other options are there? Accepting the next best theory in order to overcome the multiplicity problem is surely irrational. By assumption, there are two other theories superior to this third-best theory and it cannot be rational to adopt a knowingly worse theory in preference to a better one. Moreover, the alleged multiplicity problem could re-arise for the next best theories: perhaps there are two third-equal-best theories, and so on. Accepting no theory whatsoever in case of a tie at the top is also wrong; as with the previous response, it turns double success into failure.

Secondly, everyone faces this problem, whatever their philosophy of mathematics. In the absence of a guarantee that this situation could never arise in the empirical domain, the philosophy of science must in any case deal with this issue. Whatever the correct answer in the empirical case applies to that of a total theory. Historical attempts to claim that there could not be two equally successful yet different scientific theories—the most famous of which was the positivists' claim that two observationally equivalent theories would in fact be the same—have been unpromising.[30] If successful, such arguments would anyway presumably show that the problem does not arise for mathematics either.

Thirdly, the issue also arises for a meaning analysis. What if the two best meaning analyses of our current arithmetic conflict yet are equally supported? In other words, what if our arithmetical discourse is indeterminate between two (or more) interpretations? There is no reason to think that the alleged problem cannot arise for meaning analyses.

Finally, by way of pragmatic sop, we are unlikely to be faced with this problem for long. History suggests that whenever two rival mathematical or scientific theories are roughly on a par, one of them eventually wins out. Indeed, it teaches that the pressure for uniqueness is such that if a tie is on the horizon, our standards are likely to evolve to ensure uniqueness. This is a contingent reassurance, of course, but soothing nonetheless.

I conclude that Benacerraf has not pointed to any compelling metaphilosophical grounds for rejecting explications, and, so long as philosophy is primarily in the business of coming up with a best total theory, there do not seem to be any such grounds. And incidentally, as the reductionist sees it, the objectual reductionism articulated above is in fact hermeneutic, as explained. This completes reductionism's answer to "What Numbers Could Not Be". Reductionism is not damaged by the availability of incompatible reductions.[31]

Notes

1. Quine is best interpreted as a context-dependent objectual reductionist (1960, p. 263).
2. If concepts are distinguished from their extensions, then objectual reductionists need not claim that our arithmetical *concepts*, either pre- or post-reduction, are set-theoretic. For example, context independent objectual reductionists can maintain that the concept of zero is different from the concept of the empty set even if they have the same extension.
3. Do not confuse this with the view that set theory itself is to be interpreted structurally.
4. Did Benacerraf himself adopt structural reductionism in "Way out", the punnily titled final section of his "What Numbers Could Not Be" (1965)? It is hard to tell for several reasons. (i) That section is obscure and its argument can be interpreted in many ways. See Wetzel (1989). (ii) Benacerraf does not seem to privilege set theory over number theory, so parity of reasoning would then seem to commit him to a structural interpretation of both, which is definitely not the structuralism proposed above. (iii) As we will see, structural reductions have to choose between taking ordered tuple relations as primitive or arbitrarily reducing them to sets, an analogue of his own multiplicity objection against objectual reductionism. Benacerraf does not so much as hint at this problem, which suggests that this is not what he had in mind. (iv) Benacerraf (1998, p. 50) later signalled that "Way out" was primarily intended as meaning "crazy" rather than "solution". Thus,

though his paper inspired later structuralisms, it is hard to say with confidence that structural reductionism is Benacerraf's preferred way out.

5. For brevity, I speak here and elsewhere of numerals or numerical terms denoting sets. Of course numerical predicates, function symbols and quantifiers are also to be understood as their set-theoretic counterparts.

6. For reductionists who take the sets corresponding to the domain of arithmetical discourse to be a distinct kind, a qualitative economy of higher-level kinds is achieved via the kind inclusion of natural numbers to sets. For reductionists who do not, the kind *number* has been eliminated. For Quine's related distinction between what he calls explication and elimination, see (1960, §§53–5, esp. p. 265).

7. Reduction and/or simplification may result even if the size of the axiom set is unchanged, for example, (i) by accepting (countably many) instances of one schema rather than (countably many) instances of two schemas, (ii) by accepting only A rather than A ∧ B.

8. "A reduction clears out Plato's slum, ridding it of its 'primeval disorder'" Quine (1960, p. 267). Briefly, for the objectual reductionist, all these entities are sets and the identity claims are settled by the various chosen semantics (which could be fairly arbitrary, see below) for the theory of natural numbers, rationals, reals and complex numbers. For the structuralist, a claim such as "$2_\mathbb{N} = 2_\mathbb{Q}$" would be understood along the lines of "for any system of sets $S_\mathbb{N}$ doing the job of the natural numbers and any system of sets $S_\mathbb{N}$ doing the job of the rationals, $2_{S_\mathbb{N}} = 2_{S_\mathbb{Q}}$", which is false. Likewise this statement's negation "$\neg 2_\mathbb{N} = 2_\mathbb{Q}$" is also false under the structuralist interpretation. As the structural reductionist sees it, this is not a worry since mathematics is not firmly committed to bivalence for intra-theoretic statements of this kind. (See fn. 15 for a related point.) The case of the partial denotation reductionist is similar to that of the structural reductionist, *mutatis mutandis*.

9. Or, perhaps so long as it is at least as strong as arithmetic. In that case, the reduction of arithmetic to set theory might not be preferable to that of set theory to arithmetic.

10. Benacerraf seems to think that (confirmatory) holism is an essential part of reductionism (1998, pp. 54–56), but this is not clear. If holism is required, it need not be the strong holism promulgated by Quine that takes our total theory as the unit of confirmation.

11. It does not matter that this is biologically unrealistic.

12. Note that on the objectual reductionist view, the vexing questions of intercategorial identity turn into the question of whether different branches of mathematics have overlapping interpretations, something to be settled by pragmatic considerations.

13. Cp. Kitcher (1978). You might think that in the case of the ordered pair relation, the reduction of $\langle a, b \rangle$ to $\{\{a\}, \{a, b\}\}$ is now so standard that it represents a correct meaning analysis. However, consider the following: (i) Even if $\langle a, b \rangle$ is sometimes defined as $\{\{a\}, \{a, b\}\}$, this definition is usually acknowledged to be one several possible ones. It is rare that a teacher or text claims it as the meaning of "ordered pair". (ii) The definition is rarely given outside set theory. Mathematicians learn about ordered pairs and ordered n-tuples long before they learn about this reduction, if they ever do. (iii) The arbitrariness of reducing n-tuples for $n > 2$ remains; here, there is no conventional reduction of choice.

14. What if one tries to quantify over all ordered pair reductions? The natural way to do so would be to translate "... $\langle a,b \rangle$ ..." as "$\forall f(\Phi(f) \to ... f(a,b) ...)$" where "$\Phi(f)$" abbreviates the claim that f is an ordered pairing function. However, in first-order set theory to specify what it is for something to be a function seems to require specifying what it is to be an ordered pair, e.g. one of the conjuncts in the specification of "$f:D\to R$" is "$\forall x[x \in f \leftrightarrow \exists d{\in}D\exists r{\in}R(x=\langle d,r \rangle)]$". (This problem disappears in second-order set-theory, in which functional variables and quantification over them are given.) In any event, the translation gets the truth-conditions of ordered pair statements wrong, for example, the translation of "$\langle a,b \rangle = x$" is false for any x.

15. Field points out that if "mass" is taken to be relativistic mass, the first of the next two sentences is true and the second false; if it is taken as rest mass, the first is false and the second true: (1)"Momentum is mass times velocity"; (2) "Mass is [frame of reference] invariant". He concludes that each of the two objectual reductions is inadequate and that a partial-denotation semantics for "mass" is required (1974, p. 208). Whatever we make of this argument in the case of "mass", it has no grip on our reductionism. Any acceptable set-theoretic reduction must respect the accepted truth-values of arithmetical statements. The only cases in which they differ are mixed-theory cases such as "0 is an element of 2". However, mixed statements of this kind play nothing like the role in mathematics that (1) and (2) play in Newtonian theory. They are classic "spoils to the victor" or peripheral cases a theory ought not to be judged on. An objectual reductionist semantics is, therefore, adequate to the central uses of arithmetical language (in pure and applied mathematics). Indeed, in the physical case, the problem is that each objectual reduction only captures half the story; each is a failure. In the mathematical case, in contrast, each objectual reduction tells the full story; each is a success. The multiplicity objection arises because of double success, not double failure.

16. A related objection put to me is that psychologists and linguists might reject the reductionist interpretation as giving an incorrect account of the subjects' thoughts and language. But what reason would they have for doing so? The point of this section is to examine whether there are any such reasons. Observe in passing that the reductionist view is compatible with speakers having distinctive arithmetical concepts (as mentioned in footnote 1) and (as is evident) having distinctive arithmetical vocabulary.

17. "Arithmetical discourse" here includes pure mathematical statements such as "$3+5=8$" and applied statements such as "the number of people in this room is eight".

18. Observe that from A's point of view B does in fact utter truths—set-theoretic truths—when uttering arithmetical truths; they are simply different truths from the truths A utters with the homophonic sentences.

19. For instance, one constraint might be that the interpretation is not about concreta; another that the interpretation is not too distant from the discourse's surface form; a third that the interpretation is systematic (which might be glossed as recursively generable).

20. A related objection is that in light of the Löwenheim–Skolem theorems, reductionism leads to Pythagoreanism—the view that all of mathematics should be reduced to number theory (cf. Benacerraf 1998, fn. 28 p.57). However, this objection is unpersuasive. It assumes first-order axiomatisations of mathematical theories or non-standard semantics for higher-order axiomatisations, which

arguably do not capture their mathematical content. And it also impoverishes mathematics: if all we have is arithmetic, then we cannot use analysis or set theory since these theories' consistency cannot be proved in arithmetic. Yet, we need set theory to show that first-order analysis has an arithmetical model.

21. The equi-interpretatibility of ZF minus Infinity with Peano Arithmetic provides us with an exact measure of how much of standard set theory PA is equivalent to.

22. Quine (1992) is a brief discussion of structuralism, which is concerned more with Lewis' structuralist reduction of set theory (or subsethood) to mereology (or mereology and the singleton function) and the relation of Quine's doctrine of ontological relativity to global structuralism than with Benacerraf (1965), which Quine does not mention.

23. If it is intelligible in the first place, which Benacerraf hints that it may not be (1998, p. 57). He also imputes to the reductionist the view that it is "lacking any clear sense" to ask whether, even if sets can do the job of numbers, numbers might still exist (1998, p. 56). But he gives no reason for these strong claims.

24. I am assuming that phlogiston theory is not interpretable as oxidation theory. By "rejection" I mean theoretical rejection, not necessarily rejection in practical contexts.

25. Of course economy principles could be a facet of some deeper desideratum, for example explanatoriness.

26. Whether Quine, at least in some moods or periods, was conventionalist about ontology is another matter. To the extent that he was, we are not following him.

27. The only way to charge reductionism of conventionalism is to maintain that principles of theory choice are conventionally chosen. But if you believe that, you are a conventionalist about inquiry full stop and there is nothing specifically pro- or anti-reductionist about your view.

28. Of course, if realism is understood in such a narrow way that only countenancing *sui generis* numbers counts as arithmetical realism, then reductionism is not realist.

29. Notice that these options are analogues of the first, second and fourth reductionist semantics canvassed earlier.

30. The response Benacerraf parenthetically considers on his interlocutor's behalf is to question whether there could be theories with different contents that are as virtuous as one another (1998, p. 56). However, he does not explain why one might think that two equally virtuous theories must have the same content.

31. Thanks to audience members at the New Waves workshop in Miami, especially Chris Pincock, Hannes Leitgeb, Mary Leng, Øystein Linnebo, Roy Cook and Thomas Hofweber, members of the Birkbeck College departmental seminar, an anonymous referee for this volume, Olympia and Theodora and their parents, and Hartry Field and Penelope Maddy.

References

Benacerraf, P. (1965), "What Numbers Could Not Be", *Philosophical Review* 74, pp. 47–73.

Benacerraf, P. (1998), "What Mathematical Truth Could Not Be", in M. Schirn (ed.), *Philosophy of Mathematics Today*, Oxford: Clarendon Press, pp. 9–59.

Field, H. (1974), "Quine and the Correspondence Theory", *Philosophical Review* 83, repr. (with Postscript) in his *Truth and the Absence of Fact*, Oxford: Clarendon Press, pp. 199–221.

Field, H. (1980), *Science Without Numbers*, Princeton: Princeton University Press.

Kitcher, P. (1978), "The Plight of the Platonist", *Noûs* 12, pp. 119–136.

Quine, W.V. (1960), *Word and Object*, Cambridge, MA: Harvard University Press.

Quine, W.V. (1976), "Whither Physical Objects?", *Boston Studies in the Philosophy of Science* 39, pp. 497–504.

Quine, W.V. (1992), "Structure and Nature", *Journal of Philosophy* 89, pp. 5–9.

Wetzel, L. (1989), "That Numbers Could Be Objects", *Philosophical Studies* 56, pp. 273–292.

Part II

The Question of Realism in Mathematics

3
Mathematical Fictionalism[1]

Otávio Bueno

In this chapter, I highlight five desiderata that an account of mathematics should meet to make sense of mathematical practice. After briefly indicating that current versions of platonism and nominalism fail to satisfy all of the desiderata, I sketch two versions of mathematical fictionalism that meet them. One version is based on an empiricist view of science, and has the additional benefit of providing a unified account of both mathematics and science. The other version of fictionalism is based on the metaphysics of fiction and articulates what can be considered a truly fictionalist account of mathematics. I indicate that both versions of fictionalism satisfy all of the desiderata, and I take it that they are best developed if adopted together. As a result, mathematical fictionalism is alive and well.

1 Introduction: Platonism and nominalism

Platonism is the view according to which there are abstract entities (such as sets, functions, and numbers), and mathematical theories truly describe such objects and the relations among them.[2] Given the nature of mathematical entities—especially the fact that they are not located in space and time, and are causally inert—the postulation of such objects doesn't come lightly. Platonists are, of course, well aware of this fact. It's important, then, to highlight the benefits that immediately emerge from positing mathematical objects.

Three main *benefits* should be highlighted as follows:

(a) *Mathematical discourse can be taken at face value* (i.e. it can be taken *literally*), given that, according to the platonist, mathematical terms refer. So, when mathematicians claim that "There are infinitely many prime numbers", the platonist can take that statement literally as describing the existence of an infinitude of primes. On the platonist view, there are obvious

59

truth makers for mathematical statements: mathematical objects and their corresponding properties.

This is a major benefit of platonism. If one of the goals of the philosophy of mathematics is to provide understanding of mathematics and mathematical practice, the fact that platonists are able to take the products of that practice—such as mathematical theories—literally and do not have to rewrite or reformulate them is a significant advantage. After all, the platonist is in a position to examine mathematical theories as they are actually formulated in mathematical practice, rather than discuss a parallel discourse offered by various reconstructions of mathematics given by those who avoid the commitment to mathematical objects (the nominalists).

(b) *The platonist can also provide a unified semantics for both mathematical and scientific statements.* Again, given the existence of mathematical objects, mathematical statements are true in the same way as scientific statements are. The only difference emerges from their respective truth makers: mathematical statements are true in virtue of abstract (mathematical) objects and the relations among the latter, whereas scientific statements are true, ultimately, in virtue of concrete objects and the corresponding relations among such objects.[3]

Moreover, as is typical in the application of mathematics, there are also *mixed* statements, which involve terms referring to concrete objects and to abstract ones. The platonist has no trouble providing a unified semantics for such statements as well—particularly if platonism about mathematics is associated with realism about science. In this case, the platonist can provide a referential semantics throughout.[4]

(c) Finally, *platonists will also insist that it's possible to explain the success of the application of mathematics,* given that mathematical theories are taken to be true and mathematical terms refer to appropriate mathematical objects. So, it's not surprising that mathematical theories can be used so successfully to describe relations among physical objects. True mathematical theories correctly describe relations among mathematical objects, and suitably interpreted, such relations are then used to account for various features in the physical world.

However, that the platonist can explain the application of mathematics is actually controversial. Given that mathematical objects are abstract, it's unclear why the postulation of such entities is helpful to understand the success of applied mathematics. After all, the physical world—being composed of objects located in space and time—is not constituted by entities of the same kind as those postulated by the platonist. Hence, it's not clear why to describe correctly relations among *abstract* (mathematical) entities is even *relevant* to understand the behavior of concrete objects in the physical world involved in the application of mathematics. Just mentioning that the physical world *instantiates* structures (or substructures) described in general terms by various mathematical theories, for example, as in Shapiro (1997), is not enough. For there are infinitely many mathematical structures, and there's no way of uniquely determining which of them is actually instantiated—or even instantiated only in part—in a finite region of the physical world. There's a genuine underdetermination here, given that the same physical structure in the world can be accommodated by very different mathematical structures. For instance, according to Weyl (1928), quantum mechanical phenomena can be characterized in terms of group-theoretic structures, and according to von Neumann (1932), in terms of structures emerging from the theory of Hilbert spaces. Mathematically, such structures are very different, but there's no way of deciding between them empirically. (I'll return to this point below.) In any case, despite the controversial nature of the platonist claim, explaining the success of applied mathematics is often taken as a significant benefit of platonism.

Taken together, benefits (a)–(c) provide crucial components for platonists to make sense of mathematical practice, given that mathematical theories will not be rewritten for philosophical purposes, which allows for the many uses of mathematics, including its applications, to be understood in their own terms. Mathematical practice can be accommodated as is.

But platonism also has its *costs*: (a) Given that mathematical objects are causally inert and are not located in space and time, how exactly can *mathematical knowledge* be explained in the absence of any direct access to these objects? (b) For a similar reason, how exactly is *reference* to mathematical entities achieved? Platonists are, of course, well aware of the issue, and they have developed various strategies to address the problem. But it's still contentious how successful these strategies turn out to be.[5]

These costs motivate the development of an alternative view that doesn't presuppose the commitment to mathematical entities. According to nominalism, there are *no abstract entities*, or at least, they are *not required* to make sense of mathematics and its applications.[6]

Nominalism arguably has two main benefits. (a) Given that an ontology of mathematical entities is *not* presupposed, the *possibility of mathematical knowledge* is taken to be unproblematic. For instance, as proposed by Field (1989), mathematical knowledge is ultimately taken to be empirical or logical knowledge. (b) The same goes for *reference* to mathematical objects. Very roughly, if there are no mathematical entities, there is nothing there to be referred to! The issue simply vanishes.

But nominalism, just as platonism, also has its costs: (a) Mathematical discourse is *not* taken at *face value* (the discourse is *not* taken *literally*). Each nominalization strategy for mathematics introduces some change in either the syntax or the semantics of mathematical statements. In some cases, modal operators are introduced to preserve verbal agreement with the platonist, as in the studies by Hellman (1989). The proposal is that each mathematical statement S is translated into two modal statements: (i) If there were structures of the suitable kind, S would be true in these structures, and

(ii) It's possible that there are such structures. As a result, both the syntax and the semantics of mathematics are changed. In other cases, as can be seen in Field (1989), in order to preserve verbal agreement with the platonist despite the negation that mathematical objects exist, fiction operators (such as: "According to arithmetic...") are introduced. Once again, the resulting proposal shifts the syntax and the semantics of mathematical statements.

(b) Given that mathematical statements are not taken literally, it comes as no surprise that, on nominalist views, the semantics of science and mathematics is *not uniform*. After all, modal and fiction operators need to be introduced. But these operators have no counterparts in science given that scientific theories are taken to provide descriptions of the world, and the presence of these operators prevent us from making any such claims.

Finally, (c) it is *not* clear that the nominalist can provide an account of the *application of mathematics*. After all, if mathematical terms do not refer and mathematical theories are not true, why is it that these theories are so successful in science? *Prima facie*, it becomes mysterious exactly why that success should emerge, given the nonexistence of mathematical objects. Nominalists are, of course, aware of the issue, and some try to offer a story as to why mathematical theories can be used successfully despite being false. According to Field (1980, 1989), mathematical theories need not be true to be good, as long as they are conservative, that is, consistent with every internally consistent claim about the physical world. Field then argues that, given the conservativeness of mathematics, it's possible to dispense with mathematical objects in derivations of claims from nominalistic premises to nominalistic conclusions (that is, in claims in which mathematical terms do not occur). However, this move fails to explain the success of mathematical theories as the latter are in fact applied in mathematical practice. The fact that the nominalistic versions of certain physical theories may work in application doesn't explain how actual mathematical theories manage to work. Once again, we have an interesting philosophical proposal that produces a parallel discourse, and in terms of this discourse a particular ontological claim is made about actual practice; namely, that it need not be committed to mathematical objects, given the dispensability of the latter. But this leaves entirely open the issue as to which features *in actual mathematical practice*, if any, could be invoked to avoid commitment to mathematical objects, given that scientists don't formulate their theories in accordance with Field's nominalistic recipe.[7]

Given (a)–(c), the nominalist is ultimately unable to accommodate *mathematical practice*. Given that mathematical statements need to be rewritten, the practice cannot be taken literally. Moreover, if actual cases of the application of mathematics are not accommodated, a significant dimension of mathematical practice is left unaccounted for. An alternative proposal is thus required.

2 Motivations for fictionalism

The considerations above motivate the following question: Is it possible to develop a view that has all the *benefits* of platonism without the corresponding costs? Or, equivalently, a view that has *none* of the costs of nominalism, while keeping all of its *benefits*? In other words, what we need is a view that meets the following desiderata (each of which is independently plausible):

(1) The view explains the possibility of *mathematical knowledge*.
(2) It explains how *reference* to mathematical entities is achieved.
(3) It accommodates the *application of mathematics* to science.
(4) It provides a *uniform semantics* for mathematics and science.[8]
(5) It takes mathematical discourse *literally*.

Note that if all of these desiderata are met, the resulting view will be able to accommodate mathematical practice. After all, given that mathematical theories are taken literally, and a uniform semantics for mathematics and science is offered, there is no need for making up a parallel discourse as a replacement for the actual practice. With the desiderata in place, the resources are available to make sense of the practice in its own terms.

In this chapter, I argue that the desiderata above can all be met as long as a *fictionalist* view of mathematics is articulated, and I will sketch such a view. In fact, I'll sketch *two* such views, putting forward two different strategies to articulate a fictionalist stance, and indicating the ways in which the strategies support each other. For obvious reasons, I'll only be able to offer an outline of the views here. But hopefully enough will be said to indicate how the views look like.

3 Two fictionalist strategies

3.1 Fictionalism and nominalism

What is the difference between *fictionalism*[9] and *nominalism*? As developed here, fictionalism is an *agnostic* view; it doesn't state that mathematical objects don't exist. Rather, the issue of their existence is left open. Perhaps these objects exist, perhaps they don't. But, according to the fictionalist, we need not settle the issue to make sense of mathematics and mathematical practice. Thus, in contrast with the skeptical view offered by the nominalist—who denies the existence of mathematical objects and relations—the fictionalist provides an *agnostic* proposal. In contrast with platonism, the fictionalist is not committed to the existence of mathematical objects and relations either. In this way, and at least in temperament, fictionalism is closer to nominalism than to platonism. In fact, to highlight the connection with nominalism, the fictionalist view can also be called *agnostic nominalism*. After all, fictionalism provides a strategy to avoid commitment to the existence of mathematical entities, but without, thereby, denying their existence.

The two strategies sketched below are fictionalist in slightly different ways. The first strategy assumes a particular empiricist view about science—namely, constructive empiricism, as developed by van Fraassen (1980, 1989)—and indicates how to extend this view to make sense of mathematics. This strategy is fictionalist in a *broad* sense; it offers a way of using mathematics that is compatible with a fictionalist reading of mathematical statements. The second strategy explores the metaphysics of fiction, and what it takes to introduce a fictional object. This strategy is fictionalist in a *narrow* sense; it indicates directly the similarities between mathematical and fictional entities.

3.2   The empiricist fictionalist strategy

The empiricist strategy focuses on applied mathematics, which is the central feature of mathematics that an empiricist needs to make sense of as part of his or her account of science. Given the restriction to applied mathematics, the empiricist strategy will not offer a general account of mathematics. The second, truly fictionalist strategy, is more general—also in that it doesn't presuppose a commitment to an empiricist view of science. In turn, the second strategy doesn't have anything special to say about science. In the end, the two views are better adopted together.

3.2.1   The crucial idea

According to constructive empiricism, the aim of science is not truth, but something weaker. As van Fraassen (1980) suggests, it is empirical adequacy. Roughly, a theory is empirically adequate if it's true about the observable features of the world. As for the unobservable features, the constructive empiricist remains agnostic: it's not clear how we can know whether unobservable entities exist or not, and in case they do exist, which features they have. After all, there are incompatible accounts of unobservable phenomena that characterize the latter in drastically different ways, but which turn out to be all empirically adequate. Consider, for instance, Copenhagen and Bohmian interpretations of quantum mechanics. According to the Copenhagen view, quantum objects are such that it's not possible to measure simultaneously their position and momentum with full certainty. In contrast, according to the Bohmian conception, quantum objects can be so measured. Both interpretations are empirically adequate, and thus cannot be chosen based on empirical consideration alone. But they offer strikingly different accounts of the nature of quantum objects. Given that, empirically, we cannot choose between them, it's unclear that, in the end, we are in a position to determine the nature of quantum objects. This is, of course, a familiar underdetermination argument, and the empiricist explores it to motivate agnosticism about unobservable phenomena. So, the constructive empiricist neither denies nor asserts the existence of unobservable entities. What is offered is an agnostic stance.

The crucial idea of the empiricist fictionalist strategy to applied mathematics is to insist that (applied) mathematical theories need not be true to be good. They only need to be part of an *empirically adequate package*. This theoretical package typically involves: a scientific theory, the relevant mathematical theories (used in the formulation of the scientific theory in question), interpretations of the resulting formalism, and initial conditions. The *whole package* is never asserted to be true; it's only required to be empirically adequate, that is, to accommodate the observable phenomena.

Now, as noted, empirical adequacy is *weaker* than truth—it's truth about the observable phenomena. In particular, a theory's empirical adequacy doesn't establish the existence of *unobservable* objects—whether mathematical or physical. Given that the empirical adequacy of a theoretical package is compatible with this package being mistaken in its description of the unobservable phenomena—which includes reference to both physical and mathematical objects—the empiricist is not committed to unobservable objects when an empirically adequate package is adopted. In the end, the *existence* of unobservable objects is *not* required to make sense of scientific or mathematical practice. As a result, unobservable objects can be taken as *fictional*.

Similarly to what the constructive empiricist does in the context of science, the fictionalist can use underdetermination arguments to motivate *agnosticism* about the existence of mathematical objects. After all, it's possible to obtain the same empirical consequences of a given scientific theory using significantly different mathematical frameworks. For example, quantum mechanics can be formulated via group theory, as Weyl (1928) proposed, or via Hilbert spaces, as von Neumann (1932) did. Mathematically, these are very different formulations, which emphasize different aspects of the quantum mechanical formalism. Weyl was particularly interested in characterizing some features of quantum objects, and the use of group theory, with its transformation groups, was central to this task. In turn, von Neumann was especially concerned with offering a systematic framework to represent quantum states and introduce probability into quantum mechanics. The Hilbert space formalism was appropriate for both tasks. Despite the significant mathematical differences between the two frameworks, which emerge from the different mathematical theories that are presupposed in each case, the same empirical results about quantum phenomena are obtained.

The empiricist fictionalist will then note that the underdetermination of these two theoretical packages motivates agnosticism about the mathematical objects that are invoked in the mathematical formulation of quantum mechanics. Should the empiricist be committed to the existence of group-theoretic transformations given the success of the application of group theory to quantum mechanics? Or should the commitment go for vectors in a Hilbert space instead, given the success of the corresponding theory in quantum mechanics? Recall that each package offers a different account of what is going on beyond the observable phenomena. And given their empirical equivalence, it's unclear how to choose between them on empirical grounds. Agnosticism then emerges.

It might be argued that the empiricist should be committed to the existence of the two types of objects, given that groups and vectors have been both successfully applied in quantum mechanics. The trouble with this suggestion, however, is that we don't get a coherent picture from the adoption of both mathematical frameworks. Different features of quantum objects and their states are articulated in each case. Even if we adopted both frameworks, it wouldn't still be clear what the content of the resulting package is supposed to be. Once again, to remain agnostic seems to be the warranted option in this case.

It might be objected that the notion of empirical adequacy that is invoked here is not available to the empiricist fictionalist. After all, as formulated by van Fraassen (1980, p. 64), the concept of empirical adequacy *presupposes* abstract entities. It is, thus, a concept that the empiricist fictionalist should be agnostic about. As formulated by van Fraassen, a theory is empirically adequate if there is a model of that theory such that every empirical substructure of that model is isomorphic to the appearances (i.e. the structures that represent the outcomes of the experimental results). Given that the models involved are themselves mathematical objects, when an empiricist believes that a scientific theory is empirically adequate, she will thereby believe in the existence of abstract entities, as Rosen (1994) argues.

In response, two moves are available for the empiricist fictionalist. First, she can adopt a formulation of empirical adequacy that does not presuppose abstract entities; for instance, the characterization according to which a theory is empirically adequate as long as what it states about the observable phenomena is true, as presented in van Fraassen (1980). Second, the empiricist can use the (truly) fictionalist strategy developed below to accommodate this difficulty. After all, as will become clear shortly, the truly fictionalist strategy has the resources to accommodate structures from pure mathematics, such as mathematical models and transformations among them.

3.2.2   Meeting the desiderata

Can the empiricist fictionalist strategy accommodate the five desiderata discussed above? I think it can. Here—in very broad outline—is how this can be done.

(1) *Mathematical knowledge*: Can we expect to get any account of mathematical knowledge from a proposal that is agnostic about the existence of mathematical objects? Certainly, the proposal won't yield knowledge of the objects that, according to the platonist, make mathematical statements true. For these are precisely the objects about which the empiricist fictionalist is agnostic. So, what exactly can be expected in this case?

Consider the corresponding constructive empiricist view about knowledge of unobservable objects in science. Given the constructive empiricist's agnosticism about unobservable phenomena, it comes as no surprise that the empiricist doesn't claim to know what is going on at the unobservable level. He or she suspends the judgment about the issue. There is, however, a more positive component to the constructive empiricist approach to the problem. The fact that many incompatible, but empirically adequate, accounts of unobservable phenomena are available—such as the different interpretations of quantum mechanics—provides an important form of understanding, namely, of how the world could be if these underdetermined accounts or interpretations were true, as van Fraassen (1991, 1989) points out. Each interpretation indicates a possible way the unobservable phenomena behave in order to generate the observable features of the world that we do experience. Although we may not be able to decide which of these interpretations (if any) is true, we can still understand the conception of the unobservable world that each of them provide. For example, if Copenhagen-type quantum objects populate the world, we can understand why the position and momentum of these objects cannot be simultaneously measured with full certainty. If, however, in accordance with the Bohmian interpretation, a quantum potential exists, we can then make sense of how the position and momentum of quantum objects can be measured simultaneously. In each case, we understand how the world could be, even if we don't know how it actually is.

Similarly, in the case of the mathematical theories used in theoretical packages, it becomes clear that each of them gives us understanding. We understand how central features of quantum particles can be expressed by formulating quantum mechanics in terms of group-theoretic invariants. We can understand how quantum states can be represented in terms of suitable features of a Hilbert space. In the end, we understand how the world could be if the theoretical packages were true, despite the fact that we are unable to determine which of them (if any) is true, as van Fraassen (1991) emphasizes. And the fact that we are genuinely unable to choose between the theoretical packages on empirical grounds helps us understand why being agnostic about the existence of the corresponding objects is a perfectly acceptable stance.

Thus, in the empiricist fictionalist picture, mathematical knowledge becomes part of scientific knowledge—at least with regard to applied mathematics. We get to know mathematical results, in part, by understanding the role they play in the investigation of the empirical world. Of course, the concept of knowledge at work here is very deflationary since only the truth of the observable aspects of the theories in question is involved. This includes the observable components of applied mathematical theories. In this respect, the limited knowledge involved in applied mathematics is similar to the corresponding knowledge that the constructive empiricist recognizes in science. Whether in science or applied mathematics, knowledge and understanding go hand in hand.

(2) *Reference to mathematical entities*: How do we refer to mathematical entities on the empiricist fictionalist strategy? We refer to them in exactly the same way as we refer to unobservable objects in science. In fact, "electrons" refer to electrons just as "sets" refer to sets. Recall that the empiricist does not deny the existence of unobservable entities, whether they are mathematical or empirical. Rather than a skeptical attitude about these entities, what is offered is an agnostic one. To refer to mathematical objects or to unobservable physical entities, such as quarks or photons, all the empiricist needs is a theory that characterizes some properties of the relevant objects, even though he or she may be agnostic about whether these entities exist or not.

Although reference is often used as a success term, particularly in philosophical contexts, there is no need to assume that this is the case. After all, clearly we can refer to non-existing things, such as fictional characters (e.g. Sherlock Holmes) and non-existing posits of scientific theorizing (e.g. phlogiston). The fact that, in these cases, the corresponding objects do not exist doesn't prevent us from referring to them. We do that all the time. In the end, as we can see in Azzouni (2004), reference need not require existence of the objects that are referred to.

Thus, the mechanism of reference to mathematical entities is not different from the one in terms of which we refer to unobservable entities in science. The main difference is that, although there are mechanisms of *instrumental access* to scientific entities, there aren't such mechanisms in the case of mathematical objects.[10] This is how it should be, since we don't expect the existence of mechanisms of instrumental access to abstract entities, given that the latter are not located in space and time.

(3) *Application of mathematics*: Applied mathematics is often used as a source of support for platonism. How else but by becoming platonists can we make sense of the success of applied mathematics in science? As an answer to this question, the fictionalist empiricist will note that it's not the case that applied mathematics always works. In several cases, it doesn't work as initially intended, and it works only when accompanied by suitable empirical interpretations of the mathematical formalism. For example, when Dirac found negative energy solutions to the equation that now bears his name, he

tried to devise physically meaningful interpretations of these solutions. His first inclination was to ignore these negative energy solutions as not being physically significant, and he took the solutions to be just an artifact of the mathematics—as is commonly done in similar cases in classical mechanics. Later, however, he identified a physically meaningful interpretation of these negative energy solutions in terms of "holes" in a sea of electrons. But the resulting interpretation was empirically inadequate, since it entailed that protons and electrons had the same mass. Given this difficulty, Dirac rejected that interpretation and formulated another. He interpreted the negative energy solutions in terms of a new particle that had the same mass as the electron but opposite charge. A couple of years after Dirac's final interpretation was published Anderson detected something that could be interpreted as the particle that Dirac posited. Asked as to whether Anderson was aware of Dirac's papers, Anderson replied that he knew of the work, but he was so busy with his instruments that, as far as he was concerned, the discovery of the positron was entirely accidental.[11] For further details and references, see Bueno (2005).

The application of mathematics is ultimately a matter of using the vocabulary of mathematical theories to express relations among physical entities. Given that, for the fictionalist empiricist, the truth of the various theories involved—mathematical, physical, biological, and whatnot—is never asserted, no commitment to the existence of the entities that are posited by such theories is forthcoming. But if the theories in question—and, in particular, the mathematical theories—are *not* taken to be true, how can they be *successfully applied*? There is no mystery here. First, even in science, *false* theories can have true consequences. The situation here is analogous to what happens in fiction. Novels can, and often do, provide insightful, illuminating descriptions of phenomena of various kinds—for example, psychological or historical events—that help us understand the events in question in new, unexpected ways, despite the fact that the novels in question are not true. Second, given that mathematical entities are not subject to spatial-temporal constraints, it's not surprising that they have no active role in applied contexts. Mathematical theories need only provide a framework that, suitably interpreted, can be used to describe the behavior of various types of phenomena—whether the latter are physical, chemical, biological, or whatnot. Having such a descriptive function is clearly compatible with the (interpreted) mathematical framework not being true, as Dirac's case illustrates so powerfully. After all, as was just noted, one of the interpretations of the mathematical formalism was empirically inadequate.

(4) *Uniform semantics*: On the fictionalist empiricist account, mathematical discourse is clearly taken on a par with scientific discourse. There is no change in the semantics. Mathematical and scientific statements are treated

in exactly the same way. Both sorts of statements are truth-apt, and are taken as describing (correctly or not) the objects and relations they are about. The only shift here is on the *aim* of the research. After all, on the fictionalist empiricist proposal, the goal is not truth, but something weaker: empirical adequacy—or truth only with respect to the observable phenomena. However, once again, this goal matters to both science and (applied) mathematics, and the semantic uniformity between the two fields is still preserved.

(5) *Taking mathematical discourse literally*: According to the fictionalist empiricist, mathematical discourse is also taken literally. If a mathematical theory states that "There are differentiable functions such that...", the theory is *not* going to be reformulated in any way to avoid reference to these functions. The truth of the theory, however, is *never* asserted. There's no need for that, given that only the empirical adequacy of the overall theoretical package is required.

3.3   The fictionalist strategy

Although the two strategies described in this chapter are fictionalist, the second strategy is truly fictionalist in the sense that it explores fictionalism from the metaphysics of fiction. To distinguish these strategies, from now on I'll call the first the *empiricist strategy*, and the second the *fictionalist strategy*.

3.3.1   The crucial point

The central point of the fictionalist strategy is to emphasize that mathematical entities are like fictional entities. They have similar features that fictional objects such as Sherlock Holmes or Hamlet have. By indicating that there is nothing mysterious in the way in which we can have knowledge of fictional entities and are able to refer to them, and by arguing that mathematical entities are a particular kind of fictional entity, a truly fictionalist view can be articulated.

The fictionalist's proposal is to extend to mathematics the work on the nature of fictional characters that Amie Thomasson has developed (1999). Thomasson put forward the artifact theory of fictional objects, according to which the latter objects are *abstract artifacts*. First, fictional objects are *created* by the intentional acts of their authors (in this sense, they are *artifacts*). So, they are introduced in a particular context, in a particular time. Second, fictional objects depend on (i) the existence of copies of the artworks that describe these objects (or through memories of such works), and (ii) the existence of a community who is able to understand these works. In other words, fictional objects depend on the existence of concrete objects in the physical world (e.g. books, a community of readers, etc.). As a result, there is nothing mysterious about the way in which we refer to and obtain knowledge of such objects. By reading the story, we get information about the entities in question, and get a sense of what happened to them.

Similar points apply to mathematical entities. First, these entities are also *created*, in a particular context, in a particular time. They are *artifacts*. Mathematical entities are created when comprehension principles are put forward to describe their behavior, and when consequences are drawn from such principles.[12] Second, mathematical entities thus introduced are also dependent on (i) the existence of particular copies of the works in which such comprehension principles have been presented (or memories of these works), and (ii) the existence of a community who is able to understand these works. It's a perfectly fine way to describe the mathematics of a particular community as *being lost* if all the copies of their mathematical works have been lost and there's no memory of them. Thus, mathematical entities, introduced via the relevant comprehension principles, turn out to be *contingent*—at least in the sense that they depend on the existence of particular concrete objects in the world, such as, suitable mathematical works. The practice of mathematics depends crucially on producing and referring to these works.

In this respect, the fictionalist insists that there is nothing mysterious about how we can refer to mathematical objects and have knowledge of them. Similarly to the case of fiction, reference to mathematical objects is made possible by the works in which the relevant comprehension principles are formulated.[13] In these works, via the relevant principles, the corresponding mathematical objects are introduced. The principles specify the meaning of the mathematical terms that are introduced as well as the properties that the mathematical objects that are thus posited have. In this sense, the comprehension principles provide the context in which we can refer to and describe the mathematical objects in question. Our knowledge of mathematical objects is then obtained by examining the properties that these objects have, and by drawing consequences from the comprehension principles.

3.3.2 On the existence of mathematical objects

What about the mathematical objects that, according to the platonist, exist *independently* of any description one may offer of them in terms of comprehension principles? Do *these* objects exist on the fictionalist view? Now, the fictionalist is *not* committed to the existence of such mathematical objects, although this doesn't mean that the fictionalist is committed to the nonexistence of these objects. On the view advanced here, the fictionalist is ultimately agnostic about the issue. Here is why.

According to Azzouni (1997*b*, 2004), there are two types of commitment: *quantifier commitment* and *ontological commitment*. We incur quantifier commitment to the objects that are in the range of our quantifiers. We incur ontological commitment when we are committed to the existence of certain objects. However, despite Quine's view, quantifier commitment *doesn't* entail ontological commitment. Fictional discourse (e.g. in literature) and mathematical discourse illustrate that. Suppose that there's no way of making sense of our practice with fiction but to quantify over fictional objects, such as Sherlock Holmes or Pegasus. Still, people would strongly *resist* the claim that they are therefore *committed to the existence* of these objects. The same point applies to mathematical objects.

This move can also be made by invoking a distinction between partial quantifiers and the existence predicate.[14] The idea here is to resist reading the existential quantifier as carrying any ontological commitment. Rather, the existential quantifier only indicates that the objects that fall under a concept (or have certain properties) are less than the whole domain of discourse. To indicate that the whole domain is invoked (e.g. that every object in the domain have a certain property), we use a universal quantifier. So, two different functions are clumped together in the traditional, Quinean reading of the existential quantifier: (i) to assert the existence of something, on the one hand, and (ii) to indicate that not the whole domain of quantification is considered, on the other. These functions are best kept apart. We should use a partial quantifier (that is, an existential quantifier free of ontological commitment) to convey that only some of the objects in the domain are referred to, and introduce an existence predicate in the language in order to express existence claims.

By distinguishing these two roles of the quantifier, we also gain expressive resources. Consider, for instance, the sentence:

(*) Some fictional detectives don't exist.

Can this expression be translated in the usual formalism of classical first-order logic with the Quinean interpretation of the existential quantifier? *Prima facie*, that doesn't seem to be possible. The sentence would be contradictory! It would state that there exist fictional detectives who don't exist. The obvious consistent translation here would be: $\neg \exists x \; Fx$, where F is the predicate *is a fictional detective*. But this states that fictional detectives don't exist. Clearly, this is a different claim from the one expressed in (*). By declaring that some fictional detectives don't exist, (*) is still compatible with the existence of some fictional detectives. The regimented sentence denies this possibility.

However, it's perfectly straightforward to express (*) using the resources of partial quantification and the existence predicate. Suppose that "$\exists$" stands for the partial quantifier and "E" stands for the existence predicate. In this case, we have: $\exists x \; (Fx \wedge \neg Ex)$, which expresses precisely what we need to state.

Now, under what conditions is the fictionalist entitled to conclude that certain objects exist? In order to avoid begging the question against the platonist, the fictionalist cannot insist that only objects that we can causally interact with exist. So, the fictionalist only offers *sufficient* conditions for us to be entitled to conclude that certain objects exist. Conditions such as the following seem to be uncontroversial.[15] Suppose we have access to certain objects that is such that (i) it's *robust* (e.g. we blink, we move away, and the objects are still there); (ii) the access to these objects *can be refined* (e.g. we can get closer for a better look); (iii) the access allows us to *track* the objects in space and time; and (iv) the access is such that if the objects weren't there, we wouldn't believe that they were. In this case, having this form of access to these objects gives us good grounds to claim that these objects exist. In fact, it's in virtue of conditions of this sort that we believe that tables, chairs, and so many observable entities exist.

But recall that these are only sufficient, and not necessary, conditions. Thus, the resulting view turns out to be agnostic about the existence of the mathematical entities the platonist takes to exist—independently of any description. The fact that mathematical objects fail to satisfy some of these conditions doesn't entail that these objects don't exist. Perhaps these entities do exist after all; perhaps they don't. What matters for the fictionalist is that it's possible to make sense of significant features of mathematics without settling this issue.

Now what would happen if the agnostic fictionalist used the partial quantifier in the context of comprehension principles? Suppose that a vector space is introduced via suitable principles, and that we establish that there are vectors satisfying certain conditions. Would this entail that we are now committed to the existence of these vectors? It would if the vectors in question satisfied the existence predicate. Otherwise, the issue would remain open, given that the existence predicate only provides sufficient, but not necessary, conditions for us to believe that the vectors in question exist. As a result, the fictionalist would then remain agnostic about the existence of even the objects introduced via comprehension principles!

3.3.3 *Meeting the desiderata*

Does the fictionalist's account satisfy the five desiderata discussed in Section 2? I think it does. In very broad outline, here is why.

(1) *Mathematical knowledge*: Knowledge of mathematical entities, just as knowledge of fictional entities in general, is the result of producing suitable descriptions of the objects in question and drawing consequences from the assumptions that are made. Central in this process, we saw, is the formulation of comprehension principles, which specify and systematize the use of the relevant mathematical concepts. Note that once certain comprehension principles are introduced and a logic is adopted, it's no longer

up to us what follows from such principles. It's a matter of the relations that hold among the concepts that are introduced in the relevant comprehension principles as well as any additional assumptions that are invoked and the deductive patterns of inference that are used. In this sense, mathematical knowledge is objective, even though the objectivity in question does not depend on the existence of mathematical objects independently of comprehension principles.

The fictionalist acknowledges, of course, that we may not be able to establish some results based on certain comprehension principles; for example, the system in question may be incomplete. We may even prove that the system in question is incomplete—again, by invoking suitable comprehension principles and Gödel's incompleteness theorems. Does this mean that we can determine the truth of certain mathematical statements (say, a particular Gödel sentence) *independently* of comprehension principles? It's not clear that this is the case. After all, the comprehension principles will be needed to characterize the meaning of the mathematical terms in question; they are part of the context that specifies the concepts under consideration, as well as the objects and their relations.

(2) *Reference to mathematical entities*: How is reference to mathematical objects accommodated in the fictionalist's approach? Once again, comprehension principles play a central role here. By introducing particular comprehension principles, reference to mathematical objects is made possible. After all, the principles specify some of the properties that the objects that are introduced have, and by invoking these properties, it's possible to refer to the objects in question as those objects that have the corresponding properties. In other words, with the comprehension principles in place, it's typically unproblematic to secure reference to mathematical objects. The latter are identified as the objects that have the relevant properties.

Although this suggestion may work in most cases, some difficulties need to be addressed. Suppose, for instance, that the comprehension principles don't uniquely determine the objects that are referred to. In this case, reference won't be sharp, in the sense that exactly one object is picked out. The best we can do is secure reference to an equivalence class of objects, without uniquely identifying each member of that class. In mathematical contexts, this is typically enough, given that mathematical structures are often characterized "up to isomorphism". In other cases, the comprehension principles may turn out to be inconsistent. As a result, there won't be any consistent objects to refer to. In cases of this sort, we can still refer to inconsistent objects. This is acceptable to the fictionalist, who is not committed, of course, to the *existence* of such objects. We can refer, and we often do refer, to non-existent things. They are, ultimately, objects of thought.

In other words, similarly to knowledge of mathematical entities, reference to mathematical objects is achieved by invoking the relevant comprehension principles. As a result, reference is always contextual: it's made in the context of the comprehension principles that give meaning to the relevant mathematical terms.

(3) *Application of mathematics*: The fictionalist can make sense of crucial features of the application of mathematics, while remaining agnostic about the existence of mathematical entities. The key idea is that, for the fictionalist, the application of mathematics is a matter of using the expressive resources of mathematical theories to accommodate different aspects of scientific discourse. But this doesn't require the truth of the relevant mathematical theories. After all, whether or not mathematical theories correctly describe independently existing mathematical objects and their relations is largely irrelevant to the application of mathematics. What matters are the relations that are introduced by the mathematical theories and whether these relations can be interpreted in a physically significant way. And these issues are independent of the truth of the mathematical theories in question.

However, this still leaves the question as to why positing fictional objects can be so useful in the description of the physical world. The answer to this question depends, again, on the context. In some contexts, mathematical theories can be extremely useful. For instance, if we are interested in capturing certain structural properties of empirical phenomena—such as the representation of their speed, momentum, acceleration, or rate of growth— the mathematical vocabulary offers a rich, nuanced framework. In other contexts, however, mathematical theories are far from useful. For instance, suppose that we are interested in capturing the psychological states of some empirical phenomena; in this case, it's unclear that, in general, the use of mathematical vocabulary is of much relevance.[16] The same point applies to fictional discourse. Novels often offer insightful accounts of human psychology, but they clearly are inadequate sources of information regarding the representation of quantum states. In order to make sense of the application of mathematics, it's crucial that we are sensitive to the context in which the mathematics is in fact used.

(4) *Uniform semantics*: On the fictionalist's proposal, there is uniform semantics for scientific and mathematical statements. On this view, mathematical and scientific terms are treated in exactly the same way. There is no attempt to offer special semantic conditions for mathematical statements. Once the concept of *prime number* is introduced, the following statement: "There are infinitely many prime numbers" is true as long as there are infinitely many prime numbers. Of course, the *existence* of prime numbers is left open given the agnostic nature of the fictionalist view. But this is no change in the semantics for mathematical statements.

(5) *Taking mathematical discourse literally*: On the fictionalist proposal, mathematical discourse is also taken literally, given that the semantics of mathematical statements has not changed. The statement "There are infinitely many prime numbers" comes out true in the context, say, of

arithmetic. In this context, reference to prime numbers is achieved by the comprehension principles that provide meaning to the relevant terms.

It might be objected that the fictionalist is *not* taking mathematical discourse literally. Typically, fictionalists about mathematics (say, arithmetic) introduce a fiction operator "According to (arithmetic)". So, the syntax of mathematical statements has to be tinkered with, and as a result, mathematical discourse is not taken literally.

In response, note that the practice of mathematics always *presupposes* a given mathematical "theory":[17] certain mathematical principles (that need not be axiomatized) from which mathematicians draw their consequences. The results are always established based on such principles that provide the context for mathematical research. The fictionalist is not introducing a fiction operator to mathematical statements. The statements are used in the context of principles that characterize the properties of the relevant mathematical objects. In this sense, the fiction operator—in the form of the comprehension principles that specify a certain domain of objects— is already in place as part of mathematical practice. The fictionalist is *not* adding a new item to the language of mathematics. Properly conceptualized, the fiction operator is already there.

4   Conclusion

There is, of course, much more to be said about the two fictionalist strategies that were developed here, and I plan to expand on them in future work. My goal here has been only to sketch some of the central ideas of these proposals. If they are near the mark, it's indeed possible to be fictionalist about mathematics—making sense, in particular, of mathematical practice along the way.

Notes

1. I'd like to thank my colleagues in the Philosophy Department at the University of Miami, the participants of the New Waves in Philosophy of Mathematics Conference, and Jody Azzouni for their extremely helpful responses to earlier versions of this work.
2. For different ways of formulating platonism, see, for instance, Quine (1960), Resnik (1997), Shapiro (1997), the first part of Balaguer (1998), and Colyvan (2001).
3. This point is significantly idealized in that it assumes that, somehow, we can manage to distill the empirical content of scientific statements independently of the contribution made by the mathematics that is often used to express such statements. According to Quine (1960) and Colyvan (2001), platonists who defend the indispensability argument will insist that this is not possible to do.
4. Of course, the platonist about mathematics *need not* be a realist about science— although it's common to combine platonism and realism in this way. In principle, the platonist could adopt some form of anti-realism about science, for instance,

constructive empiricism, as developed by van Fraassen (1980). As long as the form of anti-realism regarding science allows for a referential semantics (and many do), the platonist would have no trouble providing a unified semantics for both mathematics and science.

5. See the first part of Balaguer (1998) for a critical assessment of these strategies.
6. Different versions of nominalism can be found, for example, in Field (1980, 1989), Hellman (1989), Chihara (1990), Burgess and Rosen (1997), the second part of Balaguer (1998), and Azzouni (2004). Of these views, Azzouni's is the one that comes closest to meeting all of the desiderata discussed below. But it's unclear to me that his view in fact meets them, although the point is controversial. Further details are discussed in Bueno and Zalta (2005) and Azzouni (2009).
7. Field, of course, doesn't claim that mathematical theories should be formulated in accordance with his program. He is only pointing out that, as opposed to what the platonist argues based on the indispensability argument, the successful use of mathematics in science need not force us to be committed to the existence of mathematical objects. But, as a result, the issue of making sense of the way in which actual mathematical theories are applied in science is not addressed.
8. It might be argued that we should not take the uniformity of semantics as a desideratum for a philosophical account of mathematics. After all, presumably on a more traditional, non-Quinean approach, mathematics would be taken as substantially different from empirical science—in particular, given its *a priori* character. Thus, mathematics may well correctly demand a different semantics (as well as a different epistemology). So, the uniformity of the semantics may be too contentious and substantial to be taken as a requirement here.

There's no doubt that mathematics and empirical science are importantly different. And the fact that the former is often taken to be *a priori* may well highlight a significant difference between the two domains. However, this doesn't undermine the point that the adoption of a uniform semantics would be a benefit since it won't demand a special semantic treatment for special fields. Having a uniform semantics would also simplify considerably the account of applied mathematics, given that the treatment of mixed statements—that include reference to both mathematical and non-mathematical objects—would be the same throughout applied contexts.

Moreover, the difference in epistemological status between mathematical and empirical claims, on its own, isn't sufficient to justify changing the semantics of these statements. Consider, for instance, the case of modal realism—as in Lewis (1986). For the modal realist, modal knowledge (that is, knowledge of what is possible or necessary) is *a priori*, just as mathematical knowledge is. However, the semantics for modal and mathematical statements is the same. The only difference is that, in the case of modal statements, possible worlds make them true, whereas in the mathematical case, mathematical objects and relations make the corresponding statements true. Incidentally, on the modal realist account, not even the semantics of empirical statements would be different: such statements are true in virtue of the features of the actual world. In the end, independently of the epistemological story one offers, the possibility of providing a uniform semantics should be taken as a benefit.

9. "Fictionalism" is used in dramatically different ways in the literature. In this chapter, I'll be using the expression in terms of the two strategies discussed below. I need a term that is somewhat neutral between nominalism and platonism,

but which indicates a view that is still closer to the former than to the latter. Fictionalism seems appropriate for that. So, I'll stick to it.

10. For further details, see Azzouni (1997*a*, 2004).
11. Dirac's work here clearly illustrates the point that the same mathematical formalism is compatible with multiple physically significant interpretations. Some of these interpretations, as we saw in the case of Dirac's first interpretation, may turn out to be empirically inadequate. So, the empirical success of the theoretical package involving the mathematical formalism, the relevant physical theories, and the corresponding interpretations cannot be attributed to the formalism alone. In fact, on its own, the formalism is compatible with virtually any physical outcome. Typically, at most some cardinality constraints about the size of the domain of interpretation are imposed by the mathematical formalism alone.
12. The comprehension principles in question need not be consistent. The only requirement is that they are not trivial; that is, not everything follows from them.
13. In the fiction case (e.g. in literary works), fictional objects are typically introduced simply by specifying the properties that the corresponding objects have in a much more informal way than in mathematics—even though the latter is practiced very informally too in comparison with the standards of mathematical logicians.
14. For further reference, see McGinn (2000).
15. These conditions are based on the work that Jody Azzouni has done on this issue (2004). But he put them to a very different use!
16. However, mathematical vocabulary can be extremely useful if we are interested in capturing quantitative aspects of psychological phenomena, as mathematical models used in psychology clearly show.
17. In many cases, to describe mathematical practice as presupposing a *theory* about a certain domain suggests more organization than what is actually found. Particularly in emerging mathematical fields (say, the first studies of complex numbers), the principles may be just rough attempts to systematize and describe a certain group of mathematical phenomena. Although rough, the principles are still a central feature of the practice.

References

Azzouni, J. (1997*a*). "Thick Epistemic Access: Distinguishing the Mathematical from the Empirical", *Journal of Philosophy 94*, pp. 472–484.

Azzouni, J. (1997*b*). "Applied Mathematics, Existential Commitment and the Quine-Putnam Indispensability Thesis", *Philosophia Mathematica 5*, pp. 193–209.

Azzouni, J. (2004). *Deflating Existential Consequence: A Case for Nominalism*. New York: Oxford University Press.

Azzouni, J. (2009). "Empty *de re* Attitudes about Numbers", forthcoming in *Philosophia Mathematica 17*, pp. 163–188.

Balaguer, M. (1998). *Platonism and Anti-Platonism in Mathematics*. New York: Oxford University Press.

Bueno, O. (2005). "Dirac and the Dispensability of Mathematics", *Studies in History and Philosophy of Modern Physics 36*, pp. 465–490.

Bueno, O., and Zalta, E. (2005). "A Nominalist's Dilemma and Its Solution", *Philosophia Mathematica 13*, pp. 294–307.

Burgess, J., and Rosen, G. (1997). *A Subject With No Object: Strategies for Nominalistic Interpretation of Mathematics*. Oxford: Clarendon Press.

Chihara, C.S. (1990). *Constructibility and Mathematical Existence.* Oxford: Clarendon Press.

Colyvan, M. (2001). *The Indispensability of Mathematics.* New York: Oxford University Press.

Field, H. (1980). *Science without Numbers: A Defense of Nominalism.* Princeton, NJ: Princeton University Press.

Field, H. (1989). *Realism, Mathematics and Modality.* Oxford: Basil Blackwell.

Hellman, G. (1989). *Mathematics without Numbers: Towards a Modal-Structural Interpretation.* Oxford: Clarendon Press.

Lewis, D. (1986). *On the Plurality of Worlds.* Oxford: Blackwell.

McGinn, C. (2000). *Logical Properties.* Oxford: Clarendon Press.

Quine, W.V. (1960). *Word and Object.* Cambridge, Mass.: The MIT Press.

Resnik, M. (1997). *Mathematics as a Science of Patterns.* Oxford: Clarendon Press.

Rosen, G. (1994). "What Is Constructive Empiricism?", *Philosophical Studies 74,* pp. 143–178.

Shapiro, S. (1997). *Philosophy of Mathematics: Structure and Ontology.* New York: Oxford University Press.

van Fraassen, B.C. (1980). *The Scientific Image.* Oxford: Clarendon Press.

van Fraassen, B.C. (1989). *Laws and Symmetry.* Oxford: Clarendon Press.

van Fraassen, B.C. (1991). *Quantum Mechanics: An Empiricist View.* Oxford: Clarendon Press.

Thomasson, A. (1999). *Fiction and Metaphysics.* Cambridge: Cambridge University Press.

von Neumann, J. (1932). *Mathematical Foundations of Quantum Mechanics.* (English translation, 1955.) Princeton: Princeton University Press.

Weyl, H. (1928). *The Theory of Groups and Quantum Mechanics.* (English translation, 1931.) New York: Dover.

4
Truth in Mathematics: The Question of Pluralism

*Peter Koellner**

"... before us lies the boundless ocean of unlimited possibilities."
—Carnap, *The Logical Syntax of Language*

The discovery of non-Euclidean geometries (in the nineteenth century) undermined the claim that Euclidean geometry is the one true geometry and instead led to a plurality of geometries no one of which could be said (without qualification) to be "truer" than the others. In a similar spirit many have claimed that the discovery of independence results for arithmetic and set theory (in the twentieth century) has undermined the claim that there is one true arithmetic or set theory and that instead we are left with a plurality of systems no one of which can be said to be "truer" than the others. In this chapter I will investigate such pluralist conceptions of arithmetic and set theory. I will begin with an examination of what is perhaps the most sophisticated and developed version of the pluralist view to date—namely, that of Carnap in *The Logical Syntax of Language*—and I will argue that this approach is problematic and that the pluralism involved is too radical. In the remainder of the chapter I will investigate the question of what it would take to establish a more reasonable pluralism. This will involve mapping out some mathematical scenarios (using recent results proved jointly with Hugh Woodin) in which the pluralist could arguably maintain that pluralism has been secured.

Here is the plan of the chapter. In §1, I give a brief historical overview of the emergence of pluralism and an accompanying minimalist conception of philosophy, starting with Descartes and ending with Carnap. In §2, I investigate Carnap's pluralism and his minimalist conception of philosophy, arguing that although Carnap's pluralism is defensible in some domains (for example, with respect to certain metaphysical and purely notational questions), both his pluralism in physics and his pluralism in mathematics are untenable. In the course of the argument we shall see that mathematics is rather resilient to attempts to prove that it is fleeting. I also argue that Carnap's minimalist conception of philosophy is too extreme—instead of placing philosophy before science it places philosophy after science. In contrast, I suggest that there is room (and historical precedent) for a more meaningful engagement between philosophy and the exact sciences and, moreover, that it is through such an engagement that we can properly approach the question of pluralism in mathematics. The rest of the chapter is devoted to exploring whether through such an engagement we could be led to a defensible pluralism. In §3, I lay the groundwork for this new orientation by drawing a structural parallel between physics and mathematics. Einstein's work on special relativity is given as an exemplar of the kind of meaningful engagement I have in mind. For it is with Einstein that we come to see, for reasons at the intersection of philosophy and science, that statements once thought to have absolute significance (such as "*A* is simultaneous with *B*") are ultimately relativized. Our question then is whether something similar could happen in arithmetic or set theory, that is, whether for reasons at the intersection of philosophy and mathematics—reasons sensitive to actual developments in mathematics—we could come to see that statements once thought to have absolute significance (such as the Continuum Hypothesis (CH)) are ultimately relativized. In §4, I investigate the region where we currently have reason to believe that there is no pluralism concerning mathematical statements (this includes first- and second-order arithmetic). Finally, in §5, I begin by considering some standard arguments to the effect that our current mathematical knowledge already secures pluralism in set theory (for example, with regard to CH). After arguing that this is not the case, I map out a scenario (the best to my knowledge) in which one could arguably maintain that pluralism holds (even at the level of third-order arithmetic). This scenario has the virtue that it is sensitive to future developments in mathematics. In this way, by presenting such scenarios, we will, with time, be able to test just how resilient mathematics really is.

1   The emergence of pluralism

To introduce some of our main themes let us begin by reviewing some key developments in the history of the relation between philosophy and the exact sciences, starting with Descartes (in the seventeenth century) and ending with Carnap (in the first half of the twentieth century). We shall see that under the influence of two major scientific revolutions—the Newtonian and the Einsteinian—the pendulum swung from a robust conception of mathematics and maximalist conception of philosophy (one in which philosophy is prior to the exact sciences) to a pluralist conception of mathematics and minimalist conception of philosophy (one that places philosophy after the exact sciences).[1]

1.1   Descartes

Descartes distinguished between "natural philosophy"—which concerns the corporal world—and "first philosophy" (or "metaphysics")—which concerns the incorporeal world. From the seventeenth to the nineteenth century the term "natural philosophy" was used for what we now call the exact sciences. For Descartes, first philosophy was prior to natural philosophy in that it was required to lay the rational foundation for natural philosophy.

This view of the privileged status of first philosophy with respect to natural philosophy came to be challenged by developments in physics. To begin with, this approach did not fulfill its promise of a comprehensive and unified natural philosophy. Indeed, in many cases, it led to results that were at variance with experience. For example, it led Descartes (in the *The Principles of Philosophy* (1644)) to formulate laws of impact that were at odds with observation. Huygens then found the correct laws through an ingenious blend of experience and reason (here embodied in the principle of relativity). In this way natural philosophy gained ground as an independent discipline. The true vindication of natural philosophy as an independent disciple came with Newton's *The Mathematical Principles of Natural Philosophy* (1687). Indeed it is here—in natural philosophy and not in first philosophy—that one finds the first true hope of a comprehensive and adequate account of the corporal world. Moreover, Newton's developments created additional trouble for first philosophy since there were aspects of Newton's system—such as action at a distance and absolute space—that were incompatible with the apparent results of first philosophy.

1.2   Kant

Kant rejected the pretensions of first philosophy and instead took Newtonian physics to be an exemplar of theoretical reason. On Kant's view, the task for philosophy proper is not to arrive at natural philosophy from some higher, privileged vantage point,[2] but rather to take natural philosophy as given and then articulate the necessary conditions for its possibility.[3] The Kantian solution involves the notion of the constitutive a priori: The basic laws of logic, arithmetic, Euclidean geometry, and the basic laws of Newtonian physics are not things we "find in the world", rather they are things we "bring to the world"—they are the necessary conditions of our experience of the world (in particular, they are necessary for the formulation of the Newtonian law of universal gravitation).[4]

This view of the central and secure status of both Euclidean geometry and the basic laws of Newtonian physics came to be challenged (like its predecessor) by developments in mathematics and physics. First, the discovery of non-Euclidean geometries in the nineteenth century dethroned Euclidean geometry from its privileged position and led to the important distinction between pure (formal) geometry and applied (physical) geometry. Second, many of the basic Newtonian laws were overturned by Einstein's special theory of relativity. Finally, Einstein's general theory of relativity provided reason to believe that the geometry of physical space is actually non-Euclidean.

1.3 Reichenbach

Reichenbach accommodated these developments by relativizing the notion of the constitutive a priori.[5] The key observation concerns the fundamental difference between definitions in pure geometry and definitions in physical geometry. In pure geometry there are two kinds of definition: first, there are the familiar *explicit* definitions; second, there are *implicit* definitions, that is the kind of definition whereby such fundamental terms as 'point', 'line', and 'surface' are to derive their meaning from the fundamental axioms governing them.[6] But in physical geometry a new kind of definition emerges—that of a *physical* (or *coordinative*) definition:

> The physical definition takes the meaning of the concept for granted and coordinates to it a physical thing; it is a *coordinative definition*. Physical definitions, therefore, consist in the coordination of a mathematical definition to a "piece of reality"; one might call them *real definitions*. (Reichenbach 1924, p. 8)

As an example, Reichenbach gives the coordination of "beam of light" with "straight line".[7]

Now there are two important points about physical definitions. First, some such correlation between a piece of mathematics and "a piece of physical reality" is necessary if one is to articulate the laws of physics (e.g. consider "force-free moving bodies travel in straight lines"). Second, given a piece of pure mathematics there is a great deal of freedom in choosing the coordinative definitions linking it to "a piece of physical reality".[8] So we have here a conception of the a priori which (by the first point) is constitutive (of the empirical significance of the laws of physics) and (by the second point) is relative. Moreover, on Reichenbach's view, in choosing between two empirically equivalent theories that involve different coordinative definitions, there is no issue of "truth"—there is only the issue of simplicity.[9]

Now, Reichenbach went beyond this and he held a more radical thesis—in addition to advocating pluralism with respect to physical geometry (something made possible by the free element in coordinative definitions), he advocated pluralism with respect to pure mathematics (such as arithmetic and set theory). According to Reichenbach, this view is made possible by the axiomatic conception of Hilbert, wherein axioms are treated as "implicit definitions" of the fundamental terms:

> The problem of the axioms of mathematics was solved by the discovery that they are definitions, that is, arbitrary stipulations which are neither

true nor false,[10] and that only the logical properties of a system—its consistency, independence, uniqueness, and completeness—can be subjects of critical investigation. (Reichenbach 1924, p. 3)

On this view there is a plurality of consistent formal systems and the notions of "truth" and "falsehood" do not apply to these systems; the only issue in choosing one system over another is one of convenience for the purpose at hand and this is brought out by investigating their metamathematical properties, something that falls within the provenance of "critical investigation", where there *is* a question of truth and falsehood.

This radical form of pluralism came to be challenged by Gödel's discovery of the incompleteness theorems. To begin with, through the arithmetization of syntax, the metamathematical notions that Reichenbach takes to fall within the provenance of "critical investigation" were themselves seen to be a part of arithmetic. Thus, one cannot, on pain of inconsistency, say that there is a question of truth and falsehood with regard to the former but not the latter. More importantly, the incompleteness theorems buttressed the view that truth outstrips consistency. This is most clearly seen using Rosser's strengthening of the first incompleteness theorem as follows: Let T be an axiom system of arithmetic that (a) falls within the provenance of "critical investigation" and (b) is sufficiently strong to prove the incompleteness theorem.[11] Then, assuming that T is consistent (something which falls within the provenance of "critical investigation"), by Rosser's strengthening of the first incompleteness theorem, there is a Π_1^0-sentence φ such that (provably within $T + \mathrm{Con}(T)$) both $T + \varphi$ and $T + \neg\varphi$ are consistent. However, not both systems are equally legitimate. For it is easily seen that if a Π_1^0-sentence φ is independent from such a theory, then it must be true.[12] So, although $T + \neg\varphi$ is consistent, it proves a false arithmetical statement.[13]

1.4   Carnap

Nevertheless, in full awareness of the incompleteness theorems, in *The Logical Syntax of Language* (1934), Carnap held a view that, on the face of it at least, appears to be quite similar to Reichenbach's view, both with regard to the thesis concerning the conventional element in physical geometry and the thesis concerning the purely conventional nature of pure mathematics. However, Carnap's position is much more subtle and sophisticated. Indeed I think that it is the most fully developed account of a pluralist conception of mathematics that we have and, for this reason, I will examine it in detail in the next section.

What I want to point out here is that Carnap also articulated an accompanying minimalist conception of philosophy, one according to which "*[p]hilosophy is to be replaced by the logic of science*—that is to say, by the logical analysis of the concepts and sentences of the sciences, for *the logic of science is nothing other than the logical syntax of the language of science.*"

(Carnap 1934, p. xiii) Thus the pendulum has gone full swing: We started with a robust conception of mathematics and maximalist conception of philosophy (where philosophy comes before the exact sciences) and through the influence of two scientific revolutions we were led to a pluralist conception of mathematics and minimalist conception of philosophy (where philosophy comes after the exact sciences).[14]

In the next section I will criticize both Carnap's pluralist conception of mathematics and his minimalist conception of philosophy and suggest that it is through a more meaningful engagement between philosophy and mathematics that we can properly address the question of pluralism.

2 Carnap on the foundations of logic and mathematics

In *The Logical Syntax of Language* (1934)[15] Carnap defends the following three distinctive philosophical theses: (1) The thesis that logic and mathematics are analytic and hence without content[16] and purely formal.[17] (2) A radical pluralist conception of pure mathematics (embodied in his Principle of Tolerance) according to which the meaning of the fundamental terms is determined by the postulates governing them and hence any consistent set of postulates is equally legitimate.[18] (3) A minimalist conception of philosophy in which most traditional questions are rejected as pseudo-questions and the task of philosophy is identified with the study of the logical syntax of the language of science.[19]

I will deal with these three philosophical theses in order in the following three subsections.[20]

2.1 Logic, mathematics, and content

The first philosophical thesis—that logical and mathematical truths are analytic and hence without content—involves Carnap's distinctive account of the notions of analyticity and content. So it is here that we shall begin. But first we need to make a few terminological remarks since some Carnap's terminology differs from modern terminology and, where there is overlap, his usage—most notably the term 'syntax'—is often out of step with modern usage.

2.1.1 Some key terminology

A central distinction for Carnap is that between *definite* and *indefinite* notions. A *definite* notion is one that is recursive, such as "is a formula" and "is a proof of φ". An *indefinite* notion is one that is non-recursive, such as "is an ω-consequence of PA" and "is true in $V_{\omega+\omega}$". This leads to a distinction between (i) the *method of derivation* (or *d-method*), which investigates the semi-definite (recursively enumerable) metamathematical notions, such as *demonstrable, derivable, refutable, resoluble*, and *irresoluble*, and (ii) the *method of consequence* (or *c-method*), which investigates the (typically) non-recursively enumerable metamathematical notions such as *consequence, analytic, contradictory, determinate,* and *synthetic.*

A *language* for Carnap is what we would today call a formal axiom system.[21] The rules of the formal system are definite (recursive)[22] and Carnap is fully aware that a given language cannot include its own c-notions (see Theorem 60c.1).

The *logical syntax* of a language is what we would today call metatheory. It is here that one formalizes the c-notions for the (object) language. From among the various c-notions Carnap singles out one as central, namely, the notion of (direct) consequence; from this c-notion all of the other c-notions can be defined in routine fashion.[23]

2.1.2   The analytic/synthetic distinction

We now turn to Carnap's account of his fundamental notions, most notably, the analytic/synthetic distinction and the division of primitive terms into 'logico-mathematical' and 'descriptive'. Carnap actually has two approaches. The first approach occurs in his discussion of specific languages—Languages I and II. Here he starts with a division of primitive terms into 'logico-mathematical' and 'descriptive' and upon this basis defines the c-notions, in particular the notions of being analytic and synthetic. The second approach occurs in the discussion of general syntax. Here Carnap reverses procedure: he starts with a specific c-notion—namely, the notion of direct consequence—and he uses it to define the other c-notions and draw the division of primitive terms into 'logico-mathematical' and 'descriptive'.

A. The First Approach.   In the first approach Carnap introduces two languages—Language I and Language II. It is important to note (as we have above) that here by 'language' Carnap means what we would call a 'formal system'.[24] The background languages (in the modern sense) of Language I and Language II are quite general—they include expressions that we would call 'descriptive'. Carnap starts with a demarcation of primitive terms into 'logico-mathematical' and 'descriptive'. The expressions he classifies as 'logico-mathematical' are exactly those included in the modern versions of these systems; the remaining expressions are classified as 'descriptive'. Language I is a version of PRA and Language II is a version of finite type theory built over PA. The d-notions for these languages are the standard proof-theoretic ones. So let us concentrate on the c-notions.

For Language I Carnap starts with a consequence relation based on two rules—(i) the rule that allows one to infer φ if $T \vdash \varphi$ (where T is some fixed Σ_1^0-complete formal system) and (ii) the ω-rule. It is then easily seen that one has a complete theory for the logico-mathematical fragment, that is, for any logico-mathematical sentence φ, either φ or $\neg\varphi$ is a consequence of the null set. The other c-notions are then defined in the standard fashion. For example, a sentence is *analytic* if it is a consequence of the null set; *contradictory* if its negation is analytic; and so on.

For Language II Carnap starts by defining analyticity. His definition is a notational variant of the Tarskian truth definition with *one important difference*—namely, it involves an asymmetric treatment of the logico-mathematical and descriptive expressions. For the logico-mathematical expressions his definition really just is a notational variant of the Tarskian truth definition. But descriptive expressions must pass a more stringent test to count as analytic—they must be such that if one replaces all descriptive expressions in them by variables of the appropriate type, then the resulting logico-mathematical expression is analytic, that is, true.[25] In other words, to count as analytic a descriptive expression must be a substitution-instance of a general logico-mathematical truth. With this definition in place the other c-notions are defined in the standard fashion.

The *content* of a sentence is defined to be the set of its non-analytic consequences. It then follows immediately from the definitions that logico-mathematical sentences (of both Language I and Language II) are analytic or contradictory and (assuming consistency) that analytic sentences are without content.

B. *The Second Approach.* In the second approach, for a given language, Carnap starts with an *arbitrary* notion of direct consequence[26] and from this notion he defines the other c-notions in the standard fashion. More importantly, in addition to defining the other c-notion, Carnap also uses the primitive notion of direct consequence (along with the derived c-notions) to effect the classification of terms into 'logico-mathematical' and 'descriptive'. The guiding idea is that "the formally expressible distinguishing peculiarity of logical symbols and expressions [consists] in the fact that each sentence constructed solely from them is determinate" (177).[27] He then gives a formal definition that aims to capture this idea. His actual definition is problematic for various technical reasons[28] and so we shall leave it aside. What is important for our purposes (as shall become apparent in §2.1.3) is the fact that (however the guiding idea is implemented) the actual division between "logico-mathematical" and "descriptive" expressions that one obtains as output is sensitive to the scope of the direct consequence relation with which one starts.

With this basic division in place, Carnap can now draw various derivative divisions, most notably, the division between analytic and synthetic statements: Suppose φ is a consequence of Γ. Then φ is said to be an *L-consequence* of Γ if either (i) φ and the sentences in Γ are logico-mathematical, or (ii) letting φ' and Γ' be the result of unpacking all descriptive symbols, then for every result φ'' and Γ'' of replacing every (primitive) descriptive symbol by an expression of the same genus (a notion that is defined on p. 170), maintaining equal expressions for equal symbols, we have that φ'' is a consequence of

Γ″. Otherwise φ is a *P-consequence* of Γ. This division of the notion of consequence into *L-consequence* and *P-consequence* induces a division of the notion
of demonstrable into *L-demonstrable* and *P-demonstrable* and the notion of
valid into *L-valid* and *P-valid* and likewise for all of the other d-notions
and c-notions. The terms *'analytic'*, *'contradictory'*, and *'synthetic'* are used
for 'L-valid', 'L-contravalid', and 'L-indeterminate'.

Again it follows immediately from the definitions that logico-
mathematical sentences are analytic or contradictory and that analytic
sentences are without content. This is what Carnap says in defense of the
first of his three basic theses.

2.1.3   Criticism #1: The argument from free parameters

The trouble with the first approach is that the definitions of analyticity
that Carnap gives for Languages I and II are highly sensitive to the original classification of terms into 'logico-mathematical' and 'descriptive'. And
the trouble with the second approach is that the division between 'logico-
mathematical' and 'descriptive' expressions (and hence division between
'analytic' and 'synthetic' truths) is sensitive to the scope of the direct
consequence relation with which one starts. This threatens to undermine
Carnap's thesis that logico-mathematical truths are analytic and hence
without content. Let us discuss this in more detail.

In the first approach, the original division of terms into 'logico-
mathematical' and 'descriptive' is made by stipulation and if one alters this
division one thereby alters the derivative division between analytic and synthetic sentences. For example, consider the case of Language II. If one calls
only the primitive terms of first-order logic 'logico-mathematical' and then
extends the language by adding the machinery of arithmetic and set theory,
then, upon running the definition of 'analytic', one will have the result that
true statements of first-order logic are without content while (the distinctive) statements of arithmetic and set theory have content.[29] For another
example, if one takes the language of arithmetic, calls the primitive terms
'logico-mathematical' and then extends the language by adding the machinery of finite type theory, calling the basic terms 'descriptive', then, upon
running the definition of 'analytic', the result will be that statements of
first-order arithmetic are analytic or contradictory while (the distinctive)
statements of second- and higher-order arithmetic are synthetic and hence
have content. In general, by altering the input, one alters the output, and
Carnap adjusts the input to achieve his desired output.[30]

In the second approach, there are no constraints on the scope of the direct
consequence relation with which one starts and if one alters it one thereby
alters the derivative division between 'logico-mathematical' and 'descriptive' expressions. Recall that the guiding idea is that logical symbols and
expressions have the feature that sentences composed solely of them are

determinate. The trouble is that (however one implements this idea) the resulting division of terms into 'logico-mathematical' and 'descriptive' will be highly sensitive to the scope of the direct consequence relation with which one starts.[31] For example, let S be first-order PA and for the direct consequence relation take "provable in PA". Under this assignment Fermat's Last Theorem will be deemed descriptive, synthetic, and to have non-trivial content.[32] For an example at the other extreme, let S be an extension of PA that contains a physical theory and let the notion of direct consequence be given by a Tarskian truth definition for the language. Since in the metalanguage one can prove that every sentence is true or false, every sentence will be either analytic (and so have null content) or contradictory (and so have total content). To overcome such counter-examples and get the classification that Carnap desires one must ensure that the consequence relation is (i) complete for the sublanguage consisting of expressions that one wants to come out as 'logico-mathematical' and (ii) not complete for the sublanguage consisting of expressions that one wants to come out as 'descriptive'. Once again, by altering the input, one alters the output, and Carnap adjusts the input to achieve his desired output.

To summarize: What we have (in either approach) is not a principled distinction. Instead, Carnap has merely provided us with a flexible piece of technical machinery involving free parameters that can be adjusted to yield a variety of outcomes concerning the classifications of analytic/synthetic, contentful/non-contentful, and logico-mathematical/descriptive. In his own case, he has adjusted the parameters in such a way that the output is a formal articulation of his logicist view of mathematics that the truths of mathematics are analytic and without content. And one can adjust them differently to articulate a number of other views, for example, the view that the truths of first-order logic are without content while the truths of arithmetic and set theory have content. The possibilities are endless. The point, however, is that we have been given no reason for fixing the parameters one way rather than another. The distinctions are thus not principled distinctions. It is trivial to prove that mathematics is trivial if one trivializes the claim.

2.1.4 *Criticism #2: The argument from assessment sensitivity*

Carnap is perfectly aware that to define c-notions like analyticity one must ascend to a stronger metalanguage. However, there is a distinction that he appears to overlook,[33] namely, the distinction between (i) having a stronger system S' that can *define* 'analytic in S' and (ii) having a stronger system S' that can, in addition, *evaluate* a given statement of the form 'φ is analytic in S'. It is an elementary fact that two systems S_1 and S_2 can employ the same definition (from an intensional point of view) of 'analytic in S' (using either the definition given for Language I or Language II) but *differ* on their evaluation of 'φ is analytic in S' (that is, differ on the extension of 'analytic in

S'). Thus, to determine whether 'φ is analytic in S' holds one needs to access much more than the "syntactic design" of φ—in addition to ascending to an essentially richer metalanguage one must move to a sufficiently strong system to evaluate 'φ is analytic in S'. The first step need not be a big one.[34] But for certain φ the second step must be huge.[35]

In fact, it is easy to see that to answer 'Is φ analytic in Language I?' is just to answer φ and, in the more general setting, to answer all questions of the form 'Is φ analytic in S?' (for various mathematical φ and S), where here 'analytic' is defined as Carnap defines it for Language II, *just is to answer all questions of mathematics*.[36] The same, of course, applies to the c-notion of consequence. So, when in first stating the Principle of Tolerance, Carnap tells us that we can choose our system S arbitrarily and that 'no question of justification arises at all, but *only* the question of the syntactical consequences to which one or other of the choices leads' (p. xv, my emphasis)—where here, as elsewhere, he means the c-notion of consequence—*he has given us no assurance, no reduction at all.*

2.2   Radical pluralism

This brings us to the second philosophical thesis—the thesis of pluralism in mathematics. Let us first note that Carnap's pluralism is quite radical. We are told that "any postulates and any rules of inference [may] be chosen arbitrarily" (xv); for example, the question of whether the Principle of Selection (that is, the Axiom of Choice (AC)) should be admitted is "purely one of expedience" (142); more generally,

> The [logico-mathematical sentences] are, from the point of view of material interpretation, expedients for the purpose of operating with the [descriptive sentences]. Thus, in laying down [a logico-mathematical sentence] as a primitive sentence, only usefulness for this purpose is to be taken into consideration. (142)

So the pluralism is quite broad—it extends to AC and even to Π_1^0-sentences.[37]

Now, as I argued in §1.3 on Reichenbach, there are problems in maintaining Π_1^0-pluralism. One cannot, on pain of inconsistency, think that statements about consistency are not "mere matters of expedience" without thinking that Π_1^0-statements generally are not mere "matters of expedience". But I want to go further than make such a negative claim. I want to uphold the default view that the question of whether a given Π_1^0-sentence holds is *not* a mere matter of expedience; rather, such questions fall within the provenance of theoretical reason.[38] In addition to being the default view, there are solid reasons behind it. One reason is that in adopting a Π_1^0-sentence one could always be struck by a counter-example.[39] Other reasons have to do with the clarity of our conception of the natural numbers and with our experience to date with that structure. On this basis, I would go further and maintain that for *no* sentence of first-order arithmetic is the question of whether it holds a mere matter of experience. Certainly this is the default view from which one must be moved.

What does Carnap have to say that will sway us from the default view, and lead us to embrace his radical form of pluralism? In approaching this question it is important to bear in mind that there are two general interpretations of Carnap. According to the first interpretation—the *substantive*—Carnap is really trying to *argue* for the pluralist conception. According to the second interpretation—the *non-substantive*—he is merely trying to *persuade* us of it, that is, to show that of all the options it is most "expedient".[40]

The most obvious approach to securing pluralism is to appeal to the work on analyticity and content. For if mathematical truths are without content and, moreover, this claim can be maintained with respect to an arbitrary mathematical system, then one could argue that even apparently incompatible systems have null content and hence are really compatible (since there is no contentual-conflict).

Now, in order for this to secure radical pluralism, Carnap would have to first secure his claim that mathematical truths are without content. But, as we have argued above, he has not done so. Instead, he has merely provided us with a piece of technical machinery that can be used to articulate any one of a number of views concerning mathematical content and he has adjusted the parameters so as to articulate his particular view. So he has not *secured* the thesis of radical pluralism.[41] Thus, on the substantive interpretation, Carnap has failed to achieve his end.

This leaves us with the non-substantive interpretation. There are a number of problems that arise for this version of Carnap. To begin with, Carnap's technical machinery is not even suitable for *articulating* his thesis of radical pluralism since (using either the definition of analyticity for Language I or Language II) there is no metalanguage in which one can say that two apparently incompatible systems S_1 and S_2 both have null content and hence are really contentually compatible. To fix ideas, consider a paradigm case of an apparent conflict that we should like to dissolve by saying that there is no contentual-conflict: Let $S_1 = PA + \varphi$ and $S_2 = PA + \neg\varphi$, where φ is any arithmetical sentence, and let the metatheory be $MA = ZFC$. The trouble is that on the approach to Language I, although in MT we can prove that each system is ω-complete (which is a start since we wish to say that each system has null content), we can also prove that one has null content while the other has *total* content (that is, in ω-logic, *every* sentence of arithmetic is a consequence). So, we cannot, within MT articulate the idea that there is no contentual-conflict.[42] The approach to Language II involves a complementary problem. To see this note that while a strong logic like ω-logic is something that one can apply to a *formal system*, a truth definition is something that applies to a *language* (in our modern sense). Thus, on this approach, in MT the definition of analyticity given for S_1 and S_2 is the same (since the two systems are couched in the same language). So, although in MT we can say that S_1 and S_2 do not have a contentual-conflict this is only because we have given a deviant definition of analyticity, one that is blind to the fact that in a very straightforward sense φ is analytic in S_1 while $\neg\varphi$ is analytic in S_2.

Now, although Carnap's machinery is not adequate to articulate the thesis of radical pluralism (for a given collection of systems) in a given metatheory, under certain circumstances he can attempt to articulate the thesis by changing the metatheory. For example, let $S_1 = \text{PA} + \text{Con}(\text{ZF} + \text{AD})$ and $S_2 = \text{PA} + \neg\text{Con}(\text{ZF} + \text{AD})$ and suppose we wish to articulate both the idea that the two systems have null content and the idea that $\text{Con}(\text{ZF} + \text{AD})$ is analytic in S_1 while $\neg\text{Con}(\text{ZF} + \text{AD})$ is analytic in S_2. As we have seen no single metatheory (on either of Carnap's approaches) can do this. But it turns out that because of the kind of assessment sensitivity that we discussed in §2.1.4, there are two metatheories MT_1 and MT_2 such that in MT_1 we can say both that S_1 has null content and that $\text{Con}(\text{ZF} + \text{AD})$ is analytic in S_1, while in MT_2 we can say both that S_2 has null content and that $\neg\text{Con}(\text{ZF} + \text{AD})$ is analytic in S_2. But, of course, this is simply because (any such metatheory) MT_1 proves $\text{Con}(\text{ZF} + \text{AD})$ and (any such metatheory) MT_2 proves $\neg\text{Con}(\text{ZF} + \text{AD})$. So we have done no more than (as we must) reflect the difference between the systems in the metatheories. Thus, although Carnap does not have a way of articulating his radical pluralism (in a given metalanguage), he certainly has a way of *manifesting* it (by making corresponding changes in his metatheories).

As a final retreat Carnap might say that he is not trying to persuade us of a *thesis* that (concerning a collection of systems) can be articulated in a given framework but rather is trying to persuade us to adopt a thorough radical pluralism as a *"way of life"*. He has certainly shown us how we can make the requisite adjustments in our metatheory so as to consistently manifest radical pluralism. But does this amount to more than an algorithm for begging the question?[43] Has Carnap shown us that there is no question to beg? I do not think that he has said anything persuasive in favour of embracing a thorough radical pluralism as the "most expedient" of the options. The trouble with Carnap's entire approach (as I see it) is that the question of pluralism has been detached from actual developments in mathematics. To be swayed from the default position something of greater substance is required.

2.3 Philosophy as logical syntax

This brings us finally to Carnap's third philosophical thesis—the thesis that philosophy is the logical syntax of the language of science. This thesis embodies a pluralism more wide-ranging than mathematical pluralism. For just as foundational disputes in mathematics are to be dissolved by replacing the theoretical question of the justification of a system with the practical question of the expedience of adopting the system, so too many philosophical and physical disputes are to be dissolved in a similar fashion:

> It is especially to be noted that the statement of a philosophical the- sis sometimes... represents not an *assertion* but a *suggestion*. Any dispute about the truth or falsehood of such a thesis is quite mistaken, a mere battle of words; we can at most discuss the utility of the proposal, or investigate its consequences. (299–300)

Once this shift is made one sees that

> the question of truth or falsehood cannot be discussed, but only the ques- tion whether this or that form of language is the more appropriate for certain purposes. (300)

Philosophy has a role to play in this. For, on Carnap's conception, "as soon as claims to scientific qualifications are made" (280), philosophy just *is* the study of the syntactical consequences of various scientific systems.

I want to focus on this distinction between "the question of truth or false- hood" and "the question of whether this or that form of language is the more appropriate for certain purposes" as Carnap employs it in his discussion of metaphysics, physics, and mathematics.

Let us start with metaphysics. The first example that Carnap gives (in a long list of examples) concerns the apparent conflict between the state- ments "*Numbers* are classes of classes of things" and "*Numbers* belong to a special primitive kind of objects". On Carnap's view, these material for- mulations are really disguised versions of the proper formal or syntactic formulations "Numerical expressions are class-expressions of the second- level" and "Numerical expressions are expressions of the zero-level" (300). And when one makes this shift (from the "material mode" to the "formal mode") one sees that "the question of truth or falsehood cannot be dis- cussed, but only the question whether this or that form of language is the more appropriate for certain purposes" (300). There is a sense in which this is hard to disagree with. Take, for example, the systems PA and ZFC – Infinity. These systems are mutually interpretable (in the logician's sense). It seems that there is no theoretical question of choosing one over the other—as though one but not the other were "getting things right", "carving math- ematical reality at the joints"—but only a practical question of expedience relative to a given task. Likewise with apparent conflicts between systems that construct lines from points versus points from lines or use sets ver- sus well-founded trees, and so on. So, I am inclined to agree with Carnap on such *metaphysical* disputes. I would also agree concerning theories that are mere *notational* variants of one another (such as, for example, the con- flict between a system in a typed language and the mutually interpretable system obtained by "flattening" (or "amalgamating domains")). But I think that Carnap goes too far in his discussion of physics and mathematics. I have already discussed the case of mathematics (and I will have much more to say about it below). Let us turn to the case of physics.

Carnap's conventionalism extends quite far into physics. Concerning physical hypotheses he writes:

> The construction of the physical system is not affected in accordance with fixed rules, but by means of conventions. These conventions, namely, the rules of formation, the L-rules and the P-rules (hypothesis), are, however, not arbitrary. The choice of them is influenced, in the first place, by certain practical methodological considerations (for instance, whether they make for simplicity, expedience, and fruitfulness in certain tasks). .... But in addition the hypotheses can and must be tested by experience, that is to say, by the protocol-sentences—both those that are already stated and the new ones that are constantly being added. Every hypothesis must be compatible with the total system of hypotheses to which the already recognized protocol-sentences also belong. That hypotheses, in spite of their subordination to empirical control by means of the protocol-sentences, nevertheless contain a conventional element is due to the fact that the system of hypotheses is never univocally determined by empirical material, however rich it may be. (320)

Thus, while in pure mathematics it is convention and question of expedience all the way (modulo consistency), in physics it is convention and question of expedience modulo empirical data.

I think that this conception of theoretical reason in physics is too narrow. Consider the following two examples: first, the historical situation of the conflict between the Ptolemaic and the Copernican accounts of the motion of the planets and, second, the conflict between the Lorentz's mature theory of 1905 (with Newtonian spacetime) and Einstein's special theory of relativity (with Minkowski spacetime). In the first case, the theories are empirically equivalent to within 3′ of arc and this discrepancy was beyond the powers of observation at the time. In the second case, there is outright empirical equivalence.[44] Yet I think that the reasons given at the time in favour of the Copernican theory were not of the practical variety; they were not considerations of expedience, they were solid theoretical reasons pertaining to truth.[45] Likewise with the reasons for Einstein's theory over Lorentz's (empirically equivalent) mature theory of 1905.[46]

2.3.1   Conclusion

Let us summarize the above conclusions. Carnap has not given a substantive account of analyticity and so he has not given a *defense* of his first thesis—the thesis that mathematical truths are analytic and hence formal and without content. Instead he has presented some technical machinery that can be used to formally articulate his view. In the case of his second thesis—that radical pluralism holds in mathematics—the situation is even worse. Not only has he not provided a defense of this thesis, he has not even provided technical machinery that is suitable to *articulate* the thesis (for a given collection of systems) in a given metatheory. All he can do is *manifest* his radical pluralism by mirroring the differences among the candidate systems as differences among their metatheories.

These limitations lead one to think that perhaps Carnap's first two theses were not intended as assertions—theses to be argued for—but rather as suggestions—theses to be adopted for practical reasons as the most expedient of the available options. The trouble is that he has not provided a persuasive case that these theses are indeed the most expedient. In any case, one cannot refute a proposal. One can only explain one's reasons for not following it—for thinking that it is not most expedient.

There is something right in Carnap's motivation. His motivation comes largely from his rejection of the myth of the model in the sky. One problem with this myth is that it involves an alienation of truth (to borrow an apt phrase of Tait).[47] However, the non-pluralist can agree in rejecting such a myth. Once we reject the myth and the pretensions of first philosophy, we are left with the distinction between substantive theoretical questions and matters of mere expedience. I agree with Carnap in thinking that the choice between certain metaphysical frameworks (e.g. whether we construct lines from points or points from lines) and between certain notational variants (e.g. whether we use a typed language or amalgamate types) are not substantive theoretical choices but rather matters of mere expedience. But I think that he goes too far in saying that the choice between empirically equivalent theories in physics and the choice between arbitrary consistent mathematical systems in mathematics are likewise matters of mere expedience. In many such cases one can provide convincing theoretical reasons for the adoption of one system over another. To deny the significance of such reasons appears to me to reveal a remnant of the kind of first philosophy that Carnap rightfully rejected.

Just as Kant was right to take Newtonian physics as exemplary of theoretical reason and retreat from a maximalist conception of philosophy—one that placed philosophy before science—I think that we should take Einsteinian physics as exemplary of theoretical reason and retreat from a minimalist conception of philosophy—one that places philosophy after science. Similarly, in the case of mathematics, I think we are right to take the developments in the search for new axioms seriously. But we must ensure that the pendulum does not return to its starting point. The proper balance, I think, lies in between, with a more meaningful engagement between philosophy and science. In the remainder of this chapter I shall outline such an approach. I believe that what I shall have to say is Carnapian in spirit, if not in letter.

3   A new orientation

I have embraced Carnap's pluralism with respect to certain purely metaphysical and notational questions but rejected it with regard to certain statements in physics and mathematics. For example, I agree with Carnap that there is no substantive issue involved in the choice between ZFC and a variant of this theory that uses the ontology of well-founded trees instead of sets. But I disagree in thinking the same with regard to the choice between ZFC + Con(ZFC) and ZFC + ¬Con(ZFC). Here there is a substantive difference. In slogan form my view might be put like this: Existence in mathematics is (relatively) cheap; truth is not.[48]

In the physical case I gave two examples of where I think we must part ways with Carnap. First, the historical situation of the choice between the Ptolemaic and Copernican account of the planets; second, the choice between Lorentz's mature theory (with Newtonian spacetime) and Einstein's theory of special relativity (with Minkowski spacetime). In opposition to Carnap, I hold that in each case the choice is not one of "mere expedience", but rather falls within the provenance of theoretical reason—indeed I take these to be some of the highest achievements of theoretical reason.

In addition to breaking with Carnap on the above (and other) particular claims, I think we should break with his minimalist conception of philosophy—a conception that places philosophy *after* science—and—without reverting to a conception that places it *before*—embrace a conception involving a more meaningful engagement between philosophy and science. Moreover, in doing so, I think that we can properly address the question of pluralism and thereby gain some insight into what it would take to secure a more reasonable pluralism.

In this connection, the theory of special relativity actually serves as a guide in *two* respects. First, as already mentioned, like the Copernican case, it illustrates the point that there are cases where, despite being in a situation where one has empirical equivalence, one can give theoretical reasons for one theory over another. Second, it does this in such a way that once one makes the step to special relativity one sees a kind of pluralism, namely, with respect to the class of all permissible foliations of Minkowski spacetime—the point being that it is in principle impossible to use the laws of physics to single out (in a principled way) any one of these foliations over another[49]—and so we change perspective, see each as standing on a par and bundle them all up into one spacetime: Minkowski spacetime.[50] The source of this form of pluralism is different than Carnap's—it is driven by developments at the intersection of philosophy and physics and is sensitive to developments that fall squarely within physics.

I want now to examine analogues of these two features in the mathematical case. In §4, I will argue that there are cases in mathematics where theoretical reason presents us with a convincing case for one theory over another. Indeed I will argue that theoretical reason goes quite a long way. But there are limitations to our current understanding of the universe of sets and there is a possibility that we are close to an impasse. In §5, I will investigate the possibility of being in a position where one has reason to believe that we are faced with a plurality of alternatives that cannot in principle be adjudicated on the basis of theoretical reason and, moreover, where this leads us to reconceive the nature of some fundamental notions in mathematics.

On the way toward this it will be helpful to introduce some machinery and use it to present a very rough analogy between the structure of physical theory and the structure of mathematical theory.

To a first approximation, let us take a physical theory to be a formal system with a set of coordinative definitions. We are interested in the relation of empirical equivalence between such theories. Let us first distinguish two levels of data: The *primary* data are the observational sentences (such as "At time t, wandering star w has longitude and latitude (φ, ϑ)") that have actually been verified; the *secondary* data are the observational generalizations (such as "For each time t, wandering star w has longitude and latitude $(\varphi(t), \vartheta(t))$"). The primary data (through accumulation) can provide us with inductive evidence of the secondary data. We shall take our notion of empirical equivalence to be based on the secondary data. Thus, two physical theories are *empirically equivalent* if and only if they agree on the secondary data.[51] The *problem of selection* in physics is to select from the equivalence classes of empirically equivalent theories. Some choices— for example, certain purely metaphysical and notational choices—are merely matters of expedience; others—for example, that between Lorentz's mature theory and special relativity—are substantive and driven by theoretical reason.

In the mathematical case we can be more exact. A theory is just a (recursively enumerable) formal system and as our notion of equivalence we shall take the notion of mutual interpretability in the logician's sense. The details of this definition would be too distracting to give here;[52] suffice it to say that the technical definition aims to capture the notion of interpretation that is ubiquitous in mathematics, the one according to which Poincaré provided an interpretation of two-dimensional hyperbolic geometry in the Euclidean geometry of the unit circle, Dedekind provided an interpretation of analysis in set theory, and Gödel provided an interpretation of the theory of formal syntax in arithmetic. Every theory of pure mathematics is mutually interpretable with a theory in the language of set theory. So, for ease of exposition, we will concentrate on theories in the language of set theory.[53] We shall assume that all theories contain ZFC − Infinity. As in the case of physics we shall distinguish two levels of data: The *primary* data are the Δ_1^0-sentences that have actually been verified (such as, to choose a metamathematical example, "p is not a proof of ¬Con(ZFC)") and the *secondary* data are the corresponding generalizations, which are Π_1^0-sentences (such as, "For all proofs p, p is not a proof of ¬Con(ZFC)"). As in the case of physics, in mathematics the secondary data can be definitely refuted but never definitely verified. Nevertheless, again as in the case of physics, the primary data can provide us with evidence for the secondary data. Finally, given our assumption that all theories contain ZFC – Infinity, we have the following fundamental result: Two theories are mutually interpretable if and only if they prove the same Π_1^0-sentences, that is, if and only if they agree on the secondary data. Thus, we have a nice parallel with the physical case, with Π_1^0-sentences being the analogues of observational generalizations. The *problem of selection* in mathematics is to select from the equivalence classes of mutual interpretability (which are called the interpretability *degrees*). Some choices— for example, certain purely metaphysical and notational choices—are mere matters of expedience. But in restricting our attention to a fixed language we have set most of these aside. Other choices are substantive and fall within the provenance of theoretical reason. For example, let T be ZFC – Infinity. One can construct a Δ_2^0-sentence such that T, $T + \varphi$ and $T + \neg\varphi$ are mutually interpretable.[54] The choice between these two theories is *not* a matter of mere expedience.

The question of pluralism has two aspects which can now be formulated as follows: First, setting aside purely metaphysical and notational choices, can the problem of selection be solved in a convincing way by theoretical reason? Second, how far does this proceed; does it go all the way or does one eventually reach a "bifurcation point"; and, if so, could one be in a position to recognize it? Our approach is to address these questions in a way that is sensitive to the actual results of mathematics, where here by "results" we mean results that everyone can agree on, that is, the theorems, the primary data. But while the *data* will lie in mathematics, the *case* will go beyond mathematics[55] and this is what we mean when we say that the case will lie at the intersection of mathematics and philosophy.

4   The initial stretch: First- and second-order arithmetic

There is currently no convincing case for pluralism with regard to first-order arithmetic and most would agree that given the clarity of our conception of the structure of the natural numbers and given our experience to date with that structure such a pluralism is simply untenable. So, most would agree that not just for any Π_1^0-sentence, but for *any* arithmetical sentence φ, the choice between PA + φ and PA + $\neg\varphi$ is not one of mere expedience. I have discussed this above and will not further defend the claim here. Instead I will take it for granted in what follows.

The real concerns arise when one turns to *second*-order arithmetic, *third*-order arithmetic, and more generally, the transfinite layers of the set-theoretic hierarchy.[56]

4.1 Independence in set theory

One source of the concern is the proliferation of independence results in set theory and the nature of the forms of independence that arise. Consider the statements PU (Projective Uniformization)[57] of (schematic) second-order arithmetic and CH (Cantor's continuum hypothesis), a statement of third-order arithmetic. Combined results of Gödel and Cohen show that if ZFC is consistent then these statements are independent of ZFC: Gödel invented (in 1938) the method of *inner models*. He defined a canonical and minimal inner model L of the universe of sets, V, and he showed that CH holds in L. This had the consequence that ZFC could not refute CH. Cohen invented in (in 1963) the method of *forcing* (or *outer models*). Given a complete Boolean algebra he defined a model $V^{\mathbb{B}}$ and showed that ¬CH holds in $V^{\mathbb{B}}$. This had the consequence that ZFC could not prove CH. Thus, these results together showed that CH is independent of ZFC. Similar results hold for PU and a host of other questions in set theory. In contrast to the incompleteness theorems, where knowledge of the independence of the sentences produced—Π_1^0-sentences of first-order arithmetic—actually settles the statement, in the case of PU and CH, knowledge of independence provides *no clue whatsoever* as to how to settle the statement.

4.2 The problem of selection for second-order arithmetic

How then are we to solve the problem of selection with respect to second-order arithmetic?

There are two steps in solving the problem of selection for a given degree of interpretability. The first step is to *secure the secondary data* by showing that the Π_1^0-consequences of the theories in the degree are true. One way to do this is to show (ZFC − Infinity) + $\bigcup_{n<\omega}$ Con($T \upharpoonright n$), where T is some theory in the degree and $T \upharpoonright n$ is the first n sentences of T. Having thus secured the secondary data, the second step is to *select from the degree*. For example, the degree of ZFC contains ZFC + ¬Con(ZFC), ZFC + PU, ZFC + ¬PU, ZFC + CH and ZFC + ¬CH, and many other theories. The full problem of selection for this degree would involve settling all such questions, which is a massive task. For the moment we shall concentrate on deciding between ZFC + PU and ZFC + ¬PU. In what follows I shall presuppose ZFC. I will argue that the choice between ZFC + PU and ZFC + ¬PU is not one of mere expedience; in fact, one can give strong theoretical reasons for ZFC + PU.

The case will involve the problem of selection for a much higher degree. But first a word on the structure of the hierarchy of interpretability and large cardinal axioms.

The structure of the hierarchy of interpretability is more disorderly than one might expect—it forms a distributive lattice that is neither linearly ordered nor well-founded. This is shown via the construction of non-standard theories via coding techniques. Remarkably, however, when one restricts to the natural theories that occur in mathematical practice, one finds that the theories are well-behaved—they are well-ordered under interpretability.[58]

Extensions of ZFC – Infinity via *large cardinal axioms* provide us with a canonical class of representatives within this well-ordered hierarchy in that given a theory T in the hierarchy one can generally find—via the dual techniques of inner model theory and forcing—a theory of the form (ZFC – Infinity) + LCA, where LCA is a large cardinal axiom, such that T and (ZFC – Infinity) + LCA are mutually interpretable.[59] In this way large cardinal axioms (which are (for the most part) naturally well-ordered) provide a gauge of the strength of the theories in the hierarchy of interpretability. The simplest large cardinal axioms are *reflection principles*.[60] Some notable stepping in the large cardinal hierarchy are strongly inaccessible cardinals, Mahlo cardinals, measurable cardinals, Woodin cardinals, and supercompact cardinals.

The higher degree that I shall be working with is that of the theory

$$T_1 = \text{ZFC} + \text{ there are } \omega\text{-many Woodin cardinals.}$$

There are many theories of interest in this degree. For example, the theories

$$T_2 = \text{ZFC} + \text{there is an } \omega_1\text{-dense ideal on } \omega_1,$$

$$T_3 = \text{ZFC} + \text{AD}^{L(\mathbb{R})}, \text{ and}$$

$$T_4 = \text{ZFC} + \text{AD}^{L(\mathbb{R})} + \text{MA} + \neg\text{CH.}$$

are all in the same degree as T_1, that is, T_1, T_2, T_3, and T_4 all yield the same secondary data.[61]

The first step is to secure the secondary data. As noted above it suffices to establish (ZFC – Infinity) + $\bigcup_{n<\omega} \text{Con}(T \upharpoonright n)$, where T is any one of the above theories. There is a strong case for this but it would take us too far afield to present it here. Let me just say that case rests on the intimate connection between determinacy and inner models of large cardinal axioms and that the case is so strong that set theorists who have investigated the network of theorems in this area (the primary data) are quite confident that T_1 (and hence the other theories in its degree) is consistent.[62]

The second step is to provide theoretical reasons for some of the theories in the degree. One cannot accept all of them since they are not mutually consistent (for example, T_2 and T_4 contradict one another). Nevertheless,

I think that a very strong case can be made for T_3. Moreover, T_3 implies PU and hence resolves the problem of selection for ZFC + PU versus ZFC + ¬PU. Again the case for this is based upon (but goes beyond) the primary data, namely, a large network of mathematical theorems. The case is quite involved and so we shall only give an overview. For details see Koellner (2006) and the references therein.

(1) $\text{AD}^{L(\mathbb{R})}$ is an axiom that has some degree of intrinsic plausibility.[63] But, of course, this is just a starting point. One must look to its consequences and connections with other statements to determine, for example, whether its consequences are intrinsically plausible and whether it is implied by other intrinsically plausible axioms.

(2) $\text{AD}^{L(\mathbb{R})}$ has a number of intrinsically plausible consequences—for example, that there is no paradoxical decomposition of the unit sphere using pieces that are definable (in the precise sense of being in $L(\mathbb{R})$). In fact, in addition to implying that all subsets of reals in $L(\mathbb{R})$ are Lebesgue measurable, $\text{AD}^{L(\mathbb{R})}$ implies the other regularity properties for such sets of reals, such as that they have the property of Baire and the perfect set property. In addition, $\text{AD}^{L(\mathbb{R})}$ implies that Σ_1^2-uniformization holds in $L(\mathbb{R})$. These consequences are all intrinsically plausible and these results generalize the features of Borel sets that can be established in ZFC to the level of $L(\mathbb{R})$. In short, $\text{AD}^{L(\mathbb{R})}$ has what appear to be the correct consequences for the structure theory of the sets of reals in $L(\mathbb{R})$ and this is evidence for $\text{AD}^{L(\mathbb{R})}$.

(3) In the above, there is an obvious analogy with the hypothetico-deductive method in physics, the analogue of a physical theory being $\text{AD}^{L(\mathbb{R})}$ and the analogue of observational data being the intrinsically plausible statements. One difference of course is that the notion of an intrinsically plausible statement is not very sharp; another is that judgments of intrinsic plausibility are less secure than observational statements. Remarkably, as if to compensate for this shortcoming, the mathematical case provides us with something more. To bring this out consider the concern—which arises also in the physical case—that there might be other theories with the same intrinsically plausible (cf. observational) consequences. In the physical case one can never allay this concern. Remarkably, in the mathematical case one can: $\text{AD}^{L(\mathbb{R})}$ is the *only* theory that has the above intrinsically plausible consequences, that is, the intrinsically plausible consequences themselves *imply* $\text{AD}^{L(\mathbb{R})}$ (a result of Woodin).

(4) The pattern in (2) and (3)—where one draws intrinsically plausible consequences from $\text{AD}^{L(\mathbb{R})}$ and then *recovers* $\text{AD}^{L(\mathbb{R})}$—repeats itself with respect to *other* classes of intrinsically plausible consequences. See Koellner (2006) for some examples.

(5) Let us now turn from consequences to other connections. To begin with there are other intrinsically plausible axioms—most notably large

cardinal axioms (by groundbreaking work of Martin, Steel, and Woodin)—
that imply $\mathrm{AD}^{L(\mathbb{R})}$.

(6) Large cardinal axioms and axioms of definable determinacy (such as
$\mathrm{AD}^{L(\mathbb{R})}$) spring from entirely different sources. Yet there is an intimate connection between them. It turns out that $\mathrm{AD}^{L(\mathbb{R})}$ is *equivalent* to a statement
asserting the existence of inner models of certain large cardinals.[64] We have
here a case where intrinsically plausible principles from completely different
domains reinforce one another.

(7) Not only do large cardinal axioms imply $\mathrm{AD}^{L(\mathbb{R})}$, many other theories
imply $\mathrm{AD}^{L(\mathbb{R})}$. For example, *both T_2 and T_4 imply $\mathrm{AD}^{L(\mathbb{R})}$ despite the fact that T_2
and T_4 are incompatible.*

(8) The phenomenon in (7) is quite general. Time after time it is shown
that a strong theory, which on the face of it has nothing to do with $\mathrm{AD}^{L(\mathbb{R})}$,
actually implies (through a deep result) $\mathrm{AD}^{L(\mathbb{R})}$. The technique for establishing this—the core model induction—provides evidence for the claim that
all sufficiently strong natural theories imply $\mathrm{AD}^{L(\mathbb{R})}$. In this sense $\mathrm{AD}^{L(\mathbb{R})}$
lies in the "overlapping consensus" of all sufficiently strong natural the-
ories. As one climbs the hierarchy of interpretability along any natural
path—however, remote the apparent subject matter is from determinacy—
it appears that one cannot avoid laying down something that outright
implies $\mathrm{AD}^{L(\mathbb{R})}$.

This is only a sample of the network of theorems (primary data) upon
which the case for $\mathrm{AD}^{L(\mathbb{R})}$ is based. To be sure, the case itself goes beyond
the data, as must *any* case for a new axiom, in light of the incomplete-
ness theorems. But it is remarkable that such a strong case can exist. These
arguments do not merely provide practical reasons for adopting $\mathrm{AD}^{L(\mathbb{R})}$
as a matter of expedience; they provide theoretical reasons for accepting
$\mathrm{AD}^{L(\mathbb{R})}$.[65]

Thus, we have solved the problem of selection with respect to $\mathrm{AD}^{L(\mathbb{R})}$,
an axiom that concerns the structure $L(\mathbb{R})$. Since $\mathrm{AD}^{L(\mathbb{R})}$ implies PU we
have also solved the problem of selection for PU. It turns out that in
solving the problem of selection for $\mathrm{AD}^{L(\mathbb{R})}$, we have solved the problem of selection for a host of other statements concerning $L(\mathbb{R})$. In fact,
there is a sense in which $\mathrm{AD}^{L(\mathbb{R})}$ is so central that in solving the problem of selection for it we have given an "effectively complete" solution
of the problem of selection for the entire theory of $L(\mathbb{R})$. We shall spell
this out in some detail since once we are armed with such a "complete
solution" at the level of $L(\mathbb{R})$ we will next ask how far such a solution
extends. This will form the basis of our search for a more reasonable form of
pluralism.

The axiom $\mathrm{AD}^{L(\mathbb{R})}$ appears to be "effectively complete" for the theory of
$L(\mathbb{R})$. Let us first elaborate, then quantify this. A comparison with PA is
useful here. Of course, neither PA nor $\mathrm{ZFC} + \mathrm{AD}^{L(\mathbb{R})}$ is complete because
of the incompleteness theorems. However, there are very few statements

of prior mathematical interest that are known to be independent of PA. The classic example of such a statement is the Paris-Harrington. Still, there are very few such statements and for this reason people are inclined to regard PA as "effectively complete". The case with axioms of determinacy is even more dramatic. Let PD be the restriction of the axiom $AD^{L(\mathbb{R})}$ to the domain of second-order number theory. In contrast to PA, there are many statements of prior mathematical interest that are independent of PA_2, for example, PU. But when one adds PD to PA_2 this ceases to be the case. In fact, $PA_2 + PD$ appears to be more complete than PA in that, for example, there is no analogue of the Paris-Harrington theorem. Similar considerations apply to $AD^{L(\mathbb{R})}$.

Let us try to quantify this and erase the scare quotes around "effectively complete". To do this we shall henceforth assume large cardinal axioms.[66] Our goal is to introduce a strong logic that sharpens the notion of "effective completeness" not directly of PD and $AD^{L(\mathbb{R})}$ but of the large cardinal axioms that imply these axioms. We shall use the hypothesis that there is a proper class of Woodin cardinals, which we shall abbreviate "PCWC".

The motivating result on "effective completeness" is the following:

Theorem 4.1 (Woodin). *Assume ZFC + PCWC. Suppose φ is a sentence and that $\mathbb{B}$ is a complete Boolean algebra. Then*

$$L(\mathbb{R}) \models \varphi \ \textit{iff} \ L(\mathbb{R})^{V^{\mathbb{B}}} \models \varphi.$$

Our main (and very powerful) technique for establishing independence in set theory is set forcing—the construction of such models $V^{\mathbb{B}}$. The above theorem shows that in the presence of a proper class of Woodin cardinals (which implies $AD^{L(\mathbb{R})}$) this technique *cannot* be used to establish independence with respect to statements about $L(\mathbb{R})$.[67]

The aim of the strong logic is to capture this "freezing" or "sealing" by "factoring out" the effects of forcing.

Definition 4.2 (Woodin). Suppose that T is a countable theory in the language of set theory and φ is a sentence. Then

$$T \models_\Omega \varphi$$

if for all complete Boolean algebras $\mathbb{B}$ and for all ordinals α,

$$\text{if } V_\alpha^{\mathbb{B}} \models T \text{ then } V_\alpha^{\mathbb{B}} \models \varphi.$$

A theory T is *Ω-satisfiable* if there exists an ordinal α and a complete Boolean algebra $\mathbb{B}$ such that $V_\alpha^{\mathbb{B}} \models T$.

This notion of semantic implication is robust in that large cardinal axioms imply that the question of what implies what cannot be altered by forcing:

Theorem 4.3 (Woodin). *Assume* ZFC + PCWC. *Suppose that T is a countable theory in the language of set theory and φ is a sentence. Then for all complete Boolean algebras* $\mathbb{B}$,

$$T \models_\Omega \varphi \ \ iff \ \ V^{\mathbb{B}} \models \text{"}T \models_\Omega \varphi.\text{"}$$

We are now in a position to reformulate the theorem on "freezing the theory of $L(\mathbb{R})$" in terms of Ω-logic (Theorem 4.1).

Definition 4.4. A theory T is Ω-*complete* for a collection of sentences Γ if for each $\varphi \in \Gamma$, $T \models_\Omega \varphi$ or $T \models_\Omega \neg\varphi$.

Theorem 4.5. *Assume* ZFC *and that there is a proper class of Woodin cardinals. Then* ZFC *is Ω-complete for the collection of sentences of the form* "$L(\mathbb{R}) \models \varphi$".[68]

In this sense, large cardinal axioms give an Ω-complete picture of the theory of $L(\mathbb{R})$. Furthermore, $\text{AD}^{L(\mathbb{R})}$ lies at the heart of this picture. This is how we shall quantify our success in solving the problem of selection with respect to statements of $L(\mathbb{R})$: Assuming that there is a proper class of Woodin cardinals, we have a theory that settles (in Ω-logic) *every* statement about $L(\mathbb{R})$. Our goal now is to see how far such Ω-complete pictures extend and whether we could eventually reach a "bifurcation point".

5 Bifurcation scenarios

We now turn to "bifurcation scenarios", that is, scenarios where not only have our (current) theoretical reasons have failed to settle a given question, say CH, but where we have reason to believe that no further theoretical reasons can settle the question.[69] One difficulty in articulating such scenarios is that the space of theoretical reasons one might give is not something that one can survey in advance.[70] We have to do our best and work with what we have.

5.1 An initial pass

Some have claimed that the early independence results in set theory already suffice to secure such a position. For example, it is claimed that the independence of CH with respect to ZFC shows that the choice between ZFC + CH and ZFC + ¬CH is one of mere expedience. It is maintained that although there may be practical reasons in favour of adopting one axiom over the other (say for a given purpose at hand) there are no theoretical reasons that one can give for one over the other.

Now, in response, one might argue that independence from ZFC alone cannot suffice since (under reasonable assumptions) Con(ZFC) is independent of ZFC and yet the choice between ZFC + Con(ZFC) and ZFC + ¬Con(ZFC) is hardly one of mere expedience. But to this the critic can respond the background assumption that ZFC is consistent settles the question under consideration. The critic can point out that in contrast to Con(ZFC) (and any Π_1^0-sentence for that matter), knowledge of the independence of CH does not settle CH.

However, there are Π_2^0 Orey sentences[71] in the language of arithmetic that also have this feature and yet (as we have argued above) the position that questions concerning them are questions of mere expedience is simply untenable. The critic must therefore cite some distinctive feature of the nature of the independence of CH with respect to ZFC, one that differentiates it from the case of such Π_2^0-sentences.

Perhaps the key difference is that CH is a statement of prior mathematical interest shown to be independent via set forcing. To this there are two responses. First, why should this matter? We are interested in truth not human interest. Second, PU is also such a statement and yet we have argued that the choice between ZFC + PU and ZFC + ¬PU is not one of mere expedience.[72]

So I do not think that the critic has a case here. Nevertheless, I want to see how one might respond on the critic's behalf. The goal will be to gain insight into what it would take to have theoretical reasons (driven by the primary data of mathematics) to believe that a given question of pure mathematics, say CH, is one of mere expedience.

5.2 A more promising approach

Let us begin by considering how far the case for $AD^{L(\mathbb{R})}$ and large cardinal axioms extends. Gödel had high expectations for large cardinal axioms. Indeed he went so far as to entertain a generalized completeness theorem for them:

> It is not impossible that for such a concept of demonstrability [namely, provability from true large cardinal axioms] some completeness theorem would hold which would say that every proposition expressible in set theory is decidable from the present axioms plus some true assertion about the largeness of the universe of all sets. (Gödel 1946, p. 151)

As a test case he chose CH, a statement of third-order arithmetic.

As we have seen, there has been a partial realization of this program in that large cardinal axioms provide an Ω-complete picture of second-order arithmetic and, in fact, all of $L(\mathbb{R})$. How far does this proceed? In a sense that can be made precise it holds "below CH".[73] Unfortunately, it fails at the "level of CH", namely, Σ_1^2, as follows from a series of results originating with Levy and Solovay:

Theorem 5.1.   *Assume L is a standard large cardinal axiom. Then* ZFC $+ L$ *is not* Ω-*complete for* Σ_1^2.[74]

Although large cardinal axioms do not provide an Ω-complete picture of Σ_1^2, it turns out that one can obtain such a picture provided one supplements large cardinal axioms. Remarkably, CH itself is such a statement.

Theorem 5.2 (Woodin, 1985).   *Assume* ZFC *and that there is a proper class of measurable Woodin cardinals. Then* ZFC $+$ CH *is* Ω-*complete for* Σ_1^2.

Moreover, up to Ω-equivalence, CH is the unique Σ_1^2-statement that is Ω-complete for Σ_1^2. Thus, up to Ω-equivalence, there is a *unique* Σ_1^2-sentence which (along with large cardinal axioms) provides an Ω-complete picture of Σ_1^2, namely, CH.

If one shifts perspective from Σ_1^2 to $H(\omega_2)$, there is a companion result for ¬CH, *assuming the Strong* Ω *Conjecture.*[75]

Theorem 5.3 (Woodin).   *Assume the Strong* Ω-*Conjecture.*

(1) *There is an axiom A such that*

  (i) ZFC $+ A$ *is* Ω-*satisfiable and*
  (ii) ZFC $+ A$ *is* Ω-*complete for the structure* $H(\omega_2)$.

(2) *Any such axiom A has the feature that*

$$\text{ZFC} + A \models_\Omega \text{``} H(\omega_2) \models \neg\text{CH ''}.$$

Thus, assuming the Strong Ω Conjecture, there is an Ω-complete picture of $H(\omega_2)$ and any such picture involves a *failure* of CH.

These two results raise the spectre of bifurcation at the level of CH. There are two key questions. First, are there recursive theories with higher degrees of Ω-completeness? Second, is there a unique such theory (with respect to a given level of complexity)? The answers to these questions turn on the Strong Ω Conjecture.

If the Strong Ω Conjecture holds, then one cannot have an Ω-complete picture of third-order arithmetic.

Theorem 5.4 (Woodin).   *Assume the Strong* Ω *Conjecture. Then there is no recursively enumerable theory A such that* ZFC $+ A$ *is* Ω-*complete for* Σ_3^2.

However, if the Strong Ω Conjecture fails, then such higher levels of Ω-completeness may be possible. In fact, there may be a (recursively enumerable) sequence of axioms $\vec{A}$ such that for some large cardinal axiom L the theory $ZFC + L + \vec{A}$ is Ω-complete for *all* of third-order arithmetic. Going further it could be the case that for each specifiable fragment V_λ of the universe of sets there is a large cardinal axiom L and a (recursively enumerable) sequence of axioms $\vec{A}$ such that $ZFC + L + \vec{A}$ is Ω-complete for the theory of V_λ. Moreover, it could be the case that any other theory with this feature, say $ZFC + L + \vec{B}$, agrees with $ZFC + L + \vec{A}$ on the computation of the theory of V_λ in Ω-logic. This would mean that there is a *unique* Ω-complete picture of the universe of sets up to V_λ. Furthermore, it could be the case that all of these Ω-complete pictures cohere. This would give us a unique Ω-complete picture of (the successive layers of) the entire universe of sets.

One could argue that such an Ω-complete picture of the entire universe of sets is the most that one could hope for. Should uniqueness hold that would be the end of the story from the perspective of Ω-logic for there would be nothing about the (specifiable fragments of) the universe of sets that could not be settled on the basis of Ω-logic in this unique Ω-complete picture.

Unfortunately, uniqueness *must* fail.

Theorem 5.5 (K. and Woodin). *Assume* $ZFC + PCWC$. *Suppose L is a large cardinal axiom and $\vec{A}$ is a (recursively enumerable) sequence of axioms such that*

$$ZFC + L + \vec{A} \text{ is } \Omega\text{-complete for the theory of } V_\lambda,$$

where V_λ is some specifiable fragment of the universe at least as large as $V_{\omega+2}$. Then there exists a (recursively enumerable) sequence of axioms $\vec{B}$ such that

$$ZFC + L + \vec{B} \text{ is } \Omega\text{-complete for the theory of } V_\lambda$$

but which differs from $ZFC + L + \vec{A}$ *on CH.*[76]

Thus, if there is one Ω-complete picture of such a level of the universe (and hence of arbitrarily large such levels), then there is necessarily an *incompatible* Ω-complete picture.[77]

Suppose then that there is one such theory (or a sequence of such theories for higher and higher levels, all of which extend one another). The advocate of pluralism might argue as follows: Such an Ω-complete picture is the most that theoretical reason could hope to achieve. Given that from one such picture (of the form $ZFC + L + \vec{A}$) we can generate others (say $ZFC + L + \vec{B}$) that are incompatible, and given that there is great flexibility (in the choice of $\vec{B}$)

and that we can pass from one to another (by altering $\vec{B}$), this shows that the choice is merely one of expedience. The choice of $\vec{B}$ is analogous to the choice of a timelike vector in Minkowski spacetime. The choice of such a vector in Minkowski spacetime induces a foliation of the space relative to which one can ask whether "A is simultaneous with B" but there is no absolute significance to such questions independent of the choice of a timelike vector. Likewise, the choice of the sequence $\vec{B}$ (in conjunction with ZFC + L) provides an Ω-complete picture relative to which one can ask whether "CH holds" and many other such questions but there is no absolute significance to such questions independent of the choice of $\vec{B}$. As in the case of special relativity we need to change perspective. There is no sense in searching for "the correct" Ω-complete picture, just as there is no sense in searching for "the correct" foliation. Instead of the naive picture of the universe of sets with which we started we are ultimately driven to a new picture, one that deems questions we originally thought to be absolute to be ultimately relativized.

I do not want to endorse this position. I am merely presenting it on behalf of the advocate of pluralism as the best mathematically driven scenario that I can think of where one could arguably maintain that we had been driven by the primary data of mathematics to shift perspective and regard certain questions of mathematics (such as CH) as based on choices of mere expedience.

A key virtue of this scenario is that it is sensitive to future developments in mathematics—to rule it out it suffices to prove the Strong Ω Conjecture and to establish it it suffices to find one such Ω-complete theory. In this way, by presenting mathematically precise scenarios that are sensitive to mathematical developments, the pluralist and non-pluralist can give the question of pluralism "mathematical traction" and, through time, test the robustness of mathematics.

What can we say at present concerning the above pluralist scenario? Although the scenario is an open mathematical possibility there are reasons to think that such a scenario cannot happen. For there is growing evidence for the Strong Ω Conjecture[78] and, as noted above, this conjecture rules out the existence of one (and hence many) such Ω-complete theories. Thus, one way to definitively rule out the above pluralist scenario is to prove the Strong Ω Conjecture.

Should it turn out that the Strong Ω Conjecture is true then the pluralist would have to retreat and present another scenario. Let us consider two such scenarios.

The first scenario builds on Theorem 5.2 which shows that CH is a Σ_1^2-sentence such that ZFC + L + CH is Ω-complete for Σ_1^2 (where L is a large cardinal axiom) and, moreover, that CH is the unique such sentence (up to Ω equivalence). If the Strong Ω Conjecture holds then (by Theorem 5.4) this result is close to optimal in that there is no recursively enumerable theory A such that ZFC $+ A$ is Ω-complete for Σ_3^2. However, it is an open possibility that there an axiom A such that for some large cardinal axiom L, ZFC $+ L + A$ is Ω-complete for Σ_2^2. Let us assume that this possibility is realized and consider the question of uniqueness. For each A such ZFC $+ L + A$ is Ω-complete for Σ_2^2 (where L is a large cardinal axiom) let T_A be the Σ_2^2 theory computed by ZFC $+ L + A$ in Ω-logic. The question of uniqueness simply asks whether T_A is unique. A refinement of the techniques used to prove Theorem 5.5 can be used to show that uniqueness must fail. The first pluralist scenario is this: There are incompatible Ω-complete pictures of Σ_2^2 (granting large cardinal axioms) and the choice between them is one of mere expedience.

The tenability of this scenario rests on the impossibility of giving theoretical reasons for one such T_A over another. Remarkably, from among all of the theories T_A there is *a single one* that stands out. For it is known (by a result of Woodin in 1985) that if there is a proper class of measurable Woodin cardinals then there is a forcing extension satisfying all Σ_2^2 sentences φ such that ZFC $+$ CH $+ \varphi$ is Ω-satisfiable. (See Larson, Ketchersid and Zapletal, 2008). It follows that if the question of existence is answered positively with an A that is Σ_2^2 then T_A must be this maximum Σ_2^2 theory and, consequently, all T_A agree when A is Σ_2^2.[79] So, assuming that all such T_A contain CH and that there is a T_A where A is Σ_2^2, then, although not all T_A agree (when A is arbitrary) there is one that stands out, namely, the one that is maximum for Σ_2^2 sentences.

The second scenario is based on Theorem 5.3 which shows that (granting the Strong Ω Conjecture) there is an axiom A such that ZFC $+ A$ is Ω-complete for $H(\omega_2)$ and, moreover, any such axiom has the feature that ZFC $+ A \models_{\Omega}$ "$H(\omega_2) \models \neg$CH". For each such axiom A let T_A be the theory of $H(\omega_2)$ as computed by ZFC $+ A$ in Ω-logic. Thus, the theorem shows that all such T_A agree in containing $\neg$CH. The question then naturally arises whether T_A is unique. A refinement of the techniques used to prove Theorem 5.5 can be used to answer this question negatively. And again the pluralist might use this "local bifurcation" result to ground the case for pluralism. But again, there is a T_A that stands out, namely, the maximum theory given by the axiom ($*$). (See Woodin, 1999).

In the course of this paper we have seen a number of increasingly sophisticated cases for pluralism. In each instance the case faltered for reasons that are sensitive to actual developments in mathematics. Perhaps there are deeper theorems in this vein that would lead us to embrace pluralism. The point I wish to make is that the real question of pluralism is a deep one, one that requires the combined efforts of philosophy and mathematics. It is through exploring the boundless ocean of unlimited possibilities that we can gain a sure footing in what is actual.

Notes

*I would like to thank Bill Demopoulos, Iris Einheuser, Matti Eklund, Michael Fried-man, Warren Goldfarb, Daniel Isaacson, Øystein Linnebo, John MacFarlane, Alejandro Péres, and Thomas Ricketts for helpful discussion.

1. In my overview of this historical development I am very much indebted to the work of Michael Friedman. See especially his *Dynamics of Reason*.
2. In a famous passage in the *Critique of Pure Reason* (1781/1787), Kant wrote: "With respect to the question of unanimity among the adherents of metaphysics in their assertions, it is still so far from this that it is rather a battle ground, which seems to be quite peculiarly destined to exert its forces in mock combats, and in which no combatant has ever yet been able to win even the smallest amount of ground, and to base on his victory an enduring possession. There is therefore no doubt that its procedure has, until now, been a merely random groping, and, what is worst of all, among mere concepts."
3. Thus, in the *Prolegomena to Any Future Metaphysics* (1783) the first and second parts of the transcendental problem are "How is pure mathematics possible?" and "How is the science of nature possible?".
4. There is a subdivision within these a priori truths: Logic is analytic, while arithmetic, Euclidean geometry and the basic laws of Newtonian physics are synthetic.
5. Reichenbach was one of the five students to attend Einstein's first course on the general theory of relativity in 1919. His central early works on this subject are Reichenbach (1920 and 1924).
6. Reichenbach (1924, p. 3) refers the reader to Hilbert's work on the foundations of geometry for more on the notion of implicit definition.
7. Reichenbach stresses that such definitions involve an element of idealization and that the physical notions concerned (such as "beam of light") are theory-laden. A coordinative definition is thus to be distinguished from an operational definition.
8. "... coordinative definitions are arbitrary, and "truth" and "falsehood" are not applicable to them." (Reichenbach 1924, p. 9).
9. In his discussion of Einstein's particular definition of simultaneity, after noting its simplicity, Reichenbach writes: "This simplicity has nothing to do with the truth of the theory. The truth of the axioms decides the empirical truth, and every theory compatible with them which does not add new empirical assumptions is equally true." (Reichenbach 1924, p. 11)
10. Reichenbach is extending the Hilbertian thesis concerning implicit definitions since although Hilbert held this thesis with regard to formal geometry he did not hold it with regard to arithmetic. (I am indebted to Richard Zach for discussion on this point.) Later I shall argue that this extension is illegitimate.
11. A natural choice for such an axiom system is Primitive Recursive Arithmetic (PRA) but much weaker systems suffice, for example, $I\Delta_0 + \exp$. Either of these systems can be taken as T in the argument that follows.
12. The point being that T is Σ_1^0-complete (provably so in T).
13. For the reader concerned that this argument involves the notion of truth in a problematic way, notice (as we have indicated in the parenthetical remarks) that it can be implemented in $T + \mathrm{Con}(T)$ (which is taken to fall within the provenance of "critical investigation"); that is, $T + \mathrm{Con}(T)$ proves that $T + \varphi$ and $T + \neg\varphi$ are consistent and it *also* proves φ.

14. The above historical sketch has been necessarily brief and so some qualifications and comments are in order. First, there are other views that reject first philosophy and largely place philosophy after science, for example, various forms of naturalism. Second, the above division—philosophy-before-science versus philosophy-after-science—is not intended as a complete classification of views—there are degrees. (In fact, even in the extreme case where a view attempts to place philosophy entirely after science, if it does so for philosophical reasons, then it would seem to involve first philosophy at the meta-level. To overcome such an impurity, the advocate of such a view might regard the meta-philosophical view not as a philosophical thesis but rather as a proposal. Below we shall see this as an attractive interpretation of Carnap.) Finally, it is worth mentioning that Carnap did not have an uncritical attitude toward science and he did a good deal of work that one might say lies at the intersection of philosophy and science (take, for example, his work on probability and entropy).

15. For definiteness I shall focus almost exclusively on the position held by Carnap in this work. Unless otherwise specified all references in this section are to this work.

16. See pp. xiv, 7, 41.

17. See pp. 1–2, 258.

18. See p. xv.

19. See p. 279.

20. This section is a summary of the fuller discussion in Koellner (2009), to which the reader is referred for further details and a discussion of Carnap's views during later periods. In my thinking about *The Logical Syntax of Language* I have benefited from Friedman (1999c), Gödel (1953/9), and Goldfarb and Ricketts (1992). After writing Koellner (2009), Warren Goldfarb drew my attention to Kleene's review (1939). I am in complete agreement with what Kleene has to say and there is some overlap between our discussions, though my discussion goes a good deal further.

21. See p. 167.

22. Strictly speaking Carnap does not prohibit indefinite rules (see p. 172) but in all of the cases he considers (Language I, Language II, etc.) the rules are definite.

23. See p. 168.

24. Although Carnap's usage of 'language' is somewhat misleading I will follow him in certain instances—for example, in speaking of 'Language I'—simply because the usage is well-entrenched in his writing and the secondary literature. It will always be clear from context whether I am referring to a system or a language (in the modern sense).

25. The relevant clause is DA I.C.b. on p. 111.

26. There are no conditions placed on this notion—for example, it could be a definite notion (such as "provable in T").

27. A sentence is *determinate* if either it or its negation is valid, that is, a consequence of the null set.

28. For some of these see Quine (1963), though note that Quine's discussion appears at points to mistakenly assume that the notion of direct consequence that Carnap uses is a d-notion.

29. This is more in keeping with the standard use of the term 'content'. For, in a straightforward sense, the truths of first-order logic do not pertain to a special subject matter (they are perfectly general) while those of arithmetic and set theory do.

112   *Peter Koellner*

30. For a fuller discussion—one that involves a discussion of two additional parameters—see Koellner (2009).
31. Carnap was fully aware of this sensitivity. See, for instance, the example involving $g_{\mu\nu}$ that Carnap gives (on p. 178) right after he draws the division.
32. Carnap is fully aware of such counter-examples. See p. 231 of §62 where he notes that his definitions have the consequence that the universal numerical quantifier in Whitehead and Russell's *Principia Mathematica* is really a *descriptive* symbol, the reason being that the system involves only d-rules and hence (by Gödel's incompleteness theorem) it will leave some Π_1^0-sentences undecided.
33. See, for example, pp. 1–2.
34. For example, taking S to be PA one can simply extend the language by adding a truth predicate and extend the axioms by adding the Tarskian truth axioms and allow the truth predicate to figure in the induction scheme. The resulting system S' is only minimally stronger than S. It proves Con(PA) but not much more.
35. To continue the example in the previous footnote, suppose one wishes to show that Con(ZF + AD) is analytic in S (which, as I shall argue below, it is). To do this one must move to a system that has consistency strength beyond that of "ZFC + there are ω-many Woodin cardinals".
36. For the first note that $T \vdash_\omega \varphi$ if and only if φ (where T is the fixed Σ_1^0-complete theory) and for the second note that the Tarskian truth definition has the feature that $T(\ulcorner \varphi \urcorner)$ if and only if φ (where T is the truth predicate).
37. For further evidence that Carnap's pluralism is this radical, see xv, 124 and Carnap (1939), p. 27.
38. Some readers might be tempted to interpret me as saying that there is a "fact of the matter" concerning Π_1^0-sentences. I want to resist such a formulation since I am not sure that I understand the phrase "fact of the matter" as it is often employed. I certainly understand this phrase when it is used merely as a point of contrast with "matter of mere expedience". On this reading, it means no more than that the issue is one of theoretical reason, one concerning something more than mere utility, one having something to do with the truth of one theory over another (not in some robust metaphysical sense of the "truth" but in the ordinary sense). This distinction is not as sharp as one would like but one can point to clear cases (as we have seen above and as we shall see below) and the distinction strikes me as significant. In contrast, the phrase "fact of the matter" is often used in a way that strives for something more—"thick truth", "Truth with a capital 'T'", the idea of "carving reality at the joints", and so on. I cannot think of examples where I could go along with such talk with any confidence. Moreover, it seems to me that such talk buys into the myth that there is some Archimedean vantage point from which we can survey the array of theories and compare them with "reality as it is in and of itself"—in short, a "sideways-on view" (in McDowell's apt phrase). This is something that I think Kant and Carnap were right to reject and, once we follow them in doing so, we are left with the former, thinner distinction—the one I employ in the text. (See Tait (1986) for a critical discussion of the "myth of the model in the sky". The reader might press me with the concern that the thinner distinction that I invoke (following Carnap) rests on a similar myth. I do not think that it does. But it would take us too far afield to explore the issue here.)
39. This is something that Carnap recognizes.
40. See Friedman (1999c) for the substantive interpretation and see Goldfarb and Ricketts (1992) for the non-substantive interpretation. In what follows I do not take a stance on the interpretative issue. Instead I criticize both versions of Carnap.

41. There are two other approaches that one might consider. These approaches fail as well. See Koellner (2009).
42. This is related to a point made by Michael Friedman. See p. 226 of Friedman (1999c).
43. There are many places where Carnap quite obviously begs the question in the metatheory. See, for example, §§43 and 44 where Carnap discusses intuitionism and predicativism; to people like Brouwer and Poincaré these sections would be maddening.
44. See Janssen (2002) for a discussion of the empirical equivalence of Lorentz's mature theory of 1905 and special relativity.
45. See Evans (1998), p. 412 and Wilson (1970), p. 109 for six solid reasons.
46. See DiSalle (2006), Friedman (1999a) and Janssen (2002). To be sure, the full vindication of the Copernican theory over the Ptolemaic theory (and Tychonic theory) came with Galileo (and Newton) and the full vindication of special relativity over Lorentz's mature theory came with general relativity. For my purposes it is sufficient that the reasons given before the discovery of the telescope (and Newtonian gravitation theory), in the first case, and the discovery of general relativity, in the second case, have *some* force.
47. See §72.
48. The subject requires further discussion than I can give it here. For example, *how* cheap is existence in mathematics? Does consistency suffice? Consider ZFC + CH versus ZFC + ¬CH. There is reason to believe that both are consistent. The trouble is that CH and ¬CH are existential claims and, on a straightforward reading, the objects that they assert to exist cannot coexist. I am inclined to think that existence in mathematics is "Consistency + X" but I do not know how to solve for X.
49. In the sense that the laws of physics are Lorentz invariant.
50. Some metaphysicians accept these physical limitations but are not moved by them. For example, in his entry on presentism in the *The Oxford Handbook of Metaphysics* (2003), Thomas Crisp says that we can just stipulate a preferred foliation. Well, we can do *that* but what reason do we have to think that our stipulation captures any structure (either physical or metaphysical) of our universe? We can also stipulate a preferred center or preferred direction. The possibilities are endless. But most would maintain that such stipulations are idle and do not reflect the structure of spacetime. How is the situation with simultaneity any different?
51. This is not intended as a precise, formal definition—for example, I have said nothing about what it takes to count as data. So the definition is quite flexible. Nevertheless, it is sharp enough to serve for our purposes.
52. See chapter 6 of Lindström (2003).
53. For our present purposes there is little (if any) loss of generality in this restriction since our concern now is with theoretical reason and so we do not wish to be distracted by choices between, say, ZFC − Infinity and PA.
54. Such a sentence is called an *Orey sentence* for T. See Lindström (2003) for the construction of such a sentence.
55. And necessarily so since we are not stating a theorem.
56. Recall that in our setting the various systems of arithmetic (e.g. PA and PA$_2$ (the second-order axioms of Peano Arithmetic), etc.) are cast in the language of set theory and that from the point of view of independence there is no loss of generality in this assumption.
57. This is the statement that every projective subset of the plane admits a projective choice function.

58. This is a mystery that calls for clarification.
59. There is an element of imprecision in this claim due to the lack of precision involved in both the notion of a natural theory and the notion of a large cardinal axiom.
60. Indeed ZFC is the result of supplementing our base theory ZFC – Infinity with a scheme of first-order reflection principles.
61. The main results here are due to Woodin. See Koellner (2006) for further discussion and references.
62. For a large piece of the primary data see the articles on determinacy and inner model theory in the forthcoming *Handbook of Set Theory*, in particular, Koellner and Woodin (2009b). It is noteworthy that the reasons are not merely inductively based on the fact that a contradiction has not yet been found. There is similar inductive support for the consistency of Quine's system NF but few are confident that it is consistent. The reasons for consistency are more involved and make for a very strong case. To underscore the strength of the case, at the Gödel centenary in Vienna in 2006, Woodin announced that should anyone prove one of these theories inconsistent he would resign his post and demand that his position be given to the person who established the inconsistency. (This is not an advisable strategy for securing tenure.)
63. For more on this notion, see chapter 9 of Parsons (2008).
64. See §8.1 of Koellner and Woodin (2009b).
65. A staunch skeptic could refrain from going beyond the primary data, committing to a statement only when it has been secured in the form of a theorem ("statement φ is provable in system S"). Likewise, in the physical case, a staunch skeptic could refrain from going beyond the primary data, committing to a statement only when it has been verified observationally. Each is consistent; neither is reasonable.
66. As noted above, many of these axioms are intrinsically plausible and some (though not all) of the above considerations to $AD^{L(\mathbb{R})}$ apply to them.
67. And this is not because the large cardinals are somehow throwing a wrench into the machinery of forcing. In fact, they fuel that machinery by generating more forcing extensions.
68. Although we have stated the Ω-completeness with respect to ZFC, the large cardinals are really doing the work. For this reason it is perhaps more transparent to formulate the result by saying that "ZFC + there is a proper class of Woodin cardinals" is Ω-complete for the collection of sentences of the form "$L(\mathbb{R}) \models \varphi$", noting that under this formulation the stated Ω-completeness is trivial unless our background assumptions guarantee that "ZFC + there is a proper class of Woodin cardinals" is Ω-satisfiable.
69. Compare the difference between (a) knowing that our current understanding of the physical world does not enable us to detect the luminiferous ether and (b) having reason to believe that no physical understanding will enable us to detect a luminiferous ether.
70. However, the situation in physics is similar—for example, one cannot rule out definitively the possibility that we might one day find a foliation that has fundamental physical significance and hence that the ultimate laws of physics are not Lorentz invariant.
71. Recall that φ is an *Orey sentence* for T if T, $T + \varphi$ and $T + \neg\varphi$ are mutually interpretable.

72. It would be of interest to further investigate the analogies and disanalogies between independence in arithmetic and set theory and the bearing of such results on philosophical positions in each domain.
73. See §3.3 of Koellner (2006).
74. This theorem is stated informally since the notion of a "standard large cardinal axiom" is not precise. However, one can cite examples from across the large cardinal hierarchy. For example, for L one can take "there is a measurable cardinal", "there is a proper class of Woodin cardinals", or "there is a non-trivial embedding $j : L(V_{\lambda+1}) \to L(V_{\lambda+1})$ with critical point below λ".
75. Here $H(\omega_2)$ is the set of all sets that have hereditary cardinality less than ω_2. The Strong Ω Conjecture is an outstanding conjecture in set theory. It is the conjunction of the Ω Conjecture and the statement "the AD^+ Conjecture is Ω-valid", where the Ω Conjecture is a conjectured completeness theorem for Ω-logic and the AD^+ Conjecture is the following conjecture: Suppose that A and B are sets of reals such that $L(A, \mathbb{R}) \models AD^+$, $L(B, \mathbb{R}) \models AD^+$, and the sets in $\mathscr{P}(\mathbb{R}) \cap (L(A, \mathbb{R}) \cup L(B, \mathbb{R}))$ are ω_1-universally Baire. Then either $(\underset{\sim}{\Delta}_1^2)^{L(A,\mathbb{R})} \subseteq (\underset{\sim}{\Delta}_1^2)^{L(B,\mathbb{R})}$ or $(\underset{\sim}{\Delta}_1^2)^{L(B,\mathbb{R})} \subseteq (\underset{\sim}{\Delta}_1^2)^{L(A,\mathbb{R})}$. See Woodin (1999) for definitions of the remaining terms. We caution the reader that in the existing literature one finds Theorems 5.3 and 5.4 stated with the Ω Conjecture in place of the Strong Ω Conjecture. However, Woodin recently discovered that the proofs require the latter.
76. For a more precise statement (one that spells out the notion of "specifiable fragment" (there called "robustly specifiable fragment")) see Koellner and Woodin (2009a).
77. It should be stressed that the choice of CH is just for illustration. Given one such theory one has a great deal of control in generating others.
78. See Woodin (2009).
79. A natural conjecture is that ◇ is such an A. But even if ◇ is not such an axiom A it will be in T_A.

References

Carnap, R. (1934). *Logische Syntax der Sprache*, Springer, Wien. Translated by A. Smeaton as *The Logical Syntax of Language*, London: Kegan Paul, 1937.

Carnap, R. (1939). Foundations of logic and mathematics, *International Encyclopedia*, vol. I, no. 3, University of Chicago Press, Chicago.

DiSalle, R. (2006). *Understanding Spacetime: The Philosophical Development of Physics from Newton to Einstein*, Cambridge University Press, Cambridge.

Evans, J. (1998). *The History and Practice of Ancient Astronomy*, Oxford University Press.

Friedman, M. (1992). *Kant and the Exact Sciences*, Harvard University Press, Cambridge, Massachusetts.

Friedman, M. (1999a). *Dynamics of Reason: The 1999 Kant Lectures at Stanford University*, CSLI Publications.

Friedman, M. (1999b). *Reconsidering Logical Positivism*, Cambridge University Press.

Friedman, M. (1999c). Tolerance and analyticity in carnap's philosophy of mathematics, in (Friedman 1999b), Cambridge University Press.

Gödel, K. (1946). Remarks before the Princeton bicentennial conference on problems in mathematics, in (Gödel 1990), Oxford University Press, pp. 150–153.

Gödel, K. (1953/9). Is mathematics syntax of language?, in (Gödel 1995), Oxford University Press, pp. 334–362.

Gödel, K. (1990). *Collected Works, Volume II: Publications 1938–1974*, Oxford University Press, New York and Oxford.

Gödel, K. (1995). *Collected Works, Volume III: Unpublished Essays and Lectures*, Oxford University Press, New York and Oxford.

Goldfarb, W. and Ricketts, T. (1992). Carnap and the philosophy of mathematics, in D. Bell and W. Vossenkuhl (eds), *Science and Subjectivity*, Akadamie, Berlin, pp. 61–78.

Janssen, M. (2002). Reconsidering a scientific revolution: The case of Einstein *versus* Lorentz, *Physics in Perspective* 4(4): 421–446.

Kleene, S. C. (1939). Review of *The Logical Syntax of Language, Journal of Symbolic Logic* 4(2): 82–87.

Koellner, P. (2006). On the question of absolute undecidability, *Philosophia Mathematica* 14(2): 153–188; Revised and reprinted in *Kurt Gödel: Essays for his Centennial*, edited by Solomon Feferman, Charles Parsons, and Stephen G. Simpson. Lecture Notes in Logic, 33. Association of Symbolic Logic, 2009.

Koellner, P. (2009). Carnap on the foundations of logic and mathematics. Unpublished.

Koellner, P. and Woodin, W. H. (2009a). Incompatible Ω-complete theories, *The Journal of Symbolic Logic*. Forthcoming.

Koellner, P. and Woodin, W. H. (2009b). Large cardinals from determinacy, *in* M. Foreman and A. Kanamori (eds), *Handbook of Set Theory*, Springer. Forthcoming.

Larson, P., Ketchersid, R. and Zapletal, J. (2008). Regular embeddings of the stationary tower and Woodin's Σ_2^2 maximality theorem. Preprint.

Lindström, P. (2003). *Aspects of Incompleteness*, Vol. 10 of *Lecture Notes in Logic*, second edn, Association of Symbolic Logic.

Parsons, C. (2008). *Mathematical Thought and its Objects*, Cambridge University Press.

Quine, W. V. (1963). Carnap and logical truth, in P. A. Schilpp (ed.), *The Philosophy of Rudoph Carnap*, Vol. 11 of *The Library of Living Philosophers*, Open Court, pp. 385–406.

Reichenbach, H. (1920). *Relativitätstheorie und Erkenntnis Apriori*, Springer, Berlin. Translated by M. Reichenbach as *The Theory of Relativity and A Priori Knowledge*. Los Angeles: University of California Press, 1965.

Reichenbach, H. (1924). *Axiomatik der relativistischen Raum-Zeit-Lehre*, Vieweg, Braunschweig. Translated by M. Reichenbach as *Axiomatization of the Theory of Relativity*. Los Angeles: University of California Press, 1969.

Tait, W. W. (1986). Truth and proof: The Platonism of mathematics, *Synthese* (69): 341–370. Reprinted in (Tait 2005).

Tait, W. W. (2005). *The Provenance of Pure Reason: Essays in the Philosophy of Mathematics and Its History*, Logic and Computation in Philosophy, Oxford University Press.

Wilson, C. (1970). From Kepler's laws, so-called, to universal gravitation: Empirical factors, *Archive for History of Exact Sciences* 6(2): 89–170.

Woodin, W. H. (1999). *The Axiom of Determinacy, Forcing Axioms, and the Nonstationary Ideal*, Vol. 1 of *de Gruyter Series in Logic and its Applications*, de Gruyter, Berlin.

Woodin, W. H. (2009). *Suitable Extender Sequences*. To appear.

5
"Algebraic" Approaches to Mathematics*

Mary Leng

At the turn of the twentieth century, philosophers of mathematics were predominantly concerned with the foundations of mathematics. This followed the so-called "crisis of foundations" that resulted from the apparent need for infinitary sets in order to provide a proper foundation for mathematical analysis, and was exacerbated by the discovery of both apparent and actual paradoxes in naïve infinitary set theory (most famously, Russell's paradox). Philosophers and mathematicians at this time saw their job as to place mathematics on firm, and indeed certain, axiomatic foundations, so as to provide confidence in the new mathematics being developed. Thus, the "big three" foundational programmes of logicism, formalism, and intuitionism were established, each providing a different answer to the question of the proper interpretation of axiomatic mathematical theories.

Now, 100 years on, philosophers are generally much less worried about the foundations of mathematics. Indeed, under the influence of W. V. Quine in particular, there has been a move away from "foundationalist" accounts of knowledge across the board. It is no longer assumed that there are *any* firm foundations on which to ground our beliefs. Against this backdrop, In the latter half of the twentieth century, philosophers of mathematics retreated from the question of how to establish the *certainty* of our mathematical knowledge to the question of whether we can claim to have any mathematical knowledge at all, even of our most foundational mathematical assumptions. Presented with Benacerraf's two challenges (1965, 1973) to the standard Platonist view of mathematical theories as assertions of truths about mathematical objects, it was no longer acceptable to assume without justification that we had any mathematical knowledge of such truths.

But although the debate in the philosophy of mathematics has shifted from the question of whether we can show our mathematical knowledge to be certain and secure, to the question of whether we can have mathematical knowledge at all, there is an interesting relation between contemporary approaches and an earlier debate between foundationalists. Gottlob Frege and David Hilbert clashed in their correspondence (between 1899 and 1900) over the nature of axioms (reproduced in Gabriel *et al.* (1980)). While Frege saw axioms as assertions of truths about a previously existing subject matter, Hilbert viewed axioms as contextually defining their nonlogical terms, in much the way that a system of algebraic equations in various unknowns can be thought of as contextually defining the possible values of their unknowns.

Frege's "Assertory" view is shared by traditional Platonists about mathematics, and is therefore subject to Benacerraf's worries. Moreover, it is arguable that something like Hilbert's "Algebraic" account is common to several recent responses to Benacerraf's worries, including versions of structuralism, Platonism, and fictionalism.[1] Algebraic approaches have fairly straightforward answers to Benacerraf's two challenges. Unfortunately, as I will argue, they solve these problems at the expense of introducing a new set of problems of their own.

1 "Assertory" views of mathematics and Benacerraf's problems

Traditional Platonism, according to which our mathematical theories are bodies of truths about a realm of mathematical objects, assumes that only some amongst consistent theory candidates succeed in correctly describing the mathematical realm. For platonists, while mathematicians may contemplate alternative consistent extensions of the axioms for ZF (Zermelo–Fraenkel) set theory, for example, at most one such extension can correctly describe how things really are with the universe of sets. Thus, according to Platonists such as Kurt Gödel, intuition together with quasi-empirical methods (such as the justification of axioms by appeal to their intuitively acceptable consequences) can guide us in discovering which amongst alternative axiom candidates for set theory has things right about set theoretic reality. Alternatively, according to empiricists such as Quine, who hold that our belief in the truth of mathematical theories is justified by their role in empirical science, empirical evidence can choose between alternative consistent set theories. In Quine's view, we are justified in believing the truth of the minimal amount of set theory required by our most attractive scientific account of the world.

Despite their differences at the level of detail, both of these versions of Platonism share the assumption that mere consistency is not enough for a mathematical theory: For such a theory to be true, it must correctly describe a realm of objects, where the existence of these objects is not guaranteed by consistency alone. Such a view of mathematical theories requires that we must have some grasp of the intended interpretation of an axiomatic theory that is independent of our axiomatization – otherwise inquiry into whether our axioms "get things right" about this intended interpretation would be futile. Hence, it is natural to see these Platonist views of mathematics as following Frege in holding that axioms

> . . . must not contain a word or sign whose sense and meaning, or whose contribution to the expression of a thought, was not already completely laid down, so that there is no doubt about the sense of the proposition and the thought it expresses. The only question can be whether this thought is true and what its truth rests on.
>
> FREGE to HILBERT: 27/12/1899, in Gabriel *et al.* (1980, 36)

On such an account, our mathematical axioms express genuine assertions (thoughts), which may or may not succeed in asserting truths about their subject matter. Following Geoffrey Hellman, then, we will label these Platonist views as "assertory" views of mathematics.

Assertory views of mathematics make room for a gap between our mathematical theories and their intended subject matter, and the possibility of such a gap leads to at least two difficulties for traditional Platonism. These difficulties are articulated by Paul Benacerraf (1965, 1973) in his aforementioned papers. The first difficulty comes from the realization that our mathematical theories, even when axioms are supplemented with less formal characterizations of their subject matter, may be insufficient to choose between alternative interpretations. For example, assertory views hold that the Peano axioms for arithmetic aim to assert truths about *the natural numbers*. But there are many candidate interpretations of these axioms, and nothing in the axioms, or in our wider mathematical practices, seems to suffice to pin down one interpretation over any other as the correct one. The view of mathematical theories as assertions about a specific realm of objects seems to force there to be facts about the correct interpretation of our theories even if, so far as our mathematical practice goes (for example, in the case of arithmetic), any ω-sequence would do.

Benacerraf's second worry is perhaps even more pressing for assertory views. The possibility of a gap between our mathematical theories and their intended subject matter raises the question, "How do we know that our mathematical theories have things right about their subject matter?". To answer this, we need to consider the nature of the purported objects about which our theories are supposed to assert truths. It seems that our best characterization of mathematical objects is negative: to account for the extent of our mathematical theories, and the timelessness of mathematical truths, it seems reasonable to suppose that mathematical objects are *non-physical, non-spatiotemporal* (and, it is sometimes added, *mind-* and *language-independent*) objects – in short, mathematical objects are *abstract*. But this negative characterization makes it difficult to say anything positive about how we could know anything about how things are with these objects. Assertory, Platonist views of mathematics are thus challenged to explain just *how* we are meant to evaluate our mathematical assertions – just how do the kinds of evidence these Platonists present in support of their theories succeed in ensuring that these theories track the truth?

2   "Algebraic" views and Benacerraf's problems

Frege presents his assertory view of mathematical theories in the context of a discussion with Hilbert over the nature of axioms. There, Hilbert offers a very different account of the nature of mathematical theories, according to which we should view mathematical axioms as contextually defining their subject matter. In Hilbert's view, so long as our axioms are consistent, there is no question of them *getting things wrong* about their intended interpretation: they are automatically true of *whatever* (if anything) satisfies their axioms. We can think of an axiomatic theory as analogous to a system of equations in several unknowns: together, the axioms determine what (if anything) will count as a solution to the equations (i.e., an appropriate interpretation of the primitive terms). The analogy with algebraic equations suggests a label for this kind of view of axioms: again, following Hellman, we will call views of this sort "algebraic" views of mathematical theories.

   Characteristic of algebraic views of mathematics is the notion that, while we might find some mathematical theories more interesting/worthy of study than others, insofar as a theory is to count as *mathematical*, consistency is all that matters. Thus, for Hilbert, consistency is "the criterion of truth and existence" in mathematics (HILBERT to FREGE: 29/12/1899, in Gabriel *et al.* (1980, 39–40)). In doing mathematics, according to the algebraic view, one is primarily interested in working out the consequences of consistent theoretical assumptions. No further question arises as to whether those assumptions are true of the specific objects to which they purport to refer. For one thing, one and the same theory will be true of many different interpretations:

> any theory can always be applied to infinitely many systems of basic elements. One only needs to apply a reversible one-one transformation and lay it down that the axioms shall be correspondingly the same for the transformed things. (*Ibid.*, 40–41)

Furthermore, the question of whether, as a matter of fact, there are *any* objects about which our mathematical theories can be interpreted as asserting truths takes something of a back seat: at least insofar as pure mathematics is concerned, one's interest is in working out what would have to be true in any system of objects that satisfied our axioms, without concern for whether, as a matter of fact, those axioms are true of any objects.

   Given their nonstandard understanding of the content of our mathematical theories, algebraic approaches to mathematical theories have an easier time than do assertory approaches with both of Benacerraf's two problems.

In taking an algebraic view, one does not assume that our theories are really talking about one particular interpretation of their axioms, so one does not need to choose between alternative candidates for "the referent" of a mathematical singular term. Furthermore, since in such views any logically consistent theory is seen as mathematically acceptable, the pressure to explain how our mathematical beliefs reliably reflect the mathematical realm is much reduced. From a pure mathematical perspective, what matters for a good mathematical theory is that it is consistent. And so long as we can be confident in our judgments about consistency, the question of how we can know that our theories are also *true* of a portion of mathematical reality is of reduced significance.

3 Contemporary algebraic views

Given the promise of the algebraic outlook as a solution to Benacerraf's two problems, and given the prominence of these problems in contemporary philosophy of mathematics, it is perhaps not surprising that several philosophical accounts of mathematics have arisen that contain within them elements of the algebraic view. In particular, despite their clear metaphysical differences, contemporary versions of structuralism (both Geoffrey Hellman's modal structuralism and Stewart Shapiro's *ante rem* structuralism), full-blooded Platonism, and fictionalism all share in something of the spirit of the algebraic view. In particular, for all of these accounts, logical *consistency* is what matters for a pure mathematical theory; axioms (together, perhaps, with additional implicit assumptions about appropriate interpretations) are taken to define the subject matter of such a theory, so that the question of whether the axioms are *true of* their intended subject matter does not arise; and we need not view such a theory as applicable to a unique system of objects. It is worth, then, considering these various views in turn to see how each expresses aspects of the algebraic view, and how each is able to use these aspects to respond to Benacerraf's problem.

3.1 Modal structuralism

Structuralism holds that mathematics is ultimately about the shared structures that may be instantiated by particular systems of objects. Eliminative structuralists, such as Geoffrey Hellman, try to develop this insight in a way that does not assume the existence of abstract *structures* over and above any instances. But since not all mathematical theories have concrete instances, this brings a *modal* element to this kind of structuralist view: mathematical theories are viewed as being concerned with what *would* be the case in any system of objects satisfying their axioms. In Hellman's version of the view (Hellman 1989), this leads to a reinterpretation of ordinary mathematical utterances made within the context of a theory. A mathematical utterance of the sentence *S*, made against the context of a system of axioms expressed as a conjunction AX, becomes interpreted as the claim that the axioms are logically consistent and that they logically imply S (so that, *were* we to find an interpretation of those axioms, S would be true in that interpretation). Formally, an utterance of the sentence S becomes interpreted as the claim:

$$\Diamond AX \ \& \ \Box(AX \supset S)$$

Here, in order to preserve standard mathematics (and to avoid infinitary conjunctions of axioms), AX is usually a conjunction of *second-order* axioms for a theory. The operators "$\Diamond$" and "$\Box$" are modal operators on sentences, interpreted as "it is logically consistent that", and "it is logically necessary that", respectively.[2]

This view clearly shares aspects of what we have taken to be the core of algebraic approaches to mathematics. According to modal structuralism what makes a mathematical theory *good* is that it is logically consistent. Pure mathematical activity becomes inquiry into the consistency of axioms, and into the consequences of axioms that are taken to be consistent. As a result, we need not view a theory as applying to *any* particular objects, so certainly not to one particular system of objects. Since mathematical utterances so construed do not refer to any objects, we do not get into difficulties with deciding on *the* unique referent for apparent singular terms in mathematics. The number 2 in mathematical contexts refers to *no* object, though if there were a system of objects satisfying the second-order Peano axioms, whatever mathematical theorems we have about the number 2 would apply to whatever the interpretation of 2 is in that system. And since our mathematical utterances are made true by *modal* facts, about what does and does not follow from consistent axioms, we no longer need to answer Benacerraf's question of how we can have knowledge of a realm of abstract objects, but must instead consider how we know these (hopefully more accessible) facts about consistency and logical consequence.

3.2   *Ante rem* structuralism

Stewart Shapiro's non-eliminative version of structuralism, by contrast, accepts the existence of structures over and above systems of objects instantiating those structures (Shapiro 1997). Specifically, according to Shapiro's *ante rem* view, every logically consistent theory correctly describes a structure.[3] Like Hellman, Shapiro thinks that many of our most interesting mathematical structures are described by *second-order theories* (first-order axiomatizations of sufficiently complex theories fail to pin down a unique structure up to isomorphism). Mathematical theories are then interpreted as bodies of truths about *structures*, which may be instantiated in many different *systems* of objects. Mathematical singular terms refer to the *positions* or *offices* in these structures, positions which may be occupied in instantiations of the structures by many different officeholders.

While this account provides a standard (referential) semantics for mathematical claims, the kinds of objects (offices, rather than officeholders) that mathematical singular terms are held to refer to are quite different from ordinary objects. Indeed, it is usually simply a *category mistake* to ask of the various possible *officeholders* that could fill the number 2 position in the natural number structure whether *this* or *that* officeholder is *the* number 2 (i.e., the *office*).[4] Independent of any particular instantiation of a structure, the referent of the number 2 is the number 2 *office* or *position*. And this office/position is completely characterized by the axioms of the theory in question: if the axioms provide no answer to a question about the number 2 office, then within the context of the pure mathematical theory, this question simply has no answer.

Elements of the algebraic approach can be seen here in the emphasis on logical consistency as the criterion for the existence of a structure, and on the identification of the truths about the positions in a structure as being exhausted by what does and does not follow from a theory's axioms. As such, this version of structuralism can also respond to Benacerraf's two worries. As we have seen, the question of *which* instantiation of a theoretical structure one is referring to when one utters a sentence in the context of a mathematical theory is dismissed as a category mistake. And, so long as the basic principle of structure-existence, according to which every logically consistent axiomatic theory truly describes a structure, is correct, we can explain our knowledge of mathematical truths simply by appeal to our knowledge of consistency.[5]

3.3 Full-blooded Platonism

Like *ante rem* structuralism, full-blooded Platonism (of the kind developed by Mark Balaguer (Balaguer 1998)) starts from a metaphysical assumption about the extent of the mathematical realm. But rather than holding that the mathematical realm consists of all logically consistent *structures*, full-blooded Platonism holds that the mathematical realm contains systems of objects instantiating every logically consistent theory. Balaguer's own version of full-blooded Platonism sticks with first-order theories, together with additional implicit assumptions embodying our "full conception" of the objects of a theory, to pin down a theory's interpretation, but a version that accepted full-blown second-order axiomatizations would avoid some of the difficulties of clarifying what is meant by our "full conception" of, for example, natural numbers. At any rate, given the full-blooded existence assumption, it follows that any consistent theoretical conception will describe a portion of mathematical reality. Furthermore, since on this account precisely which portion (or portions) of mathematical reality a theory describes is determined by the content of the "full conception" specifying the theory, it follows that we can know all there is to know about the objects a given theory refers to by working out the consequences of our full conception of those objects.

124   *Mary Leng*

Of course, one and the same theory may correctly describe more than one system of objects, so talk of *the referent* of (say) the number 2 does not really make sense on this account. Indeed, Balaguer embraces nonuniqueness, holding that our mathematical theories should be viewed as applying to whatever systems of objects satisfy their full conception of their objects. As such, full-blooded Platonism is in one sense a version of eliminative structuralism: the truths of a mathematical theory are truths about *all* logically possible systems of objects satisfying the theory's axioms or full conception. But since on Balaguer's view logical possibility implies mathematical actuality, this reduces to the claim that a theory's truths are those true generalizations about all systems of objects satisfying the theory's assumptions. Like modal structuralism, then, the truths of a theory are exhausted by what follows and does not follow from that theory's assumptions – again, a characteristic mark of the algebraic approach.

The fact that the apparently referential utterances of mathematical theories are best interpreted, on this account, as hidden universal generalizations (about what's true in all systems satisfying the theory's assumptions), means that Benacerraf's first worry is avoided just as it is in modal structuralism. An utterance such as, "There is a unique even prime number", can be interpreted as the claim that "In any system of objects satisfying our full conception of natural numbers, there is a unique even prime number", showing questions about the absolute identity of this unique object, independent of its identity within any particular system, to be wrongheaded. The response to Benacerraf's second worry, however, is more akin to that of the *ante rem* structuralist. Balaguer's strong existence assumptions imply that, so long as we know that our full conception characterizing a theory is consistent, it can't fail to be true of at least one (and usually many) portion of mathematical reality, so that knowledge of mathematical truths is reduced to knowledge of the consequences of consistent theories. Again, there remains a worry of how we can know that the mathematical realm contains all logically possible systems of objects, but at the very least full-blooded Platonism succeeds in providing an account of how, given its own assumptions about the nature of mathematical reality, it is reasonable to expect our mathematical beliefs to track that reality.

3.4 Fictionalism

According to fictionalists about mathematics, such as Hartry Field (1980), if we take utterances made in the context of a mathematical theory to be *assertions*, then traditional Platonists are right in thinking that these assertions aim to be about a realm of abstract mathematical objects. However, fictionalists think that we can account for the nature and value of mathematical practice without interpreting the utterances of mathematicians as genuine assertions. Rather, according to fictionalists, it is coherent to view mathematical theorizing as analogous to the utterance of sentences in fictional contexts.[6] Fictionalists view the basic assumptions of a pure mathematical theory (its axioms, or perhaps if we wish to follow Balaguer, its "full conception" of its objects) not as truths, but as generative of a fiction. Mathematical practice then becomes the practice of developing these mathematical fictions, by working out what does and does not follow from their generative assumptions. Since (classically, at least) inconsistent assumptions imply everything, only consistent fictions are mathematically interesting. So, as with other algebraic views, for fictionalism what makes a theory mathematically acceptable is consistency, and what makes a mathematical utterance appropriate in the context of mathematical theorizing is that it is a consequence of the assumptions of the theory in which it is put forward.

Benacerraf's first problem is avoided on this account since the idea that a good mathematical theory must successfully refer to *any* objects, let alone a *unique* system of objects, is abandoned. Asking which object, precisely, is the number 2 is analogous to asking who, precisely, is Hamlet. On a plausible view of fictionalist discourse,[7] the name "Hamlet" really refers to no one, and so it is wrongheaded to ask, outside of the context of the fiction, who Hamlet is. The second problem, in contrast, is avoided by simply *accepting* its conclusion: fictionalists accept that we cannot account for how we have knowledge of any abstract objects, and conclude that we do not know that mathematical utterances such as $2 + 2 = 4$ are true. This scepticism about our knowledge of mathematical utterances when construed at face value as assertions about abstracta, does not, of course, lead to an abandonment of mathematical inquiry. Rather, we are left with a shift in our understanding of the nature of mathematical knowledge. What "mathematical" knowledge we have, according to fictionalists, is really just knowledge of the consequences of our generative assumptions. That is, our mathematical knowledge turns out to be *logical* knowledge, knowledge concerning what does and does not follow from consistent mathematical assumptions.

4 Algebraic approaches: Two problems

Benacerraf's two problems can thus be solved by all four of these algebraic approaches to mathematics, and given the predominance of Benacerraf's problems in recent philosophy of mathematics, this is cause for some celebration. *But not too much.* For, it turns out that two new difficulties arise for these algebraic approaches, difficulties that are not present for assertory views of mathematics. In particular, all four algebraic approaches face a problem of modality and a problem of mixed contexts.[8]

4.1 The problem of modality

The problem of modality is most obvious for Hellman's modal structuralism. Hellman's rewriting of mathematical claims in terms of the modal operators

□ (it is logically necessary that) and ◊ (it is logically consistent that) raises two natural questions: just what do these operators *mean*, and how can we know when an utterance □*P* or ◊*P* is true?

Assertory views of mathematics have an easy answer to this, if they wish to take it.[9] We can provide reductive accounts of logical necessity and consistency in terms of model theory, analyzing these notions as truth in all models and truth in a model, respectively. Knowledge of logical necessity and consistency is then just a species of mathematical knowledge since we know about set theoretic models in just the same way as we know any facts about the sets.

Clearly, this reductive account of logical consistency and necessity is unavailable to Hellman, for whom mathematical claims concerning the existence of sets are meant to be reduced to claims about what follows logically from consistent set theoretic axioms. Further reducing consistency and logical consequence to model theory would be blatantly circular (a point which is recognized by Hellman, and stressed by Bob Hale in his (1996) critique of modal structuralism). And for fictionalists who do not think we can know any claims about the existence of *any* sets, the claim that mathematics involves working out the consequences of consistent theoretical assumptions cannot be reduced to a claim about what's true in all models of those assumptions since on such a view we would never be able to know anything about consistency and logical consequence. But it should not be thought that the ontologically weightier algebraic views, *ante rem* structuralism and full-blooded Platonism, have things any easier in accounting for the modal notions on which they rely. In fact, given that both views take the claim that the mathematical realm contains all logically consistent structures/objects to be a substantive thesis about its content,[10] rather than an almost trivial consequence of a reductive definition of consistency, the reductive account provided by assertory views is unavailable to these accounts as well. Furthermore, since these latter two accounts take the knowledge problem for the structures/objects they posit to be solved by appeal to our knowledge of logical properties such as consistency, they cannot follow assertory views in taking logical knowledge to be dependent on knowledge about the sets. If we can only know what sets there are through our knowledge of consistency, and can only have knowledge of consistency through our knowledge of the sets, mathematical knowledge would never be able to get off the ground.[11]

All four accounts, therefore, need an alternative account of the various logical concepts which does not reduce them to claims about sets. And all four accounts need to explain how we can have knowledge of the extension of these concepts which does not depend on prior knowledge of mathematical objects or structures. In fact, the main proponents of each account discussed (Hellman, Shapiro, Balaguer, and Field) are in agreement that logical consistency (and related notions) should be understood as primitive modal notions. We have, they suppose, an intuitive grasp of these notions, and can sharpen that intuitive grasp through our grasp of their inferential role (e.g., through our grasp of the validity of inferences such as: $P \supset \Diamond P$ and $\Box P \supset P$). And while our knowledge of the applicability of these notions is very often justified mathematically (e.g., we argue that $\Diamond P$ by showing that our favoured set theory implies the existence of a model of P), Field (1984) has argued that such uses of mathematics to find out about logical notions is acceptable so long as we assume that our favoured set theory is consistent, not that it is true. I will not consider the question of whether or not this kind of account is defensible here (I have done so elsewhere, in Leng (2007)), but will simply note that all four of the algebraic views under consideration depend for their feasibility on the prospects for coupling modal primitivism with a non-mathematical account of our knowledge of the logical modalities.

4.2 The problem of mixed claims

The second difficulty for algebraic views stems from the fact that all of these views are designed as accounts of the truth (or, in the fictionalist's case, the mathematical acceptability) of utterances made in the context of what I have been calling "pure" mathematical theories. By "pure" here, I do not mean to call upon the pure/applied contrast as it is usually used within mathematics – the dividing line between pure and applied theories in this sense is blurred to say the least. Rather, I use the term "pure" in contrast with "impure", as it is sometimes used in the context of set theory. Pure sets are sets which contain only sets in their transitive closures. The transitive closures of "impure" sets, in contrast, contain objects which are not themselves sets (e.g., physical objects). If we assume a reductive account of mathematical theories, according to which all mathematical objects are ultimately just sets, then we can use this pure/impure distinction to characterize a distinction between pure and impure mathematical theories: a pure mathematical theory is quantifier-committed only to pure sets, whereas an impure mathematical theory includes some impure sets (and hence some non-sets) in the range of its quantifiers. Alternatively, if we do not wish to prejudge the question of reductionism in mathematics, we can loosen this definition slightly to define pure theories as ones whose quantifiers range only over mathematical objects, and impure theories as ones whose quantifiers also range over some non-mathematical objects.

Our ordinary *empirical* theories are "impure" *mathematical* theories in this sense: they quantify over mathematical and non-mathematical objects. Formally expressed in the language of set theory with urelements, empirical theories allow the free collection of non-mathematical objects into sets. This means that cardinality claims can be made about collections of physical objects ("The number of fingers on my right hand is 5"), and that, for example, physical properties can be associated with mathematical quantities ("The mass in grams of object *a*, *m*(*a*), is 30"). Assertory views of mathematical theories can easily make sense of these claims about the relations between mathematical and non-mathematical objects using their standard semantics: on such views, only some sets exist, and in particular, the existence of impure sets will depend on what non-mathematical objects there are. Algebraic views, however, cannot make sense of these mixed claims so easily.[12]

The trouble comes from the modal aspect of the algebraic view, and can again be seen most easily with modal structuralism. As we have seen, according to modal structuralism, if *S* is a sentence uttered in the context of a pure mathematical theory with axioms *AX*, this utterance should be interpreted as the claim that

$$\Diamond AX \ \& \ \Box(AX \supset S)$$

But how, then, should a modal structuralist interpret a sentence S^{mixed}, uttered in the context of an empirical scientific theory whose laws can by expressed by the conjunction AX^{mixed}, whose quantifier commitments involve mathematical and non-mathematical objects? If we apply the same interpretation, interpreting S^{mixed} as

$$\Diamond AX^{\text{mixed}} \ \& \ \Box(AX^{\text{mixed}} \supset S^{\text{mixed}})$$

we face an immediate difficulty: such a mixed utterance will be interpreted as true just so long as it follows from our empirical theory, if that theory is logically consistent. But this standard is clearly too weak for the mixed claims of empirical science: a consistent, impure theory may be constructed which has as a consequence the claim that the moon is made of green cheese, but we would not wish to be saddled by this interpretation into accepting that this consequence is true. While neatly designed to deal with utterances made in the context of pure mathematical theories, modal structuralism needs an alternative story to deal with the semantics of mixed claims.

A similar difficulty can easily be seen for fictionalist interpretations of the mixed claims of empirical theories. If we try to interpret utterances made in the context of impure, empirical theories along the lines suggested for pure mathematical theories, we will end up saying that a sentence S^{mixed}, uttered in such a context, will be *appropriate* if and only if the hypotheses of the empirical theory in question are consistent and S^{mixed} is indeed a consequence of those hypotheses. But surely if we want to derive predictions from our empirical theories, we will want their utterances concerning empirical matters not just to be "appropriate" in the context of a given theoretical story, but also to be true? And this is something about which the fictionalist account we have sketched appears to give no guidance.

Again, one might think that the more realist aspects of full-blooded Platonism and *ante rem* structuralism allow us to avoid this difficulty since, like standard assertory Platonists, they assume that the objects of pure mathematical theories exist (either as ordinary objects or as "positions" in structures) and that therefore we can straightforwardly talk about their relations to non-mathematical objects. But the advantage is again illusory. In order to make mixed claims about the relations between mathematical and non-mathematical objects, our empirical theories introduce properly mixed objects: impure sets whose transitive closures include non-mathematical as well as mathematical objects. But as theories about the extent of the mathematical realm, full-blooded Platonism and *ante rem* structuralism are designed to tell us only which *pure* mathematical objects exist. When it comes to the nature and existence of *impure* mathematical objects, neither view has a straightforward response.

When, for example, the *ante rem* structuralist says that there is a set of fingers on my right hand whose cardinality is five, we may ask them on what structure existence principle it follows that such a set exists. The standard structure existence principle, which says that any coherent theory is true of a structure, will imply the existence of such a set within an appropriate structure, but also, in another consistent theory, a set of fingers on my right hand whose cardinality is seventeen (and, indeed, in another, a set whose sole member is a moon made out of green cheese). And similarly, if the full-blooded Platonist aims to hold that *any* logically consistent theory is true of a portion of mathematical reality, there will likewise be a portion of that reality which includes a set whose sole member is the green, cheesy moon. Structuralists and full-blooded Platonists therefore need to restrict their plenitudinous existence principles in some way, so as to ensure that only the *right* kind of mixed (impure) mathematical objects exist. For the assumption that *all* logically consistent empirical theories truly describe a realm of objects, or that all such theories correctly characterize a structure, does nothing to distinguish between merely consistent empirical theories and those which have things right about the actually existing empirical world. Mere truth of a structure, or of a portion of mathematical reality, is not enough for mixed, empirical theories: we want theories that do more than this, theories that also get things right about the physical world they purport to describe.

There are two strategies for solving this difficulty from an algebraic perspective. The first is that of Field, who takes the challenge to be to *dispense* with the problematic mixed claims in our empirical theorizing. On this account, our literally believed scientific theories can be expressed in entirely non-mathematical terms, and understood in an assertory manner as assertions that aim to be true of non-mathematical objects. *Once we have such an account*, rewriting our scientific theories so that they include no reference to pure or impure mathematical objects, we can explain the difference between between good and bad consistent "mixed" theories by showing that our good theories are not only consistent, but also consistent with the empirical facts as expressed in our literally believed non-mathematical theories. For the fictionalist, no genuinely mixed empirical theory is true, but such a theory does not have to be true to be good – it needs only to be a consistent extension of the empirical "facts" as expressed in our literally believed non-mathematical theory.

If we did have access to an entirely non-mathematical theory of the empirical world, then such an account could also be put to use by the other algebraic views we have considered to explain the difference between claims made in the context of good and bad mixed theories. For modal structuralists and full-blooded Platonists, a mixed theory will be a good one just in case it is not only consistent, but also a consistent extension of our literally believed non-mathematical theory; for *ante rem* structuralists, a mixed theory will be a good one when the structure it imposes on its non-mathematical objects corresponds to the structure posited by the non-mathematical theory of those objects. If we have access to a non-mathematical theory of the empirical world, the "goodness" of a mixed theory will come down to an inspection of the relationship between these two theories. In all cases, the claim that a mixed theory is good will amount to the claim that it conservatively extends the empirical theory we literally believe.

But it is not widely accepted that we can dispense with mathematics in our best scientific description of the non-mathematical world. In this case, what can these algebraic views say about the difference between good and bad mixed empirical theories? Hellman (1989, chapter 3) as a modal structuralist, and Balaguer (1998, chapter 7), in the context of a defense of both fictionalism and full-blooded platonism, have suggested an alternative account which does not depend on having access to a non-mathematical theory of the empirical world. Both hold that a mixed theory is a good one if its claims do not interfere with the non-mathematical *facts*, even if we have no independent theory of those facts. Extending this to *ante rem* structuralism, a mixed theory will be a good one if the structure it imposes on its non-mathematical objects respects the structural relations between those objects, *even if we have no independent non-mathematical theory of those relations.*[13] In all of these cases, the basic account of what makes a mixed theory a good one is the same as that offered under the supposition that we can dispense with mathematics: what makes the theory good is that it is correct in the picture it paints of the non-mathematical realm. However, if we do not have a non-mathematical theory of the non-mathematical realm, we will be unable to *prove* that our theory has this feature. Rather, we will simply have to hypothesize that good mixed theories are good because they are correct in their non-mathematical picture (or, as we might wish to put it, that they are *nominalistically adequate*).

Aside from worries we might have about how to characterize this notion of nominalistic adequacy (what exactly is going to count as a non-mathematical "fact"?), one major difficulty with accepting this answer is that it places all of the various algebraic views at odds with standard scientific realism. Indeed, discussions of the indispensability argument in the philosophy of mathematics recognize this to be the case for fictionalism, which, if mathematics *is* indispensable in formulating our scientific theories, cannot be combined with scientific realism standardly construed as the view that our best scientific theories are true or approximately true. It is often assumed, however, that since they interpret our mathematical theories as *truths*, the various other algebraic views considered here have an easier time with scientific realism. The argument I have presented here is that their difficulties with characterizing a suitable notion of truth for mixed claims means that these views will be similarly difficult to combine with standard scientific realism. For, whereas scientific realists think that the goodness of our best scientific theories amounts to their truth, on all of these algebraic views, what makes such a theory good is not its truth as such (which, depending on their understanding of the truth of mixed claims, may be something that is achieved quite cheaply, even for theories with empirically false consequences), but rather that it is correct in its picture of the non-mathematical realm – namely, that it is nominalistically adequate. If defenders of these various algebraic views do not wish to do the hard work of dispensing with mathematics in empirical science, then they will have to defend their account of the goodness of mixed theories over the standard account provided by scientific realism.[14]

5 Conclusion

The various algebraic approaches to mathematics show great promise in providing responses to Benacerraf's two problems. Unsurprisingly, though, they bring problems of their own. The problem of explaining our knowledge of consistency and other logical properties, and the problem of explaining what the goodness of a mixed mathematical/empirical theory amounts to, both mean that none of these accounts of mathematics can rest easy just yet. My own view is that these new problems, though difficult, are more tractable than Benacerraf's two worries. So, my hope for a new wave in the philosophy of mathematics is that it will turn to these new difficulties and show how they can be overcome.

Notes

*I am extremely grateful to participants in the New Waves in Philosophy of Mathematics workshop in Miami for their helpful discussion of this paper, and particularly to Roy Cook, Mark Colyvan, and Øystein Linnebo for thoughtful written comments.

1. The labels "Algebraic" and "Assertory" are due to Geoffrey Hellman (2003). In this volume, Thomas Hofweber uses the terminology "constitutive" versus

"descriptive" to capture a similar distinction (though Hofweber's distinction is made against the background assumption that axioms should be viewed as truths, an assumption which we will not make here).

2. More precisely, to make explicit Hellman's treatment of primitive terminology, if our theory includes primitive terms $t_1, \ldots, t_n$ (which may include relations, functions, or singular terms), treating these terms as variable expressions, the sentence $S(t_1, \ldots, t_n)$ becomes interpreted as $\Diamond \exists t_1 \ldots \exists t_n AX(t_1, \ldots, t_n)$ & $\Box \forall t_1 \ldots \forall t_n (AX(t_1, \ldots, t_n) \supset S(t_1, \ldots, t_n))$. If some of these terms are relations or functions, the relevant quantifiers are of course second-order.

3. Shapiro uses the terminology "coherent" rather than "logically consistent" in making this claim, as he reserves the term "consistent" for *deductively consistent*, a notion which, in the case of second-order theories, falls short of coherence (i.e., logical consistency), and wishes also to separate coherence from the model-theoretic notion of *satisfiability*, which, though plausibly coextensive with the notion of coherence, could not be used in his theory of structure existence on pain of circularity.

4. A special case here is the number 2 office itself: in Shapiro's view, the positions in structures (offices) are themselves to be viewed as objects, and may therefore themselves also be said to act as office-holders, occupying the positions in structures.

5. Of course, the *ante rem* structuralist has *not* shown that the basic principle of structure-existence is correct, and so we may still press the question, but *how do we know* that the mathematical realm is as they say it is? Nevertheless, *ante rem* structuralism has succeeded where traditional Platonism has failed in at least explaining how, *given its own assumptions about the mathematical realm*, it would be reasonable to expect our mathematical beliefs to reflect the facts about that realm.

6. This is not to say that it is never appropriate to interpret some *mathematicians* as genuinely intending to assert that 2 is prime when they utter the sentence "2 is prime". Rather, the fictionalist claims, one need not accept the truth of such mathematical utterances, construed as assertions, in order to account for the success of our mathematical theorizing, but may instead explain the success of mathematical discourse from a perspective which views it as, in important respects, analogous to fiction.

7. Though by no means the only view available: there are of course realists about fictional objects.

8. This is not, of course, to say that the various approaches taken individually do not face any more difficulties than these. But I take these two problems to be of particular interest because they are common to all four versions of the algebraic approach considered here.

9. Although this easy answer is available to assertory views, defenders of such views might still wonder whether this answer is the right one. As Field has argued (1989, pp. 31–32), even standard Platonists may balk at accepting the model theoretic reduction of consistency as an *analysis* of the modal notion it promises to replace. For one thing, Field argues, it seems only to be a result of an accident of first-order logic (the Löwenheim–Skolem theorem) that this reduction is extensionally adequate for first-order theories, and we have no such guarantee of extensional adequacy if we move beyond first-order logic.

10. Indeed, in the case of full-blooded Platonism, a substantive thesis is arguably (see Beall (1999)) at odds with the view's other maximalist/pluralist leanings.

11. Shapiro, while pushing the problem of logical knowledge for other algebraic views, seems to think that his *ante rem* structuralism can avoid it by taking a holistic approach which sees mathematical and logical knowledge as mutually supporting. But if this means that mathematical knowledge about the existence of structures is grounded on our logical knowledge about consistency and so on, while that logical knowledge is grounded on our knowledge of structure, it is hard to see this as anything more than viciously circular. Given the centrality of the "coherence" assumption as a requirement on the existence of structure (only coherence can tell us of the existence of anything beyond the most basic, actually instantiated, structures), it is hard to see how very much mathematical knowledge could get off the ground on this account without prior knowledge of consistency.

12. A difficulty which, arguably, was already recognized by Frege in the *Grundgesetze*, where he claimed that "It is applicability alone which elevates mathematics from a game to the rank of a science" (Geach and Black 1970, p. 187).

13. Or, as an *ante rem* structuralist might put it, a mixed theory is a good one if its non-mathematical officeholders are the way *they* would have to be to hold the offices the theory places them in.

14. I attempt such a defense at length in my (Leng 2010), but will not try to condense it here.

References

Balaguer, M. (1998). *Platonism and Anti-Platonism in Mathematics*. Oxford: Oxford University Press.

Beall, J. C. (1999). From full blooded platonism to really full blooded platonism. *Philosophia Mathematica 7*, 322–325.

Benacerraf, P. (1965). What numbers could not be. *Philosophical Review 74*, 47–73. Reprinted in Benacerraf and Putnam 1983, pp. 272–294.

Benacerraf, P. (1973). Mathematical truth. *Journal of Philosophy 70*, 661–680. Reprinted in Benacerraf and Putnam 1983, pp. 403–420.

Benacerraf, P. and H. Putnam (Eds.) (1983). *Philosophy of Mathematics: Selected Readings* (2nd ed.). Cambridge: Cambridge University Press.

Field, H. (1980). *Science Without Numbers: A Defence of Nominalism*. Princeton, NJ: Princeton University Press.

Field, H. (1984). Is mathematical knowledge just logical knowledge? *Philosophical Review 93*, 509–552. Reprinted with a postscript in Field 1989, pp. 79–124.

Field, H. (1989). *Realism, Mathematics, and Modality*. Oxford: Blackwell.

Gabriel, G., H. Hermes, F. Kambartel, C. Thiel, and A. Veraat (Eds.) (1980). *Gottlob Frege: Philosophical and Mathematical Correspondence*. Chicago: University of Chicago Press. Abridged from the German edition by Brian McGuinness.

Geach, P. and M. Black (Eds.) (1970). *Translations from the Philosophical Writings of Gottlob Frege*. Oxford: Blackwell.

Hale, B. (1996). Structuralism's unpaid epistemological debts. *Philosophia Mathematica 4*, 124–147.

Hellman, G. (1989). *Mathematics without Numbers: Towards a Modal-Structural Interpretation*. Oxford: Clarendon Press.

Hellman, G. (2003). Does category theory provide a framework for mathematical structuralism? *Philosophia Mathematica 11(2)*, 129–157.

Leng, M. (2007). What's there to know? A fictionalist account of mathematical knowledge. In Leng, Paseau, and Potter (Eds.) (2007).

Leng, M. (2010). *Mathematics and Reality*. Oxford: Oxford University Press.

Leng, M., A. Paseau, and M. Potter (Eds.) (2007). *Mathematical Knowledge*. Oxford: Oxford University Press.

Shapiro, S. (1997). *Philosophy of Mathematics: Structure and Ontology*. Oxford: Oxford University Press.

Part III

Mathematical Practice and the Methodology of Mathematics

6
Mathematical Accidents and the End of Explanation

Alan Baker

The image of mathematical sentences being true by accident is an arresting one. It is plainly repugnant to anyone who believes in a fundamentally ordered universe. That, however, is not in itself a sufficient reason to reject it. (M. Potter 1993, p. 308)

1 Introduction

A conspicuous difference between "traditional" philosophy of science and "traditional" philosophy of mathematics concerns the relative importance of the notion of *explanation*. Explanation has long featured centrally in debates in the philosophy of science, for at least two reasons. Firstly, explanation has been viewed as playing an important role in the methodology of science, principally due to the inductive character of scientific method. This has led to a focus on giving a philosophical model of scientific explanation, whose leading candidates have included Hempel's deductive–nomological model, the causal model promoted by Lewis, van Fraassen's pragmatic model, and the unification models of Kitcher and Friedman. Secondly, explanatory considerations have been an important feature of philosophical debates over scientific realism and anti-realism. This has led to a focus on inference to the best explanation and the conditions under which this mode of inference can underpin robust ontological conclusions.

By contrast, philosophical analysis of explanation in mathematics has – until very recently – been scattered and peripheral to the main debates in the philosophy of mathematics.[1] This is partly a result of the traditional philosophical emphasis on the centrality of proof in mathematics, combined with the unstated assumption that issues of explanation are irrelevant to the regimentation and evaluation of purely deductive arguments. However, there has recently been a significant increase of interest in mathematical explanation for reasons that are broadly analogous to those that have long held sway in the philosophy of science. Firstly, philosophers have begun to realize that explanatoriness is a genuine and significant feature of mathematical methodology. In particular, mathematicians frequently make a distinction between more and less explanatory proofs, and tend to value the former over the latter. Secondly, metaphysical debates concerning the existence of mathematical objects that arise from the Quine–Putnam indispensability argument have started to focus more carefully on the putative explanatory role played by mathematics in science. These two strands of philosophical interest can be thought of as focusing on mathematical explanation *in mathematics* and on mathematical explanation *in science*, respectively.[2]

One consequence of the philosophy of mathematical explanation still being in its "early adolescence" is that the philosophical terrain is still in the process of being mapped out. There is nothing like a consensus position on the correct philosophical account of mathematical explanation, or even a well-established core of basic alternative views. Even basic framework questions have been little thought about or discussed, such as whether we should look for, or expect, an account that covers both mathematical explanation and scientific explanation, or whether there is likely to be a single model of mathematical explanation as opposed to a heterogeneous collection of distinct sub-models. It may be helpful, therefore, to get some sense of the different ways in which explanatory considerations enter into philosophical analyses of mathematics. These fall fairly naturally into the following four broad categories:

1.1 Explanation in mathematics (single theory)

Explaining a given mathematical fact by drawing on results from elsewhere in the same theory. Here, the philosophical focus tends to be on *proofs*, and on comparing the relative explanatoriness of different proofs of the same result. Mancosu (2008) refers to accounts of this sort as "local" and gives as an example Steiner's account of mathematical explanation. For Steiner, an explanatory proof involves a *characterizing property*, which he defines as "a property unique to a given entity or structure within a family or domain of such entities or structures."[3] This often allows an explanatory proof to be generalized by varying the crucial characterizing property.

1.2 Explanation in mathematics (intertheoretic)

Results in one mathematical theory are explained by relating them to another, distinct mathematical theory. Sometimes the intertheoretic explanation is a proof, and sometimes it is not. In the former category is Wiles' celebrated proof of Fermat's Last Theorem, which establishes a number-theoretic result by means of a detour through elliptic curves, modular forms, and so on. Often the explanation features a second theory which expands the domain of the original theory. For an example of a non-proof-based explanation involving domain extension, consider the issue of why 1 is not considered to be a prime number. This can only be satisfactorily explained by broadening the focus from the natural numbers to the complex numbers, and in particular the Gaussian integers $a + bi$, where a and b are integers. Units are numbers which have a multiplicative inverse; among the Gaussian integers there are four units, $\{1, -1, i, -i\}$. Restricting attention to the positive integers, 1 is both the identity element and the only unit, hence these two roles are blurred together. The general definition of prime number precludes units from being prime, and this underpins the explanation for why 1 is not prime.

1.3 Mathematical explanation in science

An empirical fact about the world is explained, at least in part, by a piece of mathematics. There is a debate in the philosophical literature about whether there are in fact any genuine mathematical explanations of physical facts. Mark Colyvan claims that there are, while Joseph Melia claims that there are not.[4] I have presented in print a detailed example from the biological literature, involving the prime periods of the periodical cicada, which I argue does involve mathematics playing a genuinely explanatory role.[5] One way of thinking about this kind of explanation is by analogy with (II) above, in other words as an intertheoretic explanation where the "target" theory is scientific rather than mathematical.

1.4 Explaining the role of mathematics in science

Questions have been raised by both scientists and philosophers that demand an explanation for *why* mathematics plays such a central and important role in science. Perhaps the most famous of these challenges is from the 1960 paper, "The Unreasonable Effectiveness of Mathematics in the Natural Sciences," by Nobel prize-wining physicist Eugene Wigner. The question of why mathematics is applicable, or useful, or indispensable in science (and these are importantly distinct features) is in a sense a meta-question, not about the explanatory role of mathematics but about explaining why it plays the role that it does.[6]

For the purposes of the present chapter, I shall be restricting attention mostly to category (I) above, in other words explanation in mathematics as it occurs within a single mathematical theory. Toward the end of the chapter, I will say something about category (II) and intertheoretic explanations in mathematics. My central thesis is that philosophical understanding of the notion of explanation in mathematics can be usefully advanced by focusing on the hitherto neglected concept of an *accidental mathematical fact*, or – more briefly – a *mathematical accident*. The motivating analogy here is the distinction commonly drawn in the philosophical literature between "law" and "accidental generalization." Could there be accidental generalizations in mathematics? And, if so, what might their presence tell us about methodological issues such as confirmation, induction, and explanation?

Talk of "mathematical accidents" is apt to provoke a fairly immediate negative reaction. For one thing, this sort of terminology seems to play no role in the actual practice of mathematics. Working mathematicians are not in the habit of referring to any of the results of their investigations as "accidental." Worse still, the notion seems to be philosophically incoherent on its face. Central to our intuitive notion of accident is that accidents might have happened differently, or they might not have happened at all. Accidents, in other words, are *contingent*. But most traditional philosophical accounts of mathematics – especially those accounts according to which our core mathematical claims are true – take such claims to be necessary. If it is true that $2 + 3 = 5$, or that 17 is prime, then it is true necessarily: there is no interesting sense in which 17 might not have been prime. So we have what seems to be an incoherent notion that is irrelevant to mathematical practice. This is hardly a promising starting point for a philosophical investigation!

My plan is to address these twin worries indirectly, at least initially, by focusing not on mathematical accidents but on the related notion of *mathematical coincidence*. I shall argue that talk of coincidence does feature in mathematical practice and that it can be given a coherent philosophical analysis. Moreover, this analysis provides the groundwork for a satisfactory definition of "mathematical accident." With these worries (hopefully) allayed, I shall go on in the following section to lay out the positive case for the philosophical relevance of mathematical accidents.

2 Mathematical coincidences

Mathematicians tend to use the term "mathematical coincidence" to describe certain "surprising" low-level results. Sometimes, these results concern co-incidence in a very literal sense, for example

(1) The sequence "1828" appears twice in the first ten digits of the decimal expansion of e.

More often, the results are identities or approximate identities, such as

(2) $\pi \approx 355/113$, correct to 6 decimal places.

At this point, the philosopher of mathematics will naturally want to know more about what exactly is supposed to be meant by the term "coincidence" in this context. In their influential work on the statistical study of coincidences (in non-mathematical contexts), Diaconis and Mosteller define a coincidence as "a surprising concurrence of events, perceived as meaningfully related, with no apparent causal connection."[7] Other features that are commonly taken to be aspects of coincidences include being unpredictable, being improbable, and being inexplicable.

But how to map this onto the mathematical case? Diaconis and Mosteller talk in terms of "events" which lack "causal connection," and these are notions which presumably have no application in the context of mathematics. So, we are left with the bare criterion of "surprise," together with (potentially) related notions such as unpredictability and improbability. Arguably, these latter concepts are just as difficult to fit into the mathematical context, at least if they are construed objectively. (Making sense of subjective probabilities for mathematical claims is also a delicate issue, but it may be more tractable. See Pólya (1954) and Corfield (2003, chapter 5) for work on this topic.) Moreover, there is good reason to think that low probability and/or unpredictability in any strong sense are not essential to the notion of coincidence even in the empirical case. If the universe is deterministic, then the falling into Todd's lap of his name and phone number had a probability of 1 and was predictable in principle from the state of the world prior to the start of the football game he was attending.

The key – or so I shall argue – in importing the correct notion of coincidence into mathematics is to look for some appropriate analog of "lack of causal connection." David Owens, in his 1992 book, *Causes and Coincidences*, argues that a coincidence should be defined as an event whose constituent events are **causally independent** of one another. For example, imagine that I pray for rain tomorrow and it does in fact rain tomorrow. The atheist will claim that my prayers being answered in this case is a coincidence, which – according to Owens – is simply to claim that the causes of my praying were *independent* of the causes of the rain. Thus to say that a coincidence occurs "for no reason" is not to say that it is uncaused, but rather that there is no causal story which points to any deeper reason that explains *why* the given result should be expected to hold. This suggests in turn that the underlying essential feature here is the *inexplicability* of coincidences. But, as has already been noted, explanation is a recognized aspect of mathematical methodology. Hence this may provide the way in to defining an appropriate notion of mathematical coincidence.

Support for this approach comes from mathematicians" own reflections on the notion of coincidence. For example,

> [A] **mathematical coincidence** can be said to occur when two expressions show a near-equality that lacks direct theoretical explanation.[8]

Having pinpointed inexplicability as a crucial notion, the task now is to get clearer on what constitutes a "direct theoretical explanation." The identification of a claim as a mathematical coincidence certainly does not mean that the claim in question is *unprovable*. On the contrary, verifying the repetition of "1828" in the decimal expansion of e, or showing the accuracy of the 335/113 approximation of π, are almost trivial computational tasks.

This does not mean that proof-related factors are irrelevant, however. Consider a putative analogy between causation in the empirical cases of coincidence and proof in the mathematical cases. Empirical coincidences have causes, it is just that there is no particular link between the causes of the two coincident events.[9] Mapping this idea onto the mathematical case suggests that we characterize mathematical coincidences as claims whose separate parts require separate proofs. Take example (2) above, concerning the rational approximation of π. To prove the 6-decimal accuracy of 355/113, it is necessary and sufficient to calculate π to 6 decimal places (using one of various geometric or calculus-based methods), and to calculate 355/113 to 6 decimal places (using long division). But these two calculations are quite *distinct*. There are no parts of one calculation that are used in the other.

From a structural perspective, therefore, the crucial feature of any proof of a mathematical coincidence is its *disjointness*. It is this disjointness which prevents the explanation of the two components from constituting an explanation of the coincidence as a whole. In the case of identities (and near identities), the "components" of the coincidence are easy to pick out since they fall on each side of the identity (or approximate identity) sign. Whether there is such a clear breakdown into components for all putative cases of mathematical coincidence is unclear, but this is not an issue that I will take time to pursue here since the notion of mathematical coincidence is merely a stepping-stone on the path toward defining a broader notion of accidental mathematical fact.[10]

3   From mathematical coincidences to accidental generalizations

"Coincidence" and "accident" are closely related concepts. Frequently, one is defined in terms of the other. For example, the *American Heritage Dictionary* defines a coincidence is "an accidental sequence of events that appears to have a causal relationship." In the philosophical literature, talk of accidents is often bound up with the notion of *accidental generalization*. Accidental generalizations are taken – by those philosophers who make this distinction – to be universal, true claims that share some but not all the features of lawlike claims and hence that fall short of expressing genuine laws of nature. For example the (putative) law of nature,

>   (3) All solid spheres of enriched uranium (U235) have a diameter of less than a mile.

may be contrasted with the accidental generalization

(4) All solid spheres of gold (Au) have a diameter of less than a mile.[11]

What the key differentiating features are taken to be which accidental generalizations lack depends on what account of laws of nature is in play.

What little literature there is on accidental mathematical facts tends to follow the above pattern, although which element of lawlikeness is focused on varies from author to author. I shall begin by surveying a couple of sample approaches.

3.1 Necessity

In the context of discussing the debate between intuitionist and platonist accounts of mathematics, Michael Potter addresses the issue of whether "there may be [mathematical] sentences true accidentally." He argues that even the platonist cannot make sense of this notion, and in presenting his argument he explicitly equates being accidental with being non-necessary. In support of this analysis, it should be noted that the accidental generalization (4), does seem to lack the necessity of the lawlike claim (3). Potter frames platonism in the context of a "God's-eye" view of mathematics, and he writes

> God simply does not, under the platonist interpretation of the quantifiers on the natural numbers, have the freedom to decide their truth or falsity, whether by dice-throwing or the exercise of God's whim or anything else.[12]

I agree with Potter that making room for mathematical accidents, conceived of as matters of contingent fact, is a non-starter. The issue, then, is whether there is some other account of accidentality that may fare better.[13]

3.2 Natural kinds

David Corfield has also considered the issue of "the existence in mathematics of 'quasi-contingent' facts, i.e., facts which are shallow or 'happenstantial'."[14] Unlike Potter, Corfield argues for the existence of such facts. This difference arises mainly because Corfield focuses not on the link between lawlikeness and necessity but between lawlikeness and natural kinds. He begins by rejecting analyses that rely on modal distinctions, for reasons that are very similar to those canvassed in our earlier discussion of mathematical coincidences

> With much of the discussion of laws and necessity carried out in the meta-physical language of possible worlds, the notion that mathematical facts might vary similarly as to their lawlikeness has appeared to be hopeless.

Where I can imagine possible worlds in which very large golden balls exist, I cannot imagine a possible world in which the number denoted in the decimal system by "13" is not prime.[15]

Instead, Corfield targets the predicates used in formulating a given generalization, arguing that accidental mathematical claims are characterized by their use of "non-natural" predicates. He proposes a taxonomy of *mathematical natural kinds*, by analogy with natural kinds in empirical science. Natural mathematical predicates are those which pick out mathematical natural kinds. Consider the following example, given by Corfield, of an accidental mathematical fact:

Coining a term 'cubeprime' to characterise any natural number which is either prime or a perfect cube, we arrive at the result that:

In any base, the reversal of what we call 'thirteen' is cubeprime.[16]

I am actually quite sympathetic to Corfield's approach, and I shall return to some of the issues he raises in a later section. However, there are a couple of reasons for doubting its effectiveness as a general account of mathematical accidenthood. Firstly, it seems doubtful whether the notion of mathematical natural kind is any clearer than the notion of mathematical accident. Hence, defining the latter in terms of the former may not represent much in the way of analytical progress. In Corfield's defense, Jamie Tappenden and others have noted that invoking "naturalness" as a distinguishing mark of certain fruitful concepts and definitions is a relatively common feature of mathematical practice.[17] Nonetheless, the link between naturalness in this loose sense and natural kinds is not immediately clear. Secondly, and more seriously, there are good reasons for thinking that while featuring non-natural predicates may be a sufficient condition for a mathematical generalization to count as accident, it is not a necessary condition. Later on I shall give some *prima facie* cases of accidental mathematical generalizations which feature only non-gerrymandered, natural mathematical predicates. To partially preempt what follows, I shall argue that Corfield has the direction of dependence the wrong way around. It is not that mathematical accidents are accidental in virtue of featuring non-natural predicates; rather a predicate will count as non-natural if it features solely in accidental generalizations.

3.3   Explanation

Given my remarks in the preceding section in discussing the notion of a mathematical coincidence, it will come as little surprise that my favored analysis of accidental mathematical generalizations will be in terms of explanation. The basic idea is to treat accidental generalizations as "universal coincidences," with the only significant difference being in the number of events or phenomena that coincide. While coincidences are typically matters of particular fact, wherein two phenomena share some striking similarity, accidental generalizations are universal in form and typically involve the coinciding properties of many – perhaps even infinitely many – particular phenomena. Mapping all this into the mathematical context, accidental mathematical generalizations will be generalizations that lack any unified proof. Such generalizations, even when provable, are thus *inexplicable*, and they share this core property with mathematical coincidences. One standard way in which a mathematical generalization may lack any unified proof is for it to be verifiable only on a case-by-case basis.

Before presenting this analytical approach in more detail, I shall pause to consider a couple of putative examples. An immediate consequence of my favored definition of mathematical accident is that this designation is rarely definitive since it will usually not be possible to rule out some unified, explanatory proof of a given claim being found at some point in the future. Nonetheless, it will be useful to focus on a couple of promising – and well-known – claims that are *prima facie* candidates for being accidental mathematical generalizations.

4 Case studies

4.1 The Goldbach conjecture

In a letter to Euler written in 1742, Christian Goldbach conjectured that all even numbers greater than 2 are expressible as the sum of two primes.[18] Over the following two and a half centuries, mathematicians have been unable to prove Goldbach Conjecture (GC). However, it has been verified for many billions of examples, and there appears to be a consensus among mathematicians that the conjecture is most likely true.

Echeverria, in a recent survey article, discusses the important role played by Cantor's publication, in 1894, of a table of values of the Goldbach partition function, $G(n)$, for $n = 2$ to $1,000$.[19] The partition function measures the number of distinct ways in which a given (even) number can be expressed as the sum of two primes. Thus $G(4) = 1$, $G(6) = 1$, $G(8) = 1$, $G(10) = 2$, and so on. This shift of focus onto the partition function coincided with a dramatic increase in mathematicians' confidence in GC; however Cantor did not simply provide more of the same sort of inductive evidence since Desboves had already published, in 1855, tables verifying GC up to 10,000. To understand why Cantor's work had such an effect, it is helpful to look at the following graph (see Figure 6.1) which plots values of the partition function, $G(n)$, from 4 to 100,000.[20]

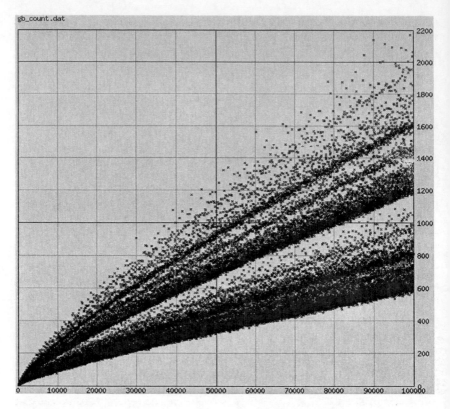

Figure 6.1 Plot, from Herkommer (2004), of the Goldbach partition function for value up to 100,000.

This graph makes manifest the close link between G(n) and increasing size of n. Note that what GC entails in this context is that G(n) never takes the value 0 (for any even n greater than 2). The overwhelming *impression* made by the above graph is that it is highly unlikely for GC to fail for some large n. At the upper end of this graph, for numbers on the order of 100,000, there is always at least 500 distinct ways to express each even number as the sum of two primes!

Reflection on the above picture reinforces the impression that GC's closest brushes with falsity occur in the first few instances. Indeed, three of the first four instances of GC have a partition function value of 1, in other words there is only a *single* way to decompose the given number into primes. My initial reaction, on looking at the graph of the Goldbach partition function, was that the fact that GC survives these first few instances intact is purely **accidental**. It is pure *happenstance* that it does not fail very early on. It was rumination on what this could possibly mean – in the mathematical context – that set me to thinking more generally about the notion of mathematical accidents.

One way to motivate the link to explanation in this case is to compare GC with the following, deliberately trivial, general claim about even numbers:

(5) All even numbers can be expressed as the sum of two odd numbers.

There is no corresponding temptation to view the truth of (5) as accidental, even though there is a sense in which (5) also "comes close" to being false for its initial few instances (there is only one way to express 2 as the sum of two odd numbers, and one way to express 4, compared with 25 ways to express 100 as the sum of two odd numbers). Partly, of course, this difference in intuition is because the truth of (5) – unlike the truth of GC – is immediately obvious. But the deeper reason, I think, is that there is a simple, general proof of (5) from which the truth of each instance straightforwardly follows.

If GC is indeed true, as most number theorists suspect, then there are several alternatives concerning its provability, including the following:

(i) No proof of GC or its negation is possible from the standard axioms.[21]

In this case, since GC has no proof, *a fortiori* it has no unified proof. So GC is an accidental mathematical fact.

(ii) GC has an elegant, unified proof, it is just that we have not found it yet.

So GC is not accidental.

(iii) There is a unified proof that all numbers greater than some specified (large) N conform to GC. The finite number of cases N and below can only be verified individually.

In this third situation, according to my analysis, GC would still count as accidental despite being provable because the best possible proof of GC is highly disjunctive and hence does not *explain* the truth of the conjecture.

4.2 The Four-Color Theorem

The Four-Color Theorem was first conjectured by a British mapmaker in the mid-nineteenth century: the claim is that, given any plane separated into regions (such as a political map of the states of a country), the regions may be colored using no more than four colors in such a way that no two adjacent regions receive the same color. Various flawed proofs were produced in the latter part of the century. Progress was made on reducing the problem during the twentieth century, culminating in Appel and Haken's 1976 proof, part of which consisted of the code for a computer program. The program was used to check through 1476 different graphs (each quite complex) on a case-by-case basis.[22]

Philosophers have worried about the Four-Color Theorem because the Appel-Haken proof makes unavoidable use of computers, and (relatedly) is unsurveyable. Mathematicians, by contrast, are dissatisfied with the Four-Color Theorem mainly because they consider the proof to be unexplanatory. The previous analysis fits nicely with this latter intuition: According to my definition, our best current evidence suggests that the Four-Color Theorem is an accidental mathematical fact. The proof is highly disjunctive: There are 1476 different sub-cases that are individually considered. Thus, the proof is very unexplanatory.[23]

5 Mathematical accidents defined

Bearing in mind the above discussion of the two case studies, I propose the following definition of the notion of a mathematical accident:

(MA) A universal, true mathematical statement is *accidental* if it lacks a unified, non-disjunctive proof.

The disjunctiveness of a proof is measured by the number of distinct subcases that need to be considered separately. Definition (MA) is coupled to the philosophical claim that – other things being equal – disjunctiveness is negatively correlated with explanatoriness in the context of proof. Hence, an accidental mathematical truth is inexplicable, or – putting the point more carefully – such a truth has only a "bottom-up" explanation of its components as opposed to a "top-down" explanation of the whole. It is important to note that (MA) is being proposed here as a sufficient condition for a mathematical claim being accidental, but it is not a necessary condition. For my underlying thesis is that accidentality is tied to lack of explanation, and there are certainly other ways than disjunctiveness in which proofs can be unexplanatory.

Though I do not have the space to pursue it here, it would be interesting to look at how the above definition might be supplemented to take into account features of the disjuncts of a proof other than simply how many there are. One candidate feature is the *naturalness* of the division of the space of possibilities into distinct subcases. Another is the *specificity* of the subcases themselves: Do they consist of individual elements (for example, the individual even numbers checked in the Goldbach case), or are they groupings of cases of a certain type (for example, the different types of map configuration checked in the Four-Color Theorem case)?

One consequence of the above definition is that accidentality is a *matter of degree*. It inherits this feature from the original definition of (non-mathematical) coincidence, based on causal independence, which inspired (MA). Causal independence may seem like an all-or-nothing matter, but strict causal independence looks to be too stringent a criterion for our ordinary notion of coincidence. Take the praying for rain example. Even for the atheist, it should seem plausible that there is *some* causal link between my praying for rain tomorrow and it raining tomorrow. My praying did not cause the rain. However, if we go back far enough, then there will presumably be some event that features as a cause both of my praying and of the rain. For example, the early conditions on Earth led to the retention of large quantities of water on its surface and in its atmosphere. Without the presence of this water I would not have been around to pray, and there would have been no raw materials for rain. And if all else fails, the Big Bang can always be cited as a common cause of any two events in the subsequent history of the universe. But neither the early atmospheric conditions of the Earth, nor the Big Bang, should undermine the claim that it raining after I prayed for rain is a *coincidence*. The reason why not, at least intuitively, is that the cited causes are very remote from the events in question. This suggests that causal independence should be taken to be a matter of degree. The longer and more circuitous the causal chain connecting the two events, the more coincidental they are. Hence, coincidence is itself a matter of degree.

The above points apply *mutatis mutandis* to the notion of mathematical accident. The optimal proof of a given result may be more or less disjunctive, and the degree of accidentality of the result corresponds to the degree of disjunctiveness of the proof. The parallel with the empirical case also provides a way to head off one potential line of objection to my account of mathematical accidents. If a mathematical claim is provable, then it is provable from the axioms of the theory in which it is embedded. So why not cite the axioms as providing an explanation of *any* provable claim? Here the axioms are analogous to the boundary conditions right after the Big Bang, and the same point about indirectness applies. Tracing inferential paths back to axioms is no more explanatory *per se* than tracing causal chains back to the Big Bang. Sometimes axioms are explanatory and sometimes they are not, but this depends on the nature of the proof and not on its bare existence.

At the beginning of the chapter, I promised an analysis of the concept of mathematical accident that would meet three benchmarks of acceptability: that the concept is coherent (i.e. that there could be mathematical accidents); that the concept has significant links to mathematical practice; and that the concept is philosophically fruitful. Hopefully it is clear enough that the first of these hurdles has already been met: It certainly seems *possible* for there to be mathematical claims whose "best" proof is highly disjunctive. What about links to mathematical practice? Some such links are already apparent, insofar as the two putative examples of mathematical accidents discussed above each concern claims (the Goldbach Conjecture and the Four-Color Theorem) that mathematicians have found interesting and significant. My analysis also respects the more general role of accidents as barriers to effective theorizing. Other things being equal, if a given fact counts as accidental according to theory A but non-accidental according to theory B, then this counts in favor of B over A. Along these lines, my account of mathematical accidents links them to a feature of proofs, namely disjunctiveness, which tends to be regarded by mathematicians as undesirable. The general preference is for "top-down" proofs rather than "brute force," case-by-case verifications. Indeed this preference is encapsulated in the – perhaps apocryphal – story of Gauss as a schoolboy taking a matter of seconds to sum the numbers from 1 to 100 while his classmates laboriously added them one by one. The early sign of Gauss's mathematical genius is here identified precisely with his ability to find a general method for immediately generating this sum, in this case by rearranging it to form fifty pairs $(1+100)$, $(2+99)$, $(3+98)$, and so on, to give a total of 5050. Not only this, but the preference for less disjunctive proofs is often expressed by citing the greater explanatory power of the "better" proof.

The third and final benchmark I promised would be met by my proposed notion of mathematical accident is that it be philosophically fruitful. Doubtless this is the hardest of the three to measure in any definitive fashion, nonetheless I hope to indicate in these concluding sections some of the various philosophical ends to which reflection on the nature and role of mathematical accidents may be put.

6   New work for a theory of mathematical accidents

6.1   The end of explanation

I have claimed as a virtue of my account of mathematical accidents that it ties accidentality to disjunctiveness of proof and disjunctiveness in turn to explanation (or lack of explanation), and that this mirrors the way in which explanatory considerations are invoked in some accounts of the law / accidental generalization distinction in empirical science. However, on closer inspection, it might seem that I am being disingenuous here. Explanation-based accounts of laws of nature typically cite the role of laws in explaining other phenomena, the implication being that accidental generalizations fail to explain their instances. By contrast, the account I have offered of mathematical accidents highlights the fact that the generalizations themselves lack explanation. So doesn't my account get the direction of inexplicability the wrong way around? This worry is further bolstered by reflection on laws of nature, for example Newton's Law of Gravitation. It seems reasonable to conclude that fundamental laws of this sort are inexplicable also. After all, to categorize them as fundamental is to imply that they do not follow from any deeper principles.

My reaction to this line of objection is to try to hang onto the motivating analogy but to subdivide it into two sorts of case. We had occasion in the previous section to talk about the role of axioms in explanatory versus non-explanatory proofs. And it is axioms that provide the most natural analogs for fundamental laws of nature. The remaining cases of "non-accidental" mathematical generalizations, in other words general mathematical claims that are provable nondisjunctively from the axioms, are analogous to "derived" laws of nature. Nor is this merely a defensive move, an unwelcome retreat forced by the previous objections, for this more fine-grained analogy has the potential to cast light on issues concerning the endpoints of chains of explanation.

In the empirical context, there seem to be two basic ways in which answers to why-questions can run out. The first kind of case involves questions that push back to *boundary conditions*. If we ask, for example, why the fauna of Australia has such little overlap with the rest of the world, then an explanation can be given in terms of Australia's historical isolation from other major land masses. If we ask how Australia came to be thus isolated, the original explanation can be extended by citing tectonic shifts and other geomorphological features. But there seems to come a point in this sequence of questions ("why were the tectonic plates in this arrangement?") where no further explanation is possible. This is just how things were: *these* were the boundary conditions. The second kind of case involves pushing back to *basic laws*. Why is X amount of energy released by the fission of amount Y of uranium 235? Because mass is converted into energy according to the equation $E = mc^2$. Why does $E = mc^2$? It just does!

One can think of these two barriers to further explanation as the *particular* and the *fundamental*. In some cases, they may come together in one and the same situation; for example, the values of some of the fundamental physical constants (the gravitational constant, Planck's constant, etc.) may perhaps best be viewed as *fundamental matters of particular fact.* As such, they are barriers to explanation in both the above senses. But when we look at more restricted domains than the cosmological, the distinction between laws and boundary conditions seems clear enough. In deterministic systems, the laws together with the boundary conditions entail all facts about the evolution of the system over time. But boundary conditions are unique to each system, indeed to each time-slice of each system, whereas laws are common across multiple systems.

What about the mathematical context? One point to bear in mind is that the axioms of a given mathematical theory come out as non-accidental, according to my definition (MA) since any axiom is trivially provable in a non-disjunctive way from itself. Hence, although axioms figure at the end of explanatory chains, they are not exactly *barriers* to explanation since each axiom explains itself! Whether there is any analog in mathematics of the law / boundary condition distinction is unclear. Certainly, it is unusual to find mathematicians using these terms in the context of pure mathematics. Perhaps one could divide up the axioms of a theory according to their logical form: universal / general axioms would correspond to laws, and existential / particular axioms would correspond to boundary conditions. Take the axioms of Peano arithmetic (PA). All but one of these axioms is universal in form: the only "boundary condition" is the axiom which states that 0 is a natural number.

I won't look in any more detail here into whether the above distinction has any mathematical significance, except to note that the role of the boundary condition axiom does seem to be different from the other Peano axioms. For one thing, it is the specification of 0 as a natural number which guarantees that the domain of natural numbers is non-empty. There is also a sense in which this is the only "non-structural" axiom of PA. Finally, it is worth mentioning that there is also another, more speculative way to map the law / boundary condition into mathematics and that is to classify *all* the axioms of a given theory as "boundary conditions", and then identify the "laws" with the rules of inference of the theory. The idea is that the evolution of a physical system from initial conditions, governed by laws, is equated with the unfolding of proofs from axioms, using rules of inference. Discussion of the philosophical merits, if any, of such a position must wait for another occasion.

6.2 Axiom choice

Mathematicians – and philosophers – have gradually moved away from the Euclidean conception of axioms as fundamental, "self-evident" truths. This traditional view has been replaced by a variety of attitudes, including the completely instrumental according to which axioms are arbitrary sets of rules, and there is no substantive sense in which one set is "better" than another. One popular view, sometimes associated with Bertrand Russell, is that axioms are justified by their consequences.[24] On this view, a mathematical theory such as arithmetic has various core claims, for example "obvious" claims such as "$2 + 2 = 4$", or "7 is a prime number." A given set of axioms is judged not according to the self-evidence of its component axioms but rather by the extent to which it allows the core claims of the theory to be deduced (and prevents the deduction of patently false claims). There is a clear analogy here with use of inference to the best explanation in the empirical sciences to justify belief in statements about theoretical posits such as electrons and black holes.

If something like inference to the best explanation (IBE) does underlie axiom choice in mathematics, doesn't this mean that we ought to pay attention to the distinction between explanatory and non-explanatory proofs? The thought is that it is not enough, as a basis for IBE, for a set of axioms simply to prove a particular core claim, the proof must also be explanatory. In other words, the optimal set of axioms for a given theory is that which (other things being equal) yields explanatory proofs of the maximum number of core claims. If we have reason to believe that amongst the core claims of a given theory – with a given set of axioms, A – there are some **mathematical accidents**, then their presence counts against A.

Of course, the explanatory power of A will not be the only criterion of evaluation, otherwise we should just keep adding axioms. There will also be desiderata such as consistency, independence, and simplicity. The most active area of debate concerning the selection of axioms is in set theory. Should provably independent axioms such as the Axiom of Choice (AC) or the Continuum Hypothesis (CH) be added to the basic ZF axioms? What about various "large cardinal" axioms? Typically, arguments in favor of adopting a particular axiom, such as AC, proceed by coming up with various powerful and useful results that can be proved only if AC is added to the core axioms. The problem – from the IBE perspective – is that there is no independent route to the *verification* of these results. What this suggests is a different route to the justification of a new axiom candidate, namely if there are important results, provable from the existing axioms in a non-explanatory way, whose proofs would be rendered much **more explanatory** if the candidate axiom were adopted.

7 Intertheoretic mathematical accidents

My discussion thus far has focused almost exclusively on the intratheoretic case, in other words on mathematical coincidences and mathematical accidents where the "coinciding" facts all lie within a single mathematical theory, typically number theory. We already know, from Gödel's incompleteness theorems, that any consistent axiomatization of number theory will fail to prove certain arithmetical truths. Of course the Gödel results *per se* tell us nothing about whether any of these unprovable truths are mathematically *interesting*. My presumption has been that there may well be mathematical truths, such as the Goldbach Conjecture, that are both mathematically interesting and are accidental either because they are unprovable or because their best proof is highly disjunctive. There may, in other words, be interesting mathematical truths that are true for no reason.

What happens when we broaden our focus to encompass multiple mathematical theories? Intertheoretic considerations raise some new possibilities. The first is when a result that is naturally expressed in mathematical theory X has no non-disjunctive proof in X (and perhaps no proof of any kind in X), but it does have a non-disjunctive proof in some stronger background theory Y. Relatively well-known examples of this sort include the Paris–Harrington theorem and Goodstein's theorem, both of which are arithmetically expressible claims which can be shown to be unprovable in first-order Peano arithmetic but which are provable in stronger theories such as ZF set theory. I am not aware of any examples where the background theory reduces the disjunctiveness of a proof in the weaker theory, but there seems no reason in principle why this should not occur. How should such examples be classified on the accidental / non-accidental spectrum? One idea is to conceive of accidentality as a theory-relative notion. Thus, one and the same result might be both accidental from a number-theoretic perspective and explicable from a set-theoretic perspective.

A second sort of possibility is where the coincidence – or apparent coincidence – itself concerns items from more than one theory. One notorious example along these lines concerns the so-called "Monstrous Moonshine." The (so-called) *j*-function is connected to the parameterization of elliptic curves, and it has the following Fourier expansion in $q = \exp(2\pi i \tau)$:

$$j(\tau) = 1/q + 744 + 196884q + 21493760q^2 + \ldots$$

Mathematician John McKay was the first to notice that the third coefficient, 196884, was the same as the sum of the dimensions of the two smallest irreducible representations of the Monster finite group. Yet, the two theories in which these numbers appear – elliptic curves and group theory – seem to be almost completely unrelated to one another. The initial reaction of many mathematicians to McKay's observation was that it was purely a **coincidence** that this number appeared in both theories. Yet it soon became clear that all the coefficients of the *j*-function can be expressed as linear combinations of irreducible representations of the Monster group, and this prompted mathematicians to start searching for a connection between these two theories. A connection was eventually found by Richard Borcherds, who won a Fields Medal in 1998 for his work. As Corfield summarizes it, Borcherds "managed to spin a thread from the *j*-function to the 24-dimensional Leech lattice, and from there to a 26-dimensional space-time inhabited by a string theory whose vertex algebra has the Monster as its symmetry group."[25]

Although McKay's initial observation spanned two distinct theories, the analysis of mathematical coincidence that I sketched earlier in the chapter seems to apply. Recall that coincidence was there characterized as "lack of direct theoretical explanation" and this was cashed out in turn in terms of *disjointness* of any proof of the result. If the Monstrous moonshine were a genuine coincidence, then the only way to prove it would be to give a proof in each theory of the respective occurrence of 196884. Borcherds result gives a "theoretical explanation" of the result by connecting the two halves of the apparent coincidence together in a single proof. Hence, the Monstrous moonshine turns out not to be a coincidence after all.

A less well-known, and more recent, example of a putative intertheoretic mathematical coincidence concerns some surprising correspondences that have been discovered between the theory of certain ordinary differential equations and particular integrable lattice models and quantum field theories in two dimensions. This phenomenon is sometimes referred to as the *ODE/IM correspondence*.[26] Unlike in the Monstrous moonshine case, the consensus seems to be that the ODE/IM has not (yet) been adequately explained.

The above episodes raise a question that has more general significance for the status – and existence – of mathematical coincidences and mathematical accidents, namely what attitude mathematicians ought to take to putative examples of coincidences. Whatever the answer to this normative question, it appears that as a matter of fact mathematicians are more inclined to accept genuine coincidences within one theory than they are between two or more theories. For instance, nearly all of the dozens of examples that appear on the Wikipedia page devoted to "Mathematical Coincidences" concern results within a single theory. Where the single theory is arithmetic, this acceptance of coincidences might be motivated by appeal to Gödel's incompleteness results. But if we insist on coincidences being "surprising," "significant," or "interesting," then there is no guarantee that any strictly unprovable arithmetical claims will be coincidental.

By contrast, when a striking feature occurs in two distinct mathematical theories, usually the presumption is that there is some substantive connection to be unearthed. In this regard, the following views expressed by mathematician Philip Davis seem quite typical:

> I cannot define coincidence. But I shall argue that coincidence can always be elevated or organized into a superstructure which performs a unification along the coincidental elements. The existence of a coincidence is strong evidence for the existence of a covering theory.[27]

Elsewhere in the same paper, Davis is even more definitive, writing that "the existence of a coincidence implies the existence of an explanation."[28] As was pointed out in our earlier discussion, there are good reasons to rule out mathematical coincidences if one takes a modal perspective according to which coincidence implies contingency. What is striking about Davis's view is that he seems to be following the sort of explanation-based account of mathematical coincidence that I have been arguing for, and yet still concludes that there are no (intertheoretic) coincidences in mathematics. The only general grounds I can see for adopting this kind of position is via some version of Leibniz's "Principle of Sufficient Reason" (PSR) for mathematics. As Leibniz formulates this principle, it states that "nothing happens without a reason why it should be so, rather than otherwise."[29] Traditionally, PSR has only been taken to apply to *contingent* matters of fact, and Leibniz himself thought that a "principle of contradiction" was sufficient to found arithmetic and geometry. Quite apart from the historical incongruity of applying PSR to mathematics, there is also the pressing issue of how – if at all – PSR itself is to be justified.

8 Conclusions

There are many ways to carve up the space of mathematical truths, and many purposes toward which such a carving-up might be put. We might distinguish the pure from the applied, the universal from the particular, the provable from the unprovable, and so on. My proposal has been to add another way of carving up this space, namely by distinguishing the *accidental* from the *non-accidental* mathematical truths. In so doing, I am placing together on one side of the divide mathematical truths which are unprovable and mathematical truths whose only proofs are highly disjunctive (as well as truths whose "best" proofs are unexplanatory in other ways). This is my category of **mathematical accidents**. This unorthodox suggestion can best be judged – or so I claim – by the extent to which it helps our philosophical understanding of mathematical methodology. And here there are, I think, at least two significant areas of benefit:

8.1   Means of justification

We may care about whether there exists – in some abstract sense – a proof of some given conjecture, C. But we may care even more whether *we can in fact* formulate a proof of C. Some provable claims are not provable by us. Furthermore, there are also provable claims which we can only prove with the help of computers. The category of mathematical accidents encompasses both of these categories. Once a proof is disjunctive enough then, as a matter of practical necessity, our only way of formulating and checking the proof is by harnessing the power and speed of electronic computers. As their degree of disjunctiveness increases, proofs become inaccessible even to computers. In this latter case, our only way of gathering evidence for the truth of the result is by enumerative induction based on verification of some of its instances.

From a methodological point of view, a unifying feature of mathematical accidents is that they are amenable to (and may *only* be amenable to) investigation using "experimental" methods. The newly emerging field of **experimental mathematics** harnesses the power of computers to discover plausible-looking conjectures and to gather evidence of their truth.[30] If the conjecture in question is a mathematical accident, then this may be the best that can be done.

8.2   Explanatory basis

A second unifying feature of mathematical accidents is the barrier they present to explanation. Mathematical proofs often function as explanations of the results which they prove, but this link between proof and explanation is broken for results whose only proofs are highly disjunctive. Nothing in what has been said above rules out the possibility that an accidental mathematical truth may function as an explanation of some other mathematical fact. However, even if this were to be the case, the mathematical accident would itself be inexplicable. Mathematical accidents are at best the endpoints of explanatory chains and at worst more-or-less completely isolated from the broader explanatory framework of mathematical theories. Nor is there any plausible way to view these endpoints as "self-explanatory," along the lines (perhaps) of axioms. Mathematical accidents are brute facts but they are not fundamental in any significant sense.[31]

Notes

1. As Tappenden (2008, p. 4) puts it, "the study of mathematical explanation is still in early adolescence."
2. Mancosu (2008) makes roughly this distinction between two facets of mathematical explanation.
3. Steiner (1978, p. 144).
4. See Colyvan (2001); Melia (2000, 2002).
5. Baker (2005). See also Baker (forthcoming) for further discussion of the cicada case study.
6. See Pincock (2007).
7. Diaconis and Mosteller (1989, p. 853).
8. "Mathematical Coincidences," *Wikipedia*.
9. One option, terminologically speaking, would be to reserve the label "mathematical miracle" for those mathematical statements within a given theory which are true yet unprovable from the axioms of that theory. Thus, unprovability would here correspond to the lack of causation for miracles in the empirical context.
10. In fact example (1), considered earlier, may not be straightforwardly analysable in terms of disjointness since the calculation of the 5th to 8th decimal digits of e presupposes the calculation of the 1st to 4th digits.
11. See van Fraassen (1989, p. 27).
12. Potter (1993, p. 308).
13. Also cf. Davis (1981, p. 320): "A Platonic philosophy of mathematics might say that there are no coincidences in mathematics because all is ordained."
14. Corfield (2005, p. 33).
15. *op. cit.*, p. 31.
16. *op. cit.*, pp. 34–35. The reversal of a number consists of the digits of that number taken in reverse order. Thus, the reversal of "125" is "521." The reversal of the representation of 13 in base 10 is "31," but in base 5 (for example), 13 is represented as "23," so its reversal is "32."
17. Tappenden (2008).
18. In fact, Goldbach made a slightly more complicated conjecture which has this as one of its consequences.
19. *op. cit.*, pp. 29–30.
20. Of course the number of results displayed here is orders of magnitude beyond Cantor's own efforts, but the qualitative impression is analogous. This graph is taken from Mark Herkommer's "Goldbach Conjecture Research" website at http://www.petrospec-technologies.com/Herkommer/goldbach.htm.
21. Note that the undecidability of GC would entail its truth, because if it were false then it would fail for some number, n, and hence the negation of GC would be provable.
22. For more on the mathematical details of the Four-Color Theorem, see Thomas (1998) and Wilson (2002).
23. For discussion of some philosophical considerations arising from the Four-Color Theorem, see Tymoczko (1979), and McEvoy (forthcoming).
24. See, for example, Russell (1973, p. 282).
25. Corfield (2003, p. 126).
26. Here "ODE" refers to "ordinary differential equations", and "IM" refers to "integrable models." See Dorey (2007) for more details.
27. Davis (1981, p. 311).

28. *op. cit.*, p. 320.
29. Leibniz (1956, L.II.1).
30. See Baker (forthcoming).
31. I would like to thank audiences at the New Waves in Philosophy of Mathematics Conference at the University of Miami, at Wayne State University, and at Temple University for helpful questions and comments on earlier versions of this paper, and Chris Pincock for valuable suggestions on improving the penultimate draft.

Bibliography

Baker, A. (2005) "Are There Genuine Mathematical Explanations of Physical Phenomena?" *Mind*, 114, 223–238.

Baker, A. (2007) "Is There a Problem of Induction for Mathematics?" in *Mathematical Knowledge*, ed. M. Leng, A. Paseau, and M. Potter, Oxford University Press: Oxford, 59–73.

Baker, A. (2008) "Experimental Mathematics," *Erkenntnis*, 68, 331–344.

Baker, A. (forthcoming) "Mathematical Explanation in Science," *British Journal for the Philosophy of Science*.

Colyvan, M. (2001) "The Miracle of Applied Mathematics," *Synthese*, 127, 265–278.

Corfield, D. (2003) *Towards a Philosophy of Real Mathematics*, Cambridge: Cambridge University Press.

Corfield, D. (2005) "Mathematical Kinds, or Being Kind to Mathematics," *Philosophica*, 74, 30–54.

Davis, P. (1981) "Are There Coincidences in Mathematics?" *American Mathematical Monthly*, 88(5), 311–320.

Diaconis, P. and Mosteller, F. (1989) "Methods for Studying Coincidences," *Journal of the American Statistical Association*, 84, 853–861.

Dorey, P. (2007) "The ODE/IM Correspondence," *Journal of Physics A: Mathematical and Theoretical*, 40, R205–R283.

Echeverria, J. (1996) "Empirical Methods in Mathematics. A Case-Study: Goldbach's Conjecture," in *Spanish Studies in the Philosophy of Science*, ed. G. Munevar, Kluwer: Dordrecht/Boston/London, 19–55.

Frege, G. (1980) *The Foundations of Arithmetic*, transl. J. L. Austin, Evanston: Northwestern University Press.

Gannon, T. (2006) "Monstrous Moonshine: the First Twenty-five Years," *Bulletin of the London Mathematical Society*, 38, 1–33.

Herkommer, M. (2004) "Goldbach Conjecture Research," online at http://www.petrospec-technologies.com/Herkommer/goldbach.htm.

Leibniz, G. (1956) *The Leibniz-Clarke Correspondence*, Manchester: Manchester University Press.

McEvoy, M. (forthcoming) "The Epistemological Status of Computer-Assisted Proofs," *Philosophia Mathematica*.

Mancosu, P. (2008) "Mathematical Explanation: Why It Matters," in P. Mancosu (ed.) (2008), 134–150.

Mancosu, P. (ed.) (2008) *The Philosophy of Mathematical Practice*, Oxford: Oxford University Press.

Melia, J. (2000) "Weaseling Away the Indispensability Argument," *Mind*, 109, 458–479.

Melia, J. (2002) "Response to Colyvan," *Mind*, 111, 75–79.

Owens, D. (1992) *Causes and Coincidences*, Cambridge: Cambridge University Press.

Pincock, C. (2007) "A Role for Mathematics in the Physical Sciences," *Nous*, 41, 253–275.

Pólya, G. (1954) *Mathematics and Plausible Reasoning: Patterns of Plausible Inference, Vol. 2*, Princeton: Princeton University Press.

Potter, M. (1993) "Inaccessible Truths and Infinite Coincidences," in *Philosophy of Mathematics: Proceedings of the 15th International Wittgenstein Symposium*, ed. J. Czermak, Verlag Hölder-Pichler-Tempsky: Vienna, 307–313.

Russell, B. (1973) *Essays in Analysis*, London: Allen & Unwin.

Steiner, M. (1978) "Mathematics, Explanation and Scientific Knowledge," *Nous*, 12, 17–28.

Tappenden, J. (2008) "Mathematical Concepts and Definitions," in P. Mancosu (ed.) (2008), 256–275.

Thomas, R. (1998) "An Update on the Four-Color Theorem," *Notices of the AMS*, 45, 848–859.

Tymoczko, T. (1979) "The Four-Color Problem and Its Philosophical Significance," *Journal of Philosophy*, 76, 57–83.

Van Fraassen, B. (1989) *Laws and Symmetry*, Oxford: Clarendon Press.

Wigner, E. (1960) "The Unreasonable Effectiveness of Mathematics in the Natural Sciences," *Communications in Pure and Applied Mathematics*, 13, I, 1–14.

Wilson, R. (2002) *Graphs, Colourings and the Four-colour Theorem*, Oxford: Oxford University Press.

7
Applying Inconsistent Mathematics

*Mark Colyvan**

At various times, mathematicians have been forced to work with inconsistent mathematical theories. Sometimes, the inconsistency of the theory in question was apparent (e.g. the early calculus), while at other times, it was not (e.g. pre-paradox naïve set theory). The way mathematicians confronted such difficulties is the subject of a great deal of interesting work in the history of mathematics but, apart from the crisis in set theory, there has been very little philosophical work on the topic of inconsistent mathematics. In this chapter, I will address a couple of philosophical issues arising from the applications of inconsistent mathematics. The first is the issue of whether finding applications for inconsistent mathematics commits us to the existence of inconsistent objects. I then consider what we can learn about a general philosophical account of the applicability of mathematics from successful applications of inconsistent mathematics.

1 Introduction

Inconsistent mathematics has a special place in the history of philosophy. The realisation, at the end of the nineteenth century, that a mathematical theory—naïve set theory—was inconsistent prompted radical changes to mathematics, pushing research in new directions and even resulted in changes to mathematical methodology. The resulting work in developing a consistent set theory was exciting and saw a departure from the existing practice of looking for self-evident axioms. Instead, following Russell (1907) and Gödel (1947), new axioms were assessed by their fruits.[1] Set theory shook off its foundationalist methodology. This episode is what philosophers live for. Philosophers played a central role in revealing the inconsistency of naïve set theory and played pivotal roles as new set theories took shape. This may have been philosophy's finest hour.[2]

Despite the importance of the crisis in set theory, inconsistent mathematics has received very little attention from either mathematicians or philosophers. Looking for inconsistency so that it might be avoided seems to be the extent of the interest. But inconsistent mathematics holds greater interest than merely providing an impetus for finding new, consistent theories to replace the old. Indeed, there are many reasons for taking inconsistent mathematical theories seriously and to be worthy of study in their own right. For example, there has been work on non-trivial, inconsistent mathematical theories such as finite models of arithmetic (Meyer, 1976; Meyer and Mortensen, 1984; Priest, 1997, 1998). While such mathematical theories might seem like mere curiosities, that's not the case. Chris Mortensen (1997, 2004) has argued that the best way to model inconsistent pictures (such as Penrose triangles and figures from Escher and Reutersvärd) is to invoke inconsistent geometry.[3] This work provides an interesting application for inconsistent mathematics. Although applications of inconsistent mathematics is the main theme of this chapter, I want to focus on other, more mundane applications of inconsistent mathematics—applications in modelling bits of the actual world (as opposed to Escher worlds). In particular, I will argue that there are a couple of puzzles arising from the applications of inconsistent mathematics, and in both cases, the puzzles have wider implications for philosophy of mathematics.

2 Indispensability of inconsistent mathematical objects

The first puzzle concerns ontology. In particular, it seems that an indispensability argument can be mounted for inconsistent (mathematical) objects. To see this, it will be useful to recall a particular, inconsistent mathematical theory, namely the early calculus.

The early calculus was inconsistent in at least two ways. First, infinitesimals were taken to be zero and non-zero. Moreover, they were taken to be zero at one place and non-zero at another place within the same proof. When dividing by infinitesimals, they were taken to be non-zero (for otherwise the division was illegitimate) and at other times, they were taken to be equal to zero (for example, when an infinitesimal appeared as a term in a sum). Newton, at least, tried to address such concerns by giving an interpretation of infinitesimals (or fluxions) as changing quantities. But, alas, this interpretation was itself inconsistent. After all, if an infinitesimal, δ, is a changing quantity, it cannot appear in equations such as

$$a = a + \delta$$

where a is a constant. Why? Well, the term on the right is changing (since δ is changing) so it cannot equal anything fixed, such as a constant a. Yet early calculus required equations like the one above to hold.[4]

As it turns out, the calculus could be put on a firm basis, but that didn't come until the nineteenth century, when Bolzano, Cauchy, and Weierstrass developed a rigorous theory of limits and the ϵ-δ notation.[5] Now the puzzle is that in the interim—over 150 years—the calculus was widely used, both in mathematics and elsewhere in science. Indeed, it is hard to imagine a more widely used and applicable theory. This presents a problem for those (like me) who take indispensability to science to be a reason to believe in the entities in question.[6]

According to this line of thought, we should be committed to the existence of all and only the entities that are indispensable to our best scientific theories and, and yet, for over 150 years, inconsistent mathematical entities—infinitesimals—were indispensable to these theories. This leads to the conclusion that we ought to have believed in the existence of inconsistent entities in the period between the late seventeenth century (by which time the calculus was finding widespread applications) and the middle of the nineteenth century (when the calculus was finally placed on a firm foundation). It seems that if one subscribes to the indispensability argument, there's a rather unpalatable conclusion beckoning: sometimes we ought to believe in the existence of inconsistent objects (Colyvan, 2008a, 2008b; Mortensen, 2008).[7] Indeed, it seems that the case for inconsistent mathematical objects (in the eighteenth century) was every bit as good as the case for believing in consistent mathematical objects.

A couple of comments on drawing ontological conclusions from inconsistent theories. Take any inconsistent theory along with classical logic and everything is derivable, including every other contradiction and the existence of all kinds of inconsistent objects. So what do we take to be the ontological commitments of an inconsistent theory? It is clear, and well-known, that in inconsistent settings like this, a paraconsistent logic is required.[8] With such a logic in place, triviality is avoided and we can make sense of specific inconsistent objects and conclusions being entailed by the theory in question.[9] Of course, Quine would have no truck with inconsistency and paraconsistent logics, but, nevertheless, what I'm arguing for here does seem to be a very natural extension of the Quinean approach to ontology. More importantly, it is at least plausible that scientists, when working with inconsistent theories, implicitly invoke a paraconsistent logic. Of course, most working scientists (even mathematicians) don't explicitly invoke a particular logic at all. The usual story that they all use classical logic is a rational (and heavily theory-laden) reconstruction of the practice. But it is interesting to note that when contradictions arise, working mathematicians do not derive results using the familiar C.I. Lewis proof,[10] even though such proofs are classically valid. This suggests, at least, that mathematical practice might be more appropriately modelled using a paraconsistent logic, in which such proofs are invalid (disjunctive syllogism is invalid in paraconsistent logics). Of course, there are other ways of explaining the practice. All I'm claiming here is that invoking paraconsistency is not as radical a move as it might first seem; it might be thought to be already implicit in mathematical practice.

On an historical note, it is interesting that both Newton and Leibniz believed that the methods of the calculus were in need of justification, and both sought geometric justifications. Newton took the justification task to be that of providing a geometric proof in place of each calculus proof—calculus for discovery, but geometry for justification. Leibniz, however, took the task to be that of providing a general justification of the methods of the calculus, then business as usual (Gaukroger, 2008). Although both Newton and Leibniz were thinking in terms of justification, they can also be seen to be offering two quite different anti-realist strategies in response to the indispensability argument I just presented. Newton was advocating a kind of eliminativist strategy, whereas Leibniz was seeking a non-revisionary account. Indeed, Leibniz's quest for a general justification of the methods of the calculus has a modern-day fellow traveller in Hartry Field (1980). Leibniz sought a general geometric limit account that would ensure that the calculus, despite being inconsistent, always gave the right answers on other matters. With a bit of massaging, we can see Leibniz as seeking something like a conservativeness proof: a demonstration that the calculus was a conservative extension of standard mathematics.[11]

I have argued elsewhere Colyvan (2008b), that it is not clear what to make of this argument for the existence of inconsistent objects. Does it tell us that consistency should be an overriding constraint in such matters? If so, why?[12] I have also (tentatively) suggested that the apparently unpalatable conclusion should be accepted: There are times when we ought to believe in inconsistent objects. But before you dismiss such thoughts as madness or perhaps as a *reductio* of the original indispensability argument, it is important to make sure that other accounts of ontological commitments do not also fall foul of inconsistent objects. Both mathematical realists and anti-realists alike have always assumed the consistency of the mathematics in question. Considering inconsistent mathematical theories adds a new wrinkle to the debate over the indispensability argument, and the ontology of mathematics, more generally.

3 A philosophical account of applied mathematics

There is another, perhaps more disturbing, conclusion beckoning. If our only theories of space and time need to invoke inconsistent mathematical theories (as they did in the eighteenth century), this might be thought to give us reason to be realists about not just the inconsistent mathematical objects, but about the inconsistency of space and time themselves.[13] But putting such disturbing thoughts aside for the moment, let's assume that the world itself is consistent. Now there is a puzzle about how inconsistent mathematical models can be applied to the world. Again this is a new twist on an old problem.

The general problem is that of providing a philosophical account of the applicability of mathematics. This debate had its origins in the indispensability debate but has taken on a life of its own. The problem, in a nutshell, is as follows: how is it that mathematical structures can be so useful in modelling various aspects of the physical world.[14] The obvious answer is that when some piece of mathematics is applied to a physical system, the mathematics is applicable because there are structural similarities between the mathematical structure and the structure of the physical system. So, for example, there's no surprise that $\mathbb{R}^3$ is useful in modelling physical space, for the two are isomorphic (putting aside relativistic curvatures). But in general, isomorphism is not the appropriate structural similarity—there is usually either more structure in the world or more in the mathematics. This is where things get interesting. We need to explain how non-isomorphic structures can be used to model one another and although there are several proposals around (Batterman, forthcoming; Bueno and Colyvan, forthcoming; Leng, 2002, 2008; Pincock, 2004, 2007), none of these is complete. The realisation that sometimes the mathematics in question is inconsistent, changes the way we might approach the problem. Assuming that the world is consistent, the problem is that of explaining how an inconsistent mathematical theory can be used to model a consistent system. This seems much tougher than explaining cases where there's simply no isomorphism, and some of the proposals do not seem well-suited to dealing with this tougher problem.

Let me make a couple of suggestions about how this problem might be solved. First, note that although the early calculus was inconsistent, it was eventually put on a firm foundation. Indeed, even when calculus was first developed, it might be argued that the consistent version existed, even though the existence of the latter wasn't known at the time. It might be further argued that this is all that's required; the inconsistent seventeenth century calculus is useful in applications because of its similarity to a consistent latter-day calculus.[15] The idea here is that what matters in applying mathematics is whether or not the mathematical model is capturing the salient features of the empirical phenomena in question. The model can achieve this irrespective of the knowledge of the modeller. An example might help here.

Early electrical theory had it that when there was a potential difference across a conductor, positively charged particles moved from the higher potential to the lower. This, it turns out, is wrong in a couple of ways. First, it's negatively-charged particles (electrons) that move, and in the opposite direction to that of the proposed positive particles. Second, electrons do not move very far in a conductor and they tend to oscillated—they certainly don't flow. The electrical current is the result of small movements of the electrons compounding to a net drift. So why was the original theory, which had all this wrong, so useful? It was useful because, for many purposes, these details are unimportant. The old, incorrect theory had a correct cousin— even though the latter was not known—and that's all that matters. Electrons do not care whether electricians know about them or not. There are electrons and there is (known or unknown) a correct theory of them, so all that matters is that any useful theory of electricity resembles the correct electron theory in certain respects. Clearly, positive particles flowing in one direction as opposed to negative particles flowing in the other, does not matter (unless one is specifically interested in the direction of particle movement), nor does it matter whether the particles in question flow or merely oscillate to ensure a net drift in one direction.

Returning to the case of the inconsistent calculus, we can see how the earlier suggestion might be fleshed out. It doesn't matter that the early calculus was inconsistent; it was, as a matter of fact, very similar to a consistent theory of calculus and it is this that explains the usefulness of the former. Indeed, on the account I'm proposing here, the usefulness of the calculus in itself suggests that there is a consistent theory in the offing. And this makes good sense of several key episodes in the history of mathematics where applications helped legitimise some questionable pieces of mathematics.[16] "It works" does seem like a very good response to suspicions about a new piece of mathematics.

This seems a promising start but there are some questions to be addressed. How can a consistent theory be similar to an inconsistent one? After all, it might appear that any given consistent theory is more like an arbitrary consistent theory than any inconsistent theory. And relatedly, it might seem that there is only one inconsistent theory since an inconsistent theory is trivial. The second worry is easily dealt with so let me tackle it first. In classical (and other explosive logics, such as intuitionistic logic), there is a sense in which there is only one inconsistent theory, namely, the trivial theory, where every proposition is true. But when dealing with inconsistency—or even potential inconsistency—we have already seen that we need to adopt a paraconsistent logic. Once this has been done, good sense can be made of *different* inconsistent theories. Seeing this also helps address the first question. Once we realise that we can discriminate between inconsistent theories, we can also determine which of these theories are similar to each other and to their consistent neighbours. It is not the case that all consistent theories are more like one another than they are to any inconsistent theory. Indeed, it is hard to see what would motivate such a thought, apart from the aforementioned mistake that there is only one inconsistent theory, namely, the trivial theory, and that this is radically unlike any consistent theory. There is still the difficult problem of how we compare theories, and nothing I've said here sheds any light on that more general problem. All I'm arguing for here is that inconsistent theories (in the context of a paraconsistent logic) can be compared in just the same way—whatever that is—to other theories, consistent and inconsistent.

The account just given seems right, in broad brush strokes, but further details will depend on the particular theory of applied mathematics adopted. So let me finish up by saying just a little about how some of the details might look in an account of the applications of mathematics I've recently developed with Otávio Bueno (forthcoming): *The Inferential Conception of Applied Mathematics*. The full details of this theory are not important for present purposes; the basic idea is that there are three separate stages of applying mathematics. First, there's the *immersion* step where a empirical set up is represented mathematically. The mathematics must be chosen in order to faithfully represent the parts of the empirical set up that are of interest. We do not require that the mathematics is isomorphic to the empirical set up. In general, the mathematical model and the empirical set up will not be isomorphic, but some structural features will be preserved in the mathematics. The second step is the inferential step where the mathematical model is investigated and various consequences of the model are revealed. The final step is the interpretation step where the results of the inferences conducted in the mathematical model in step two are interpreted back into the empirical set up. It is important to note that the interpretation step is constrained by the immersion step—mathematically representing some physical quantity in a specific way means that one must interpret the mathematics in question as a representation of the physical quantity—but the interpretation is not just the inverse of the immersion. For instance, at the interpretation step one is free to interpret more than what was initially represented in the immersion step. It is this feature of modelling that allows the mathematics to deliver novel phenomenon for investigation. It is one of the strengths of the inferential conception of applied mathematics that it is able to make sense of this important role mathematisation plays in science.[17]

With this framework in place, we can see how inconsistent mathematical theories, such as the early calculus, might be applied. First, we might explicitly treat the world as being inconsistent in the limit. That is, we treat the world as approximately inconsistent. (This is similar to when we make other idealisations, such as treating a fluid as being approximately incompressible and model it as such, despite holding that it really is compressible.) The inconsistent mathematics is then invoked to model this inconsistent picture of the world. Indeed, the inconsistent mathematics is essential here. No consistent mathematics could model the inconsistent limit being envisaged *as an inconsistent limit*. The inferential steps, of course, would need to be conducted using a paraconsistent logic, then the results of the inferences would be interpreted as being part of the nearest consistent story (if there is one).[18] There may be more than one consistent story in the neighbourhood. If there is, all such theories would need to be considered and the question of which theory to prefer would be settled by consideration of their theoretical virtues—simplicity, unificatory power, and so on.

Alternatively, we might only discover the (implicit) inconsistent assumptions about the world after the immersion and derivation steps. So, for example, we might have an implicitly inconsistent theory of instantaneous change. But the inconsistencies in this theory might not be apparent until the theory is represented using calculus, and some of the consequences of the theory are revealed. Once it is realised that the theory is inconsistent, we have no reason to force the interpretation to be consistent (as we did in the case just considered). After all, in this case it was a discovery of the mathematisation process that the underlying theory is inconsistent, so it should come as no surprise that some inconsistent results are delivered. The inconsistency here is more serious than in the previous case, and may prompt further work on developing a consistent theory. In the meantime, however, we can continue using the inconsistent theory (after employing a paraconsistent logic). Again, the inconsistent mathematics is essential here. The original theory of the empirical set up was (implicitly) inconsistent—indeed, inconsistent in ways of interest (or so we are assuming here). In order to faithfully represent the theory of the empirical set up, inconsistent mathematics is needed. Any consistent mathematics used for the immersion would not only hide the inconsistencies, but would render the mathematicised theory consistent. This would make it more difficult to discover the inconsistency of the original theory. But it might seem that that's preferable. At the end of the day, we are seeking a consistent theory, so using consistent mathematics might seem like a good way to facilitate this. But this is to misunderstand the role of discovering the inconsistency. We are seeking a consistent theory, but we are not seeking *any* consistent theory. The problem with this suggestion is that using consistent mathematics to model an inconsistent theory simply renders the inconsistent theory consistent; it does not reveal the inconsistency and it does not allow for careful reflection on how best to resolve the inconsistency. It just papers over the problem. Recognising an inconsistent theory as having a specific inconsistency is an important step in securing the appropriate consistent theory.[19]

There is much more work to be done before we have an adequate philosophical account of the applications of mathematics. Although it might be tempting to ignore cases of inconsistency—both inconsistent mathematics and inconsistent empirical theories—when considering the applications of mathematics, this would be a serious mistake. As I have just shown, considering applications of inconsistent mathematics forces attention onto the nature of the structures in question and the relevant notion of similarity in a way that is enlightening—we must be able to make sense of similarity between consistent and inconsistent structures, for example. Considering the applications of inconsistent mathematics, it seems, will help shed light on the general problem. And it is worth stressing that the inconsistent cases are not mere test cases either. A great deal of one of the most important periods in the history of science—the late seventeenth century to the mid-nineteenth century—relied heavily on inconsistent mathematics. During this period, most scientists were working with an inconsistent mathematical theory and the theory was used almost everywhere. Ignoring inconsistent mathematics in a general account of applied mathematics would simply be negligence.

4 Conclusion

I have discussed just two of the many philosophical issues that arise in connection to inconsistent mathematics. Both the issues discussed in this chapter revolve around applications of inconsistent mathematics. The first concerned drawing conclusions about ontological commitments from the indispensability of mathematics. When we find ourselves forced to admit the indispensability of inconsistent mathematical theories, a counterintuitive conclusion looms: sometimes we ought to believe that inconsistent objects exist. The second issue concerns the provision of an adequate account of applied mathematics—one that provides an adequate account of applications of inconsistent mathematics.

Apart from the much-discussed crisis in set theory, there has been very little work in philosophy on inconsistent mathematical theories, presumably because such mathematics is thought not to occupy a central position in mathematics itself; inconsistent mathematics is thought to be at best a curiosity or a pathological limiting case, and at worst something to be avoided at all costs. I hope this chapter has gone some way to establishing that inconsistent mathematics is interesting in its own right and that including it in our stock of examples will help shed light on major issues in mainstream philosophy of mathematics.[20]

Notes

*Department of Philosophy, University of Sydney, Sydney, NSW, 2006, Australia. *Email:* mcolyvan@usyd.edu.au.

1. Russell, for example, suggests that "[w]e tend to believe the premises because we can see that their consequences are true, instead of believing the consequences because we know the premises to be true" [1907, p. 273].
2. See Giaquinto (2002) for a nice account of this episode and its fallout.
3. The consistent treatments of such figures (Penrose, 1991; Penrose and Penrose, 1958) do not do justice to the cognitive dissonance one experiences when viewing the figures in question *as inconsistent*.
4. In modern calculus, we'd say that the limit of $a + \delta$, as δ goes to zero, is a. But in the early days of calculus, a rigorous theory of limits was not available. Newton and Leibniz were stuck with equations like the one above. It's also worth noting that inconsistency is usually thought to be a property of formal theories and the

early calculus was a long way from anything that would count as a formal theory. To claim that the early calculus was inconsistent, then, also involves some substantial claims about the interpretation of that theory *as inconsistent*. This is a big issue and much more needs to be said in order to establish beyond doubt that the early calculus was inconsistent. (For example, the early practitioners may have been groping towards one of the modern consistent interpretations of the calculus.) But it does seem that *prima facie*, at least, both the natural interpretation and Newton's changing quantity interpretations of the early calculus were inconsistent. See Mortensen (1995) for more on this.

5. Later in the 1960s work on non-standard analysis (and infinitesimals) by Robinson (1966) provided a separate consistent interpretation of the calculus, and, arguably, one closer to the spirit of the original. A little later Conway (1976) provided yet another way to rehabilitate infinitesimals.

6. See Colyvan, 2001a, 2008a; Putnam, 1971 and Quine, 1981 for details of the indispensability argument.

7. Throughout this chapter, I will take an inconsistent object to be an object that has inconsistent properties assigned to it by the theory positing it.

8. This is a logic where there is some Q such that $P \wedge \neg P \nvdash Q$. That is, in paraconsistent logics not everything follows from a contradiction.

9. And we can also deal with the related worry that there would seem to be only one inconsistent theory. As we shall see shortly, in a paraconsistent setting, we can make sense of different inconsistent theories.

10. For example, since an infinitesimal $\delta \neq 0$, it follows that either $\delta \neq 0$ or the fundamental theorem of calculus holds. But since $\delta = 0$, by disjunctive syllogism, we have the fundamental theorem of calculus.

11. Of course, it's hard to think about conservativeness when inconsistent theories are in the mix. If we take a theory Δ to be a conservative extension of Γ, then conservativeness amounts to (roughly) that any statement formulated in the vocabulary of Γ and derivable from $\Delta + \Gamma$ is derivable from Γ alone. But if Δ is inconsistent, then it can never be conservative, so long as the logic in question is explosive (i.e. supports *ex contradictione quodlibet*). But sense can be made of conservativeness in such settings, if the logic is paraconsistent.

12. You might think that in order for a theory to count as one of our best theories (and thus relevant to the indispensability argument), it needs to be consistent. This would rule out such cases as I'm considering here right from the start. It is hard to motivate such a privileged position for consistency, though (Bueno and Colyvan, 2004, Priest, 1998). Consistency is one among many virtues theories can enjoy, but it does not seem to trump all other virtues in the way this response would require. Indeed, if I am right that scientists take inconsistent theories seriously, anyone wishing to argue that such theories are never candidates for our best theories (so no ontological conclusions can be drawn from them) would seem to be at odds with scientific practice and thereby flying in the face of philosophical naturalism. See Colyvan (2008b) for further objections and responses to the indispensability argument I've outlined here.

13. There are interesting connections here with debates about ontological vagueness (or vagueness in the world) and Russell's dismissal of it as "the fallacy of verbalism" (Colyvan, 2001c). There are also related debates about whether vagueness might give us reason to believe that the world is inconsistent (Beall and Colyvan, 2001).

14. There is also a related puzzle, often called the unreasonable effectiveness of mathematics (Colyvan, 2001b; Steiner, 1995, 1998; Wigner, 1960), of understanding

how an apparently *a priori* discipline such as mathematics can provide the tools so often required by empirical science.

15. Of course, nothing I've said here tells us why the latter is so useful, but the strategy here is to deal with any special issues arising from the inconsistency.

16. I'm thinking here of the role of applications in helping legitimise the Dirac delta function, the early complex numbers, and, of course, the calculus (Kline, 1972).

17. See Bueno and Colyvan (forthcoming) for more details and a defence of the account.

18. The situation here is not unlike using continuous mathematics to model discrete phenomena (e.g. differential equations in population ecology). The discrete phenomena are treated as continuous in the limit, modelled using continuous mathematics, then inferences drawn in the mathematics, and the results interpreted discretely.

19. Something like this may have been going on with at least some of the applications of the early calculus: the underlying (unmathematised) theories of change, for example, were inconsistent in precisely the ways revealed when these theories were represented using the inconsistent calculus.

20. I'd like to thank Otávio Bueno, Stephen Gaukroger, Øystein Linnebo, and an anonymous referee for comments on earlier drafts or for discussions that helped clarify my thinking on the issues addressed in this paper. I am also grateful to audiences at the Baroque Science Workshop at the University of Sydney in February 2008, at the conference on New Waves in Philosophy of Mathematics at the University of Miami in April 2008, and at the 4th World Congress of Paraconsistency at the University of Melbourne in July 2008. Finally, I'd like to thank Chris Mortensen for impressing on me the significance of inconsistent mathematics. Work on this paper was funded by an Australian Research Council Discovery Grant (grant number DP0666020).

References

Batterman, R. forthcoming. "On the Explanatory Role of Mathematics in Empirical Science".

Beall, Jc and Colyvan, M. 2001. "Looking for Contradictions", *The Australasian Journal of Philosophy*, 79(4): 564–569.

Bueno, O. and Colyvan, M. 2004. "Logical Non-Apriorism and the Law of Non-Contradiction", in G. Priest, J.C. Beall, and B. Armour-Garb (eds.), *The Law of Non-Contradiction: New Philosophical Essays*. Oxford: Oxford University Press, pp. 156–175.

Bueno, O. and Colyvan, M. forthcoming. "An Inferential Conception of the Application of Mathematics".

Colyvan, M. 2001a. *The Indispensability of Mathematics*. New York: Oxford University Press.

Colyvan, M. 2001b. "The Miracle of Applied Mathematics", *Synthese*, 127: 265–278.

Colyvan, M. 2001c. "Russell on Metaphysical Vagueness", *Principia*, 5(1–2): 87–98.

Colyvan, M. 2008a. "Who's Afraid of Inconsistent Mathematics?", *Protosociology*, 25: 24–35.

Colyvan, M. 2008b. "The Ontological Commitments of Inconsistent Theories", *Philosophical Studies*, 141: 115–123.

Colyvan, M. 2008c. "Vagueness and Truth", in H. Dyke (ed.), *From Truth to Reality: New Essays in Logic and Metaphysics*. London: Routledge, pp. 29–40.

Conway, J.H. 1976. *On Numbers and Games*. New York: Academic Press.

Field, H. 1980. *Science Without Numbers: A Defence of Nominalism*. Oxford: Blackwell.

Gaukroger, S. 2008. "The Problem of Calculus: Leibniz and Newton On Blind Reasoning", paper presented at the Baroque Science Workshop at the University of Sydney in February 2008.

Giaquinto, M. 2002. *The Search for Certainty: A Philosophical Account of Foundations of Mathematics*. Oxford: Clarendon Press.

Gödel, K. 1947. "What Is Cantor's Continuum Problem?", reprinted (revised and expanded) in P. Benacerraf and H. Putnam (eds.), *Philosophy of Mathematics Selected Readings*, second edition. Cambridge: Cambridge University Press, 1983, pp. 470–485.

Kline, M. 1972. *Mathematical Thought from Ancient to Modern Times*. New York: Oxford University Press.

Leng, M. 2002. "What's Wrong With Indispensability? (Or, The Case for Recreational Mathematics)", *Synthese*, 131: 395–417.

Leng, M. 2008 *Mathematics and Reality*. Oxford: Oxford University Press.

Meyer, R.K. 1976. "Relevant Arithmetic", *Bulletin of the Section of Logic of the Polish Academy of Sciences*, 5: 133–137.

Meyer, R.K. and Mortensen, C. 1984. "Inconsistent Models for Relevant Arithmetic", *Journal of Symbolic Logic*, 49: 917–929.

Mortensen, C. 1995. *Inconsistent Mathematics*. Dordrecht: Kluwer.

Mortensen, C. 1997. "Peeking at the Impossible", *Notre Dame Journal of Formal Logic*, 38(4): 527–534.

Mortensen, C. 2004. "Inconsistent Mathematics", in E.N. Zalta (ed.), *The Stanford Encyclopedia of Philosophy* (Fall 2004 edition), URL= <http://plato.stanford.edu/archives/fall/2004/entries/mathematics-inconsistent/>.

Mortensen, C. 2008. "Inconsistent Mathematics: Some Philosophical Implications", in A.D. Irvine (ed.), *Handbook of the Philosophy of Science Volume 9: Philosophy of Mathematics*. North Holland: Elsevier.

Penrose, L.S. and Penrose, R. 1958. "Impossible Objects, a Special Kind of Illusion", *British Journal of Psychology*, 49: 31–33.

Penrose, R. 1991. "On the Cohomology of Impossible Pictures", *Structural Topology*, 17: 11–16.

Pincock, C. 2004. "A New Perspective on the Problem of Applying Mathematics", *Philosophia Mathematica (3)*, 12: 135–161.

Pincock, C. 2007. "A Role for Mathematics in the Physical Sciences", *Noûs*, 41: 253–275.

Priest, G. 1997. "Inconsistent Models of Arithmetic Part I: Finite Models", *Journal of Philosophical Logic*, 26(2): 223–235.

Priest, G. 1998. "What Is So Bad About Contradictions?", *The Journal of Philosophy*, 95(8): 410–426.

Priest, G. 2000. "Inconsistent Models of Arithmetic Part II: The General Case", *Journal of Symbolic Logic*, 65: 1519–1529.

Putnam, H. 1971. *Philosophy of Logic*. New York: Harper.

Quine, W.V. 1981. "Success and Limits of Mathematization", *Theories and Things*. Cambridge MA.: Harvard University Press, pp. 148–155.

Robinson, A. 1966. *Non-standard Analysis*. Amsterdam: North Holland.

Russell, B. 1907. "The Regressive Method of Discovering the Premises of Mathematics", reprinted in D. Lackey (ed.), *Essays in Analysis*. London: George Allen and Unwin, 1973, pp. 272–283.

Steiner, M. 1995. "The Applicabilities of Mathematics", *Philosophia Mathematica (3)*, 3: 129–156.

Steiner, M. 1998. *The Applicability of Mathematics as a Philosophical Problem*. Cambridge MA: Harvard University Press.

Wigner, E.P. 1960. "The Unreasonable Effectiveness of Mathematics in the Natural Sciences", *Communications on Pure and Applied Mathematics*, 13: 1–14.

8
Towards a Philosophy of Applied Mathematics

Christopher Pincock

Most contemporary philosophy of mathematics focuses on a small segment of mathematics, mainly the natural numbers and foundational disciplines like set theory. Although there are good reasons for this approach, in this chapter, I will examine the philosophical problems associated with the area of mathematics known as applied mathematics. Here mathematicians pursue mathematical theories that are closely connected to the use of mathematics in the sciences and engineering. This area of mathematics seems to proceed using different methods and standards when compared to much of mathematics. I argue that applied mathematics can contribute to the philosophy of mathematics and our understanding of mathematics as a whole.

I

If there is any trend in contemporary philosophy of mathematics worthy of the label "New Wave", it is surely the call to turn our attention to the practices, priorities and developments that are prized by working mathematicians. Most actual mathematicians, unsurprisingly, have little interest in the questions that have dominated the philosophy of mathematics since the 1960s. Few mathematicians, for example, are likely to be troubled by Benacerraf's argument in "What Numbers Could Not Be" that numbers are not objects for the simple reason that they typically do their mathematics without worrying about what numbers might be. Similarly, the extended debates between platonists and nominalists, and the associated epistemological worries about our knowledge of ordinary mathematics, have little impact on what is usually called "mathematical practice".

In different ways, two philosophers of mathematics have recently argued that this disengagement is a serious problem for the philosophy of mathematics. Starting from Quine's conception of naturalized epistemology, Maddy argues that the only questions that philosophers of mathematics should be asking are questions about the methodology of working mathematicians. The philosophers need not accept these methods uncritically, but can only criticize them on the basis of internal mathematical criteria, as opposed to external philosophical tests. In particular, when set theorists consider adopting new axioms to resolve open questions like the Continuum Hypothesis, the philosopher should seek to understand the set theorists on their own terms, and not criticize their practice for philosophical reasons.

A further call to focus on mathematical practice is found in Corfield's *Towards a Philosophy of Real Mathematics* (2003). He argues that mathematical practice beyond the foundations of mathematics should inform the philosophy of mathematics. This may be because, like Maddy, he thinks that traditional philosophical questions about mathematics are illegitimate. Alternatively, he may simply prefer to broaden the list of questions pursued by philosophers of mathematics. For Corfield, real mathematics includes things like the Riemann hypothesis, algebraic topology and higher-dimensional algebra. Exactly what gets into the category of real mathematics is never really clarified, but he does emphasize a turn to "interpretational issues interior to branches of mathematics in such a way as to provide us with insight into reasonably large portions of mathematics" (Corfield 2003, p. 20).

I am uncomfortable with the claim that mathematicians should be the ultimate arbiters of philosophical interest. At the same time, I believe that the turn to practice will inject new vitality into the philosophy of mathematics. This belief is reinforced by the appearance of the volume *The Philosophy of Mathematical Practice*, edited by Mancosu. In his introduction, Mancosu explains that "What is distinctive of this volume is that we integrate local studies with general philosophy of mathematics, contra Corfield, and we also keep traditional ontological and epistemological topics in play, contra Maddy" (Mancosu 2008b, p. 19). It is this spirit which animates the present chapter with its focus on the part of mathematics known loosely as applied mathematics. I aim to relate a local study from applied mathematics to some broader philosophical issues. Neither Maddy nor Corfield has pressed their turn to practice in this direction.[1] My questions are as follows: What relevance might applied mathematics, thought of as a kind of mathematics, have for the philosophy of mathematics? Are there new philosophical questions that we can fruitfully pursue about applied mathematics? Do traditional philosophical issues get new life based on an examination of applied mathematics? I will argue that the answer to these last two questions is "yes".

II

A series of new philosophical questions about applied mathematics is not difficult to formulate. What is applied mathematics, after all? Is it a distinct kind of mathematics? How does applied mathematics relate to the rest of mathematics, if it is a distinct kind? And, what mathematical interest does the study of applied mathematics offer? Where might we find the parts of applied mathematics that have some intrinsic mathematical value?

It is very hard to answer these questions, and it seems to me that philosophers of mathematics can help to answer them in a more systematic way than we are likely to find by asking the mathematicians themselves. As a first step, we can begin with the clear sociological and disciplinary divisions between pure and applied mathematics.[2] We have special courses in applied mathematics along with associated textbooks, journals, professional organizations and prizes. In addition to these practices, there are various reflections by mathematicians on the distinction between pure and applied mathematics, and whether it is a good or a bad thing. These divisions can serve as our initial data in the attempt to distinguish pure and applied mathematics, although, as with all philosophical problems, we must be prepared to reject some or all of this data if they prove to be ungrounded.

Two influential textbooks provide a natural starting point. They are both introductory surveys of applied mathematics designed for the undergraduate mathematics major. These collections are most useful for our purposes as it is presumably here that a student can find out what they are getting themselves into if they continue to pursue applied mathematics as a specialization. Volume 1 of the SIAM series "Classics in Applied Mathematics",[3] for example, is Lin and Segel's *Mathematics Applied to Deterministic Problems in the Natural Sciences*. Lin and Segel's (1988, p. vii) book "is concerned with the construction, analysis, and interpretation of mathematical models that shed light on significant problems in the natural sciences". Part A of the book discusses "Deterministic Systems and Ordinary Differential Equations", "Random Processes and Partial Differential Equations" and "Superposition, Heat Flow, and Fourier Analysis". Part B continues with several chapters under the heading of "Some Fundamental Procedures Illustrated on Ordinary Differential Equations". These procedures include dimensional analysis, perturbation theory and stability analysis. Finally, there is Part C, called "Introduction to Theories of Continuous Fields", with discussions on the wave equation, potential theory and various kinds of fluid flow (inviscid, viscous and compressible). A similar approach is taken in Fowler's *Mathematical Models in the Applied Sciences*, volume 17 in the Cambridge Texts in Applied Mathematics series.[4] Here, the first four chapters set out methods such as nondimensionalization, asymptotics and perturbation theory. These chapters are followed by four on "Classical models", eight on "Continuum models" and four on "Advanced models". Each model discussed is of a specific physical system, ranging from "Heat transfer" and "Electromagnetism" through "Spruce budworm infestations" and "Frost heave in freezing soils".

The organization of both volumes suggests that applied mathematics is thought of primarily as the development of a limited number of mathematical theories and techniques with an eye on formulating and analyzing acceptable mathematical models of specific sorts of physical systems. This proposal finds some confirmation in the programmatic remarks that begin each volume. Lin and Segel (1988, p. 5) say

> The purpose of applied mathematics is to elucidate scientific concepts and describe scientific phenomena through the use of mathematics, and to stimulate the development of new mathematics through such studies. The process of using mathematics for increasing scientific understanding can be conveniently divided into the following three steps:
>
>   (i) The *formulation* of the scientific problem in mathematical terms.
> (ii) The *solution* of the mathematical problems thus created.
> (iii) The *interpretation* of the solution and its empirical verification in scientific terms.

Later, Lin and Segel (1988, p. 7) add "(iv) The generation of scientifically relevant new mathematics through creation, generalization, abstraction, and axiomatic formulation". In a similar way, Fowler (1997, p. 3) divides the practice of applied mathematics into the following six stages: problem identification, model formulation, reduction, analysis, computation and model validation. For good measure, he adds

> Applied mathematicians have a procedure, almost a philosophy, that they apply when building models. First, there is the phenomenon of interest that one wants to describe or, more importantly, explain. Observations of the phenomenon lead, sometimes after a great deal of effort, to a hypothetical mechanism that can explain the phenomenon. The purpose of a model is then to formulate a description of the mechanism in quantitative terms, and the analysis of the resulting model leads to results that can be tested against the observations. Ideally, the model also leads to predictions which, if verified, lend authenticity to the model.[5]

Both Lin and Segel's and Fowler's programmatic remarks fit with the details of their later discussions. In each case, it is not sufficient just to learn the mathematics. An applied mathematician, we are told, must learn how to judge which areas of mathematics are likely to be useful in formulating a given scientific problem as well as how to interpret the results of their analysis in physical terms.[6]

If we take these textbooks to be accurate indications of the difference between pure and applied mathematics, then we can draw the following sociological conclusions. While both pure and applied mathematicians are concerned with the development of mathematical theories, the applied mathematician has different priorities and tests for the adequacy of their work. An applied mathematician aims to have her mathematics available to provide mathematical formulations for scientifically motivated problems.

Furthermore, even if her mathematics succeeds in the formulation task, it will only be judged adequate if she can also solve or otherwise analyze the resulting mathematical problem. An even more challenging requirement is that she must remain available to assist the scientist in interpreting her solutions to mathematical problems. That is, it must be possible to assess the links between the simplified or analyzed mathematical model and whatever experimental data can be collected.

In contrast, the task of the pure mathematician seems different. The tests for adequate pure mathematics are more internal to pure mathematical practice. The pure mathematician either takes on a problem that has been previously identified by mathematicians as mathematically important, or else formulates new conjectures and problems of intrinsic mathematical interest. What is expected seems more fixed and amenable to sustained treatment. This is not meant to suggest that achieving these goals in pure mathematics is any easier than reaching the different goals of the applied mathematician. Still, the applied mathematician seems constantly left to the whims of scientists and their theoretical deficiencies and so the goals here are often less well-defined.

III

In starting with a sociological approach to the pure/applied mathematics distinction, I have followed the paper by Vincenti and Bloor (2003), "Boundaries, Contingencies and Rigor: Thoughts on Mathematics Prompted by a Case Study in Transonic Aerodynamics". Here Bloor, the well-known advocate of the strong program in the sociology of science, is ably assisted by Walter Vincenti, a pioneer in the treatment of transonic aerodynamics from both a mathematical and an experimental perspective. While the case study they discuss is too involved to develop here, a prior episode in the history of fluid mechanics will suffice to investigate in more detail some additional philosophical questions raised by applied mathematics. This is Prandtl's development of what is known as boundary layer theory. Its significance for our understanding of fluid mechanics is hard to exaggerate, and a contemporary textbook calls it "perhaps the greatest single discovery in the history of fluid mechanics" (Kundu and Cohen 2008, p. 851).[7] As I hope to explain, though, it is also of great significance as a piece of applied mathematics, as the techniques pioneered by Prandtl can be applied quite widely beyond the equations he was concerned with. Boundary layer theory, then, refers not only to a part of fluid mechanics, but also to a general technique of solving otherwise intractable mathematical problems.

To start to understand what Prandtl's theory amounts to, consider the flow of a fluid like air around an object.[8] See Figure 8.1.

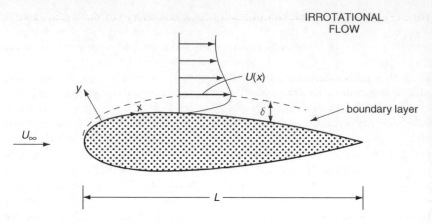

Figure 8.1   Flow around an object. From Kundu and Cohen 2008, p. 340

Here, the vector U_∞ represents the left-to-right incoming velocity of the fluid. The general problem is to determine how the velocity and pressure values are arrayed around the object and the associated forces impressed on the object. Depending on the incoming velocity, the shape of the object and the properties of the fluid, the object may experience both an upward force (lift) and a left-to-right force (drag). The lines labeled "$U(x)$" show the velocity in the x-direction as we approach the object for a particular value of y. Crucially, the velocity first increases and then dramatically decreases to 0.

We focus on steady flow, so none of the functions in question are time-dependent. We aim to discover for a given flow the velocity functions u(x, y), v(x, y) and the pressure function p(x, y) where u is the velocity component in the x-direction and v is the velocity component in the y-direction. As shown in Figure 8.1, the x and y coordinates are not the regular Euclidean axes, but are chosen so that the x-direction is always tangential to the surface of the body and the y-direction is always normal to the surface of the body.

Prior to Prandtl, there were two approaches to this sort of system.[9] First, there were the Navier–Stokes equations. For our simplified situation, the Navier–Stokes equations are

$$u\, \partial u/\partial x + v\, \partial u/\partial y = -1/\rho\, \partial p/\partial x + \nu(\partial^2 u/\partial x^2 + \partial^2 u/\partial y^2) \qquad (1)$$
$$u\, \partial v/\partial x + v\, \partial v/\partial y = -1/\rho\, \partial p/\partial y + \nu(\partial^2 v/\partial x^2 + \partial^2 v/\partial y^2) \qquad (2)$$

In addition, conservation laws tell us that

$$\partial u/\partial x + \partial v/\partial y = 0 \qquad (3)$$

For the case under consideration, additional boundary conditions are that

$$u(0, y) = U_\infty, \quad v(0, y) = 0 \tag{4}$$

ρ here represents the density of the fluid. We treat ρ as a constant, which is only appropriate when the fluid is incompressible and homogeneous, that is the density is not changed by the pressure and the fluid's features do not vary. Neither assumption, of course, is correct for air, but for speeds well below the speed of sound, these assumptions are widely employed. The v in (1) and (2), represents the kinematic viscosity of the fluid.[10] The viscosity of a fluid measures its tendency to absorb the momentum of adjacent fluid elements. A highly viscous fluid resists internal circulation whereas a non-viscous fluid easily propagates internal disturbances. It may be represented by $v \equiv \mu/\rho$ or μ, known simply as viscosity. As with density, we treat viscosity as a constant. The units of v and μ are cm^2/s and $g/cm \cdot s$, respectively, and typical values are

	v	μ
water	10^{-2}	10^{-2}
air	$15 \cdot 10^{-2}$	$2 \cdot 10^{-4}$

(Segel 2007, p. 87).

The appearance of the viscosity term makes the Navier–Stokes equations very difficult to solve, except for highly artificial contexts. The reason is that the viscosity terms are the only place where the second-order partial differential operators occur.

The second approach employed the Euler equations. The Euler equations result from the assumption that terms preceded by v are small compared to the other terms. Neglecting these terms gives us

$$u \, \partial u/\partial x + v \, \partial u/\partial y = -1/\rho \, \partial p/\partial x \tag{1E}$$

$$u \, \partial v/\partial x + v \, \partial v/\partial y = -1/\rho \, \partial p/\partial y \tag{2E}$$

The conservation law and boundary conditions are unaffected:

$$\partial u/\partial x + \partial v/\partial y = 0 \tag{3E}$$

$$u(0, y) = U_\infty, \quad v(0, y) = 0 \tag{4E}$$

We can motivate this simplification by calculating a rough estimate of the relative size of the terms in (1). As we can see, $u\partial u/\partial x$ is roughly of order U_∞^2/L, where L is the length of the body. By contrast, $v\partial^2 u/\partial x^2$ is roughly of order vU_∞/L^2. The ratio between the two terms, then, is

$(U_\infty^2/L)/(\nu\, U_\infty/L^2) = U_\infty L/\nu$. If $U_\infty = 50\,$km/h, $L = 10\,$cm and $\nu = 0.15\,$cm^2/s (the value for air), then the u $\partial u/\partial x$ term is about 100,000 times larger than the $\nu\partial^2 u/\partial x^2$ term (Segel 2007, p. 106). So, the approximation procedure seems eminently reasonable.

We can make this reasoning more rigorous if we deploy the technique of scaling. As Lin and Segel explain,

> in the process of scaling one attempts to select intrinsic reference quantities so that each term in the dimensional equations transforms into the product of a constant dimensional factor which closely estimates the term's order of magnitude and a dimensionless factor of unit order of magnitude (Lin and Segel 1988, p. 211).

The procedure is not an algorithm as there is no way to be assured in advance that an adequate set of scales has been found. See Appendix A1 for the choice of scales and how they result in the Euler equations.

The crucial assumption here is that these are the right scales for the whole domain under consideration, including the region close to the object. The failure of this assumption is revealed by d'Alembert's paradox: For this sort of situation, the Euler equations predict that no drag will be felt by the suspended object. This is easily refuted by experiment. Another way to see the limitations of the Euler equations is to note that they cannot be reconciled with a plausible and experimentally motivated boundary condition:

$$u(x, 0) = 0, \quad v(x, 0) = 0 \tag{4B}$$

That is, at the surface of the object, the fluid should obey a "no slip" condition. This requires that the velocity of the fluid elements decrease as we consider elements closer and closer to the object, in line with Figure 8.1. By contrast, the Euler equations require that the velocity increase as we approach the object, as depicted in Figure 8.2.

Mathematically speaking, then, there is something wrong with our scaling procedure. It has transformed our presumably correct original equations (1)–(3) into the demonstrably incorrect Euler equations (1E)–(3E).

The flaw in the procedure was the assumption that the scales would lead to terms of comparable magnitude in the *entire* domain. What we have just seen tells us that in some parts of the domain, presumably in close proximity to the object, the magnitudes of the terms dramatically change and our scaling becomes inappropriate. Clearly, it is the terms originally paired with ν that grow in relative significance. Must we then return to the intractable Navier–Stokes equations, or is there another alternative?

Prandtl's boundary layer theory opts to combine the Euler equations, restricted to the outer region where they work, with some new equations called the boundary layer equations, restricted to the thin region near the

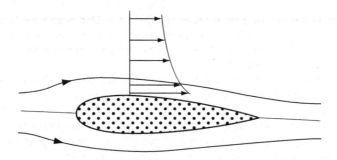

Figure 8.2 Solution to Euler equations. From Kundu and Cohen 2008, p. 166

object called the boundary layer. These new equations are obtained by a different set of scales. These scales are picked using a new quantity δ, which we can think of as the width of the boundary layer. This reflects our recognition that when it comes to the y-direction, effects operate on a much smaller scale than L. See Appendix A2 for the choice of the scales and how they lead to the boundary layer equations:

$$u^* \, \partial u^*/\partial x^* + v^* \, \partial u^*/\partial y^* = -\partial p^*/\partial x^* + \partial^2 u^*/\partial y^{*2} \qquad (1B)$$

$$\partial p^*/\partial y^* = 0 \qquad (2B)$$

While (1B) still remains second-order, we have removed one of the problematic terms. In conjunction with (3), the original constraint (4) and some additional boundary constraints, the problem becomes mathematically tractable.

These constraints arise from the general requirement that the flow within the boundary layer match the flow in the outer region along the upper edge of the boundary layer. By (2B), we know that the pressure at the upper edge of the boundary layer remains constant in the y-direction and must agree with the pressure in the outer region. In addition to (4B), we impose the conditions

$$u^*(x, \infty) = U_\infty \qquad (6B)$$

$$u^*(x_0, y) = u(x_0, y)$$

where x_0 is some point upstream which can be used to determine the flow downstream in the boundary layer by matching it with the outer region.

What are we to make of this approach? We can divide it into two stages. First, there is the choice of scales. Discussions of scaling typically emphasize the empirical character of the choice of scales and, as we have seen, an apparently well-motivated choice of scales can prove inappropriate for certain purposes. In their discussion of scaling, for example, Lin and Segel offer six ways to arrive at the correct scales that include "Utilize experimental or observational evidence", "Make certain order of magnitude assumptions merely because the concomitant neglect of terms renders to problem tractable", "Use a trial and error approach" and "Employ the results of computer calculations" (1988, p. 222). Even after reviewing the manifestly successful example of boundary layer theory, Segel concedes that "Formulation of the boundary layer problem is hardly a straightforward matter, and the critical reader may well feel uneasy at this point" (2007, p. 114).[11] While scaling is clearly a mathematical transformation of a given set of equations, its adequacy in a given case is not something accessible to the pure mathematician.

The second stage is the elimination of terms preceded by a small factor, as with $(1/Re)$ $(\partial^2 u'/\partial x'^2 + \partial^2 u'/\partial y'^2)$ in (1″) and $(1/Re)$ $\partial^2 u^*/\partial x^{*2}$ in (1**). See Appendix A. A mathematically more sophisticated way to motivate this elimination is to think of it as an application of what is known as perturbation theory. The basic idea of perturbation theory is that terms involving a small parameter, like $(1/Re)$, can be rewritten as a series expansion, where the terms of the series decrease quickly. This allows an approximate evaluation of the contribution of the problematic term through an evaluation of the beginning of the series. The cogency of the procedure, as Lin and Segel (1988, p. 230) suggests, depends on the assumption of asymptotic validity: "the exact solution and the approximation can be made to differ by an arbitrarily small amount for the entire range of the independent variable provided the parameter is taken sufficiently small". When this assumption is correct, the situation is one where what is called regular perturbation theory is adequate. The failure of the Euler equations shows that the assumption is not correct in our case as asymptotic validity fails in the region near the object.

To deal with the failure of this assumption, applied mathematicians have developed a theory known as singular perturbation theory. One part of singular perturbation theory employs two series expansions that are said to be asymptotic in their respective regions of the domain and then matches them in a common, overlap region. This helps to account for the peculiar procedure employed above when we forced the two sets of equations to agree along the "edge" of the boundary. From our current perspective, what we are really doing is taking two series expansions and claiming that matching them yields an asymptotically valid representation of the original Navier–Stokes version of the problem. The mathematical techniques here provide considerable flexibility and can be deployed to manage a wide range of problems where regular perturbation theory fails. For example, when Fowler explains these singular perturbation methods in chapter 4 of his book, he notes that "They provide the analytic platform upon which much of the later discussion in this book is based" (1997, p. 45). Lin and Segel motivate their discussion in the following way:

It might seem that the practitioner of singular perturbation theory would find relatively few important problems to work on, but this is not at all the case. The abundance of such problems is at least partially explained by the fact that singular perturbation techniques seem to be appropriate for examining the behavior of solutions near their "worst" singularities. "Taking the bull by the horns" and examining functions near their worst singularities is the best way to obtain information about the qualitative behavior, and it is the elucidation of this behavior that is very often the goal of a theoretician. In this lies the point of a remark once made about a distinguished applied mathematician, "All he can do is singular perturbation problems. But of course he can turn all problems into singular perturbation problems!" (Lin and Segel 1988, p. 277).

Here, then, is a powerful technique whose domain of application extends far beyond the fluid mechanics case that I have focused on.[12]

This example seems to have all the ingredients that we saw are used to distinguish applied mathematics from pure mathematics. In particular, the invocation of a thin boundary layer was motivated by the need to recover the details of a physical phenomenon, and not by any intrinsic mathematical interest. It is, of course, a mathematical discovery that the Navier–Stokes equations can be simplified in this way, but what made the innovation so important was its success in resolving various scientific problems having to do with fluid flow. Coinciding with these non-mathematical priorities, we find a failure of mathematical rigor. The disconnect between boundary layer theory as a practical tool for science and engineering and its status as a mathematical theory is well summarized by Nickel in his 1973 survey article "Prandtl's Boundary-Layer Theory from the Viewpoint of a Mathematician" (pp. 405–406):

> During the first 50 years of boundary-layer theory the fundamental mathematical questions could not be answered. It was not possible to establish a sound mathematical connection to the Navier–Stokes differential equations. There was no evidence of the existence, uniqueness, and well-posedness of a solution.[13]

Similarly, Segel (2007, pp. 77–78) quotes Carrier's 1972 evaluation of Prandtl's contributions:

> they have provided the basic foundations for the heuristically-reasoned, spectacularly successful treatment of many important problems in science and engineering. This success is probably most surprising to rigor-oriented mathematicians (or applied mathematicians) when they realize that there still exists no theorem which speaks to the validity or the accuracy of Prandtl's treatment of his boundary layer problem; but seventy

years of observational experience leave little doubt of its validity and its value.

While the situation had improved up to the 1970s (and has no doubt improved since then), the failure of boundary layer theory as a rigorous mathematical theory is significant. It shows that the standards for admission as a tool of applied mathematics are different from what a pure mathematician would hope for. A sacrifice in intrinsic mathematical clarity is typically made if a consequent reward in tractability can be achieved.

IV

Let us return to our main topic and see what the implications of our example for philosophical questions about the respects in which pure and applied mathematics are different. To begin with, there appear to be epistemic differences. That is, I can come to know some piece of applied mathematics based on different kinds of justifications than are available for pure mathematics, for example, that this set of equations agrees with some other set of equations in some region. A second dimension to explore is semantic. Here the issue is whether the contents of sentences found in applied mathematics have some different character than the contents expressed by sentences of pure mathematics. Finally, one might insist on a metaphysical difference. Here the question is whether the subject matter of applied mathematics is different from pure mathematics, and if they are different, how they might be related. I will take each of these issues in turn. It seems to me that a strong case can be made that significant epistemic, semantic and metaphysical consequences result from reflecting on applied mathematics, but I can only begin to make that case here.

The epistemic differences between pure and applied mathematics are brought out clearly by our example from Prandtl. What, in the end, justifies the claim that the relevant instances of the Navier–Stokes equations can be simplified by the boundary layer equations? As Nickel and Carrier note, the justification was not wholly mathematical; although for certain sorts of situations, this sort of wholly mathematical justification was later made available. Instead, it seems clear that experiments and observations played a crucial role in justifying this technique in applied mathematics. It was only after various experiments were done, and the results found to agree with the predictions of the solutions to the boundary layer equations, that the approach could be judged to be sound. Here, I think, we have a clear difference with pure mathematics. Additional sources of justification become available for applied mathematics. This point is not meant to deny that some empirical sources of justification also exist in pure mathematics, but their motivation is, I think, quite different.[14]

This empirical justification of applied mathematics fits with the common assumption that applied mathematics is less rigorous than pure mathematics. For in testing this or that instance of the boundary layer technique, we may not gain a good understanding of why the theory works for this case and where it might break down for others. To take a relatively unsurprising example, the approach pioneered by Prandtl succeeded for subsonic fluid flows, but was not easily extended for flows near the speed of sound or into the transonic range. The reasons for this are not hard to appreciate. Our whole discussion began with the assumption that the fluid was incompressible, that is, its density is not affected by pressure. Beyond roughly Mach 0.3, however, the flow of fluids like air shows effects inconsistent with incompressibility. This is because the pressure changes become much more significant, and so the density of the fluid changes as it flows around the object. The difficulties in understanding the resulting flow, and how they were solved so that, for example, supersonic flight became possible is the focus of Vincenti and Bloor's paper.

Here we have another role for empirical justification which distinguishes pure from applied mathematics. The incompressibility of the fluid corresponds to a simplification of the full Navier–Stokes equations, and consequently to the applicability of the further simplifications achieved by positing the boundary layer. The mathematics by itself is, of course, not able to tell us when these simplifications will fail. But recall that it was part of the job of the applied mathematician to interpret the results of their mathematical models and so advise the scientist and engineer on when their models would fail to fit a given physical situation. So, again, we see the need for the applied mathematician to draw on experimental data. For it is only by looking to experiments that we can come to any clear decision about when the necessary approximations are acceptable.

We can press things in a different direction by turning to the semantic difficulties that applied mathematics engenders. Here I have in mind the extended discussion by Mark Wilson in his book *Wandering Significance: An Essay on Conceptual Behavior* (2006). Wilson is the philosopher who has made the strongest case that applied mathematics is important to philosophy.[15] His view is essentially that general facts about language and its relation to the world are revealed most clearly when we consider the details of applied mathematics. This includes Prandtl's boundary layer theory, so here we can use our example to illustrate the more general semantic conclusions that Wilson aims to draw.

A characteristic overview of Wilson's book is the following:

a strong and unverified faith in classical physics' guaranteed axiomatizability generally stems from a *false picture* of how its admirable stock of predicates gather their descriptive utilities: there are important alternatives – including my facades – that have been overlooked. And that mistake, in microcosm, encapsulates many of the basic mechanisms responsible for the other ur-philosophical difficulties we explore in this book. Anytime we blithely presume that the "conceptual contents" attached to a passel of predicates behave in the simple manner sketched within the classical picture, we are in danger of building ourselves up for an awful letdown, as Fred Astaire once put it: some unfortunate ur-philosophical muddle may lie in the offing. That warning of optimism-induced error represents the chief message of this book, which we will examine from various vantage points throughout the book (2006, p. 192).

The classical picture criticized here is the view that words like "pressure", "velocity" and "viscosity" can be easily attached to genuine physical properties possessed by the objects that we encounter in our experience. Most contemporary philosophy of language and mind assumes the classical picture. Wilson also opposes what he thinks of as an equally confused alternative to the classical picture which he calls "hazy holism". On this view, the semantic contents of predicates are determined entirely by the broader theoretical context in which they are embedded.

Between the classical picture and hazy holism, we have Wilson's conception of scientific theories as facades. A facade is composed of several patches. What distinguishes one patch from another are the rules that guide the application of predicates within that patch. A central motivating example for this approach is the boundary layer theory that we have described. After outlining it, Wilson explains "the general problem of variable reduction" as a problem of moving from a "high dimensional phase space" representation of a given system that is completely unworkable to "a set of *reduced variables* that can efficiently capture the *main features* of our swarm's complicated behavior" (2006, p. 186). In our case, we start with the relatively realistic Navier–Stokes equations, but these are not able to contribute to a representation we can actually work with.[16] To achieve variable reduction, we divide the fluid flow into two patches. In the upper region, we can have the Euler equation because viscosity is neglected. In the boundary layer, we include viscosity, but the high Reynolds number (see Appendix A1), allows us to manage the resulting complexity. These two patches thus correspond to inconsistent representations of the fluid flow.[17] We make the two representations consistent along their edges by stipulating that the relevant physical magnitudes agree along that line, for example, pressure.

Given this reconstruction of this representation, the problem remains to say exactly how the world has to be in order for the representation to be true. One concern is the reference of the predicates across patches. If the rules governing each patch are so different, then how might we make sense of the claim that the predicates appearing in each patch pick out the same physical property? Here we have an instance of Wilson's main challenge to the classical picture of the predicate–property relationship. Similarly, what status does the edge between the two regions really have? Consider the claim that "Around a typical airplane wing it [the thickness of the boundary layer] is of order of a centimeter" by Kundu and Cohen (2008, p. 341). This seems to mean that the edge represents a genuine physical difference between the boundary layer and the outer region. Still, there is no real physical variation in the fluid that we can easily match to this edge, for example, it is not as if the fluid suddenly becomes non-viscous when it gets a certain distance from the object. Kundu and Cohen's estimate clearly comes from the value of δ that was adopted in the scaling procedure (see Appendix A2).[18] But this choice of δ seemed to be motivated more by mathematical considerations than physical considerations. The crucial difference between the outer region and the boundary layer is that the velocity decreases when we enter the boundary layer, as in Figure 8.1. But the physical causes of this change are not given by our representation, and this complicates the determination of a location where the boundary layer "really" begins. In our brief discussion of singular perturbation theory, we also saw that the relationship between the outer region and the boundary layer is more complicated than it might initially appear.[19] According to Wilson, what we have here is really an instance of what he calls "physics avoidance" whereby the representation omits any reference to the complex physical interactions between two patches. Without physics avoidance and other associated methods of variable reduction, usable scientific representations would no longer be possible. It follows that the content of the boundary layer representation is much more nuanced than an initial presentation might suggest.

Without endorsing Wilson's complicated proposal, it seems clear that there is a genuine philosophical problem associated with representational content and applied mathematics. Extended investigation is needed to determine what consequences this has for scientific representation and for representation more generally. For our purposes here, though, the main result from Wilson's argument is the possibility that the semantic link between predicate and property might work quite differently for pure and applied mathematics. As I said above, Wilson takes these cases from applied mathematics to undermine the classical picture across the board, and he may wish to extend the lessons from reflection on boundary layer theory to number theory, set theory and other areas of pure mathematics.[20] If this was right, then mathematical terms could not be assigned a definite referent across contexts. This would complicate arguments about the metaphysical commitments of mathematical language. Despite this, it remains possible that the classical picture is correct for pure mathematics if it turns out that these complications do not arise when we use mathematical language in pure contexts. That is, there may be a semantic difference between pure and applied mathematics. But it will only be when philosophers of mathematics turn their focus to applied mathematics in its own right that these differences, if they exist, will be found and clarified.[21]

V

I close with some brief remarks about the metaphysical problems raised by applied mathematics. The main metaphysical issue is the relationship between the entities studied in applied mathematics and the entities studied in more ordinary pure mathematics. I want to suggest that with applied mathematics we have an open-endedness that seems to be different from what we find in ordinary pure mathematics. As a result, the study of applied mathematics can call into question certain overly simplified pictures of what the metaphysics of mathematics amounts to.

To start with, I assume that the well-established domains of mathematics like analysis and algebra have their own entities like the natural numbers, real numbers, groups and vector spaces. These theories are sometimes axiomatized and studied in a fully rigorous fashion. By contrast, applied mathematics proceeds in a more tentative and non-axiomatic fashion. New techniques and innovations continually project the applied mathematician into new domains where peculiar entities are encountered whose genuine metaphysical status is often unclear. When these developments in applied mathematics are successful, I would insist that we are studying new kinds of mathematical entities. These new entities are related in different ways back to ordinary mathematical entities, and it is in this way that applied mathematics can shed new light on ordinary pure mathematics. The most well-known example of this phenomena is the Dirac delta function (Urquhart 2008, pp. 429–431). In our case study, we have a newly introduced transformation between two sets of equations. This new entity contributes to a successful applied mathematical practice, but its mathematical features remain opaque. For example, it is not clear when the transformation is valid and when it will fail to preserve the relevant aspects of the solutions at issue.

On the resulting picture, mathematics has an open-ended character in at least two directions. In one direction, there is the link between the demands of the sciences, in all of their theoretical, experimental and applied aspects, and the discovery of new entities in applied mathematics.[22] In the other direction, there is the attempt to axiomatize, or make rigorous by other means, the study of the new entities found in applied mathematics. Here, the task is to relate the entities from applied mathematics back to the better understood subject-matter of pure mathematics. I would not insist that it is necessary that we do this, and the successful practice of applied mathematics seems to require that it go through periods where its standards of rigor are less than what a pure mathematician might prefer. The interaction between the sciences, applied mathematics and pure mathematics, then, could bring us to countenance new kinds of entities that stand in novel metaphysical relationships to the entities already found in our ontology.

Somewhat surprisingly, this reveals a respect in which foundational disciplines like set theory are similar to applied mathematics. In set theory, there are widely accepted axioms, like ZFC, but the active research in set theory often involves considering new axioms and their consequences. For this reason, it is naïve to think that sets are adequately characterized by the ZFC axioms. The foundations of mathematics, then, has an open-ended metaphysical character that is more apparent than in ordinary mathematics. Still, set theory is not an isolated mathematical discipline, and so whatever open-endedness it has should inform our understanding of mathematics more generally. Among other things, it is well-known that stronger and stronger axioms of set theory resolve questions about other mathematical objects, for instance the natural numbers, that are independent of their usual axiomatizations. The metaphysical significance of this relationship remains unclear and the subject of continuing philosophical reflection.

We arrive at a picture, then, where ordinary mathematical practice sits at the center of a network of disciplines that includes both applied mathematics and foundations. The philosopher of mathematics needs to study not only the pure mathematics at the heart of this network, but also the interactions between pure mathematics and its neighbors. This proposal receives some independent support from the striking cases assembled by Urquhart in his "Mathematics and Physics: Strategies of Assimilation" (2008). He relates how innovative techniques from physics have not only been put on a rigorous foundation by mathematicians, but have also led to mathematical developments of independent interest. Urquhart closes by

> urging logicians and philosophers to look beyond the conventional boundaries of standard mathematics for mathematical work that is interesting but non-rigorous, with a view to making it rigorous. The examples...all show that some of the best mathematics can result from this process. Similarly, philosophers can surely find fruitful areas for their studies in areas lying beyond the usual set-theoretical pale (p. 437).

I agree. It is only in this way that we can hope to understand what mathematical objects really are and how they might relate to other kinds of objects. While it is premature to claim that the study of applied mathematics will prove as philosophically important as the study of foundations, I hope it is clear why I think more philosophers of mathematics should turn their attention in this new direction.[23]

Appendix A. Two scaling procedures

A1. From the Navier–Stokes equations to the Euler equations

For our simplified situation our starting equations are

$$u\,\partial u/\partial x + v\,\partial u/\partial y = -1/\rho\,\partial p/\partial x + \nu(\partial^2 u/\partial x^2 + \partial^2 u/\partial y^2) \tag{1}$$

$$u\,\partial v/\partial x + v\,\partial v/\partial y = -1/\rho\,\partial p/\partial y + \nu(\partial^2 v/\partial x^2 + \partial^2 v/\partial y^2) \tag{2}$$

$$\partial u/\partial x + \partial v/\partial y = 0 \tag{3}$$

$$u(0,y) = U_\infty, v(0,y) = 0 \tag{4}$$

We select the following scales:

$$x' = x/L, y' = y/L, u' = u/U_\infty, v' = v/U_\infty, p' = p/\rho U_\infty^{\,2} \tag{5E}$$

The scale for pressure may seem ad hoc, but it is chosen based on the units of pressure and so that the resulting $\partial p'/\partial x'$ term is of the same order of magnitude as the other terms. Making the substitution produces

$$(U_\infty^{\,2}/L)u'\,\partial u'/\partial x' + (U_\infty^{\,2}/L)\,v'\,\partial u'/\partial y' = -(U_\infty^{\,2}/L)\partial p'/\partial x' \tag{1'}$$
$$+ (\nu U_\infty/L^2)\,(\partial^2 u'/\partial x'^2 + \partial^2 u'/\partial y'^2)$$

$$(U_\infty^{\,2}/L)u'\partial v'/\partial x' + (U_\infty^{\,2}/L)v'\,\partial v'/\partial y' = -(U_\infty^{\,2}/L)\partial p'/\partial x' \tag{2'}$$
$$+ (\nu U_\infty/L^2)\,(\partial^2 v'/\partial x'^2 + \partial^2 v'/\partial y'^2)$$

which simplifies to

$$u'\,\partial u'/\partial x' + v'\,\partial u'/\partial y' = -\partial p'/\partial x' + (\nu/U_\infty L)(\partial^2 u'/\partial x'^2 + \partial^2 u'/\partial y'^2) \tag{1''}$$

$$u'\,\partial v'/\partial x' + v'\,\partial v'/\partial y' = -\partial p'/\partial x' + (\nu/U_\infty L)(\partial^2 v'/\partial x'^2 + \partial^2 v'/\partial y'^2) \tag{2''}$$

Here we see the presence of $\nu/U_\infty\,L$, which is a dimensionless quantity that is characteristic of a given flow. Its reciprocal is known as the Reynold's number:

$$\mathrm{Re} \equiv U_\infty L/\nu$$

For plausible values, we saw that this quantity is large, so its inverse is small. If we have chosen the right scales, then all the terms of the equations involving our functions are of the same order of magnitude. The $\nu/U_\infty L = 1/\mathrm{Re}$ in front of one of these terms tells us that the product is of a different order of magnitude, and so it can be safely dropped.

A2. From the Navier–Stokes equations to the boundary layer equations

We work with a new quantity δ, which we can think of as the width of the boundary layer. We can estimate δ by first applying the following set of scales, and then seeing what it must be comparable to for the resulting terms to balance:

$$x^* = x/L, y^* = y/\delta, u^* = u/U_\infty, v^* = v/(U_\infty\delta/L), p^* = p/\rho U_\infty^2 \qquad (5B)$$

This results in the equations:

$$(U_\infty^2/L)\,u^*\,\partial u^*/\partial x^* + (U_\infty^2/L)\,v^*\,\partial u^*/\partial y^* = -(U_\infty^2/L)\,\partial p^*/\partial x^* + \qquad (1^*)$$
$$(U_\infty^2 v/L)\,\partial^2 u^*/\partial x^{*2} + (U_\infty^2 v/\delta^2)\,\partial^2 u^*/\partial y^{*2}$$

$$(U_\infty^2\,\delta/L^2)\,u^*\partial v^*/\partial x^* + (U_\infty^2\,\delta/L^2)\,v^*\partial v^*/\partial y^* = -(U_\infty^2/\delta)\partial p^*/\partial y^* + \qquad (2^*)$$
$$(U_\infty\,\delta v/L^3)\partial^2 v^*/\partial x^{*2} + (U_\infty\,v/\delta L)\partial^2 v^*/\partial y^{*2}$$

Looking to (1*), we want at least one of the v terms to be of the same order of magnitude as the other terms and only $(U_\infty^2 v/\delta^2)\,\partial^2 u^*/\partial y^{*2}$ has our so far undetermined scale δ. We can fix δ by requiring

$$(U_\infty^2/L) = (U_\infty^2 v/\delta^2)$$

which simplifies to

$$\delta = L/\sqrt{Re} \qquad (\delta)$$

In turn, this allows us to rewrite (1*) and (2*) as

$$u^*\,\partial u^*/\partial x^* + v^*\,\partial u^*/\partial y^* = -\partial p^*/\partial x^* + 1/Re\,\partial^2 u^*/\partial x^{*2} + \partial^2 u^*/\partial y^{*2} \qquad (1^{**})$$
$$(1/Re)u^*\,\partial v^*/\partial x^* + (1/Re)v^*\,\partial v^*/\partial y^* = -\partial p^*/\partial y^* + \qquad (2^{**})$$
$$(1/Re^2)\partial^2 v^*/\partial x^{*2} + (1/Re)\partial^2 v^*/\partial y^{*2}$$

As Re is much larger than 1, it follows that $1/Re$ and $(1/Re^2)$ are much smaller than 1 so the corresponding terms can be dropped. This results in the boundary layer equations:

$$u^*\,\partial u^*/\partial x^* + v^*\,\partial u^*/\partial y^* = -\partial p^*/\partial x^* + \partial^2 u^*/\partial y^{*2} \qquad (1B)$$
$$\partial p^*/\partial y^* = 0 \qquad (2B)$$

Notes

1. While Maddy discusses applications of mathematics in Maddy (2007), chapter IV.2, her discussion does not focus on applied mathematics as a discipline within mathematics.
2. Historical methods would also be useful, but I have not employed them here.
3. http://www.siam.org/books/series/cl.php
4. http://www.cambridge.org/series/sSeries.asp?code=CTAM&srt=P
5. The notion of mechanism here remains unclear. As the example given below indicates, not all adequate mathematical models represent fundamental causal mechanisms. (This passage is partly quoted in Chihara 2004, p. 263.)
6. The remaining books in these series seem to fit this framework as they are either further elaborations of the mathematical theories introduced in the survey textbooks or else detailed investigations of particular kinds of mathematical models of physical systems.
7. See Heidelberger (2006) for an important discussion on this case and its relevance for the ongoing debates in the philosophy of science about modeling.
8. My exposition follows most closely the presentation of Segel (2007, chapter 3). This is the companion volume to Lin and Segel (1988), where chapters 6, 7 and 9 provide an instructive background. Kundu and Cohen's discussion is also excellent, but their summary of the link between boundary layer theory and perturbation theory is less detailed. See also Fowler (1997, chapter 4).
9. Here, I offer only a simplified "textbook" history. An insightful historical reconstruction of Prandtl's discovery can be found in Darrigol (2005, chapter 7).
10. The viscosity ν should not be confused with the velocity v.
11. See also Fowler (1997, p. 20).
12. It is worth emphasizing that the boundary layer approach is only one part of singular perturbation theory. Fowler (1997, p. 49), for example, also discusses interior layer techniques. Batterman has discussed this sort of reasoning from the more general perspective of "asymptotics of the second kind". See Batterman (2002, p. 56) for an explicit link to singular perturbation theory.
13. Cited by Vincenti and Bloor (2003, p. 500, fn. 28).
14. See Baker (2008) for an important discussion of experimental mathematics.
15. A pioneering effort, in a somewhat different direction, was made by Steiner (1998). Several philosophers of science have made similar claims involving applied mathematics, but they do not always focus on the mathematics itself. I am most sympathetic to the work being done along these lines by Batterman. See his 2002 and 2005. Wilson (2000) presents a criticism of Cartwright's philosophy of science based upon her ignoring applied mathematics' current understanding of the reasoning steps involved in commonplace physical procedures. Sheldon Smith (2007) also has made strides in the "philosophy of applied mathematics".
16. Wilson (2006, p. 185) notes that "utilizing our highest capacity computers, for example, a smooth N–S [Navier–Stokes] solution can be projected about 1/5 of a second into the future, after which accumulated roundoff error completely swamps the validity of our results".
17. See Colyvan's essay in Chapter 7 for further discussion of inconsistency.
18. Using the numbers considered earlier, $\delta = 0.03$ cm.
19. Furthermore, taking the boundary layer too seriously leads to a breakdown in the representation of the fluid at the point of initial separation between the boundary layer and outer region. See Stewartson (1981, p. 312) as well as Birkhoff (1960, p. 40, fn. 21).

20. There is little discussion of pure mathematics in *Wandering Significance*, but the 2000 paper seems to extend the point to some areas of pure mathematics.
21. Cf. Maddy's discussion of Wilson (2007, chapter II.6).
22. I would distinguish theoretical physics, for example, from applied mathematics based on the fact that the techniques developed in applied mathematics extend beyond physics to other sciences. The links here are complicated, as is the link between applied mathematics and what Fowler terms the "applied sciences".
23. An earlier version of this paper was presented at the conference *New Waves in Philosophy of Mathematics* at the University of Miami in April, 2008. I would like to thank the organizers of the conference and the audience for their helpful suggestions. In addition, Alan Baker, Robert Batterman, Roy Cook, Øystein Linnebo, Paolo Mancosu and Mark Wilson offered generous comments and encouragement.

References

Baker, Alan (2008), "Experimental Mathematics", *Erkenntnis* 68: 331–344.

Batterman, Robert (2002), *The Devil in the Details: Asymptotic Reasoning in Explanation, Reduction, and Emergence*, Oxford University Press: New York, NY.

Batterman, Robert (2005), "Critical Phenomena and Breaking Drops: Infinite Idealizations in Physics", *Studies in History and Philosophy of Modern Physics* 36: 225–244.

Birkhoff, Garrett (1960), *Hydrodynamics: A Study in Logic, Fact and Similitude*, Revised edition, Princeton University Press: Princeton, NJ.

Chihara, Charles (2004), *A Structuralist Account of Mathematics*, Oxford University Press: New York, NY.

Corfield, David (2003), *Towards a Philosophy of Real Mathematics*, Cambridge University Press: Cambridge, UK.

Darrigol, Olivier (2005), *Worlds of Flow: A History of Hydrodynamics from the Bernoullis to Prandtl*, Oxford University Press: New York, NY.

Fowler, A.C. (1997), *Mathematical Models in the Applied Sciences*, Cambridge Texts in Applied Mathematics, Cambridge University Press: Cambridge, UK.

Heidelberger, Michael (2006), "Applying Models in Fluid Dynamics", *International Studies in the Philosophy of Science* 20: 46–67.

Kundu, Pijush K. and Ira M. Cohen (2008), *Fluid Mechanics*, Fourth edition, Academic Press: Boston, MA.

Lin, C.C. and L.A. Segel (1988), *Mathematics Applied to Deterministic Problems in the Natural Sciences*, Classics in Applied Mathematics, Vol. 1, SIAM: Philadelphia, PA.

Maddy, Penelope (2007), *Second Philosophy*, Oxford University Press: New York, NY.

Mancosu, Paolo (ed.) (2008a), *The Philosophy of Mathematical Practice*, Oxford University Press: New York, NY.

Mancosu, Paolo (2008b), "Introduction", in Mancosu (ed.) (2008a), 1–21.

Nickel, Karl (1973), "Prandtl's Boundary-Layer Theory from the Viewpoint of a Mathematician", *Annual Review of Fluid Mechanics* 5: 405–428.

Segel, L.A. (2007), *Mathematics Applied to Continuum Mechanics*, Classics in Applied Mathematics, Vol. 52, SIAM: Philadelphia, PA.

Smith, Sheldon (2007), "Continuous Bodies, Impenetrability, and Contact Interactions: The View from the Applied Mathematics of Continuum Mechanics", *British Journal for the Philosophy of Science* 58: 503–538.

Stewartson, Keith (1981), "D'Alembert's Paradox", *SIAM Review* 23: 308–343.

Steiner, Mark (1998), *The Applicability of Mathematics as a Philosophical Problem*, Harvard University Press: Cambridge, MA.

Urquhart, Alasdair (2008), "Mathematics and Physics: Strategies of Assimilation", in Mancosu (ed.) (2008a), 417–440.

Vincenti, Walter G. and David Bloor (2003), "Boundaries, Contingencies and Rigor: Thoughts on Mathematics Prompted by a Case Study in Transonic Aerodynamics", *Social Studies of Science* 33: 469–507.

Wilson, Mark (2000), "The Unreasonable Uncooperativeness of Mathematics in the Natural Sciences", *Monist* 83: 296–314.

Wilson, Mark (2006), *Wandering Significance: An Essay on Conceptual Behavior*, Oxford University Press: New York, NY.

Part IV

Mathematical Language and the Psychology of Mathematics

9
Formal Tools and the Philosophy of Mathematics

Thomas Hofweber

1 How can we do better?

In this chapter, I won't try to defend a particular philosophical view about mathematics, but, in the spirit of *New Waves*, I would instead like to think about how we can hope to make real progress in the field. And to do that it seems best to think about what in the field as it is now is holding up progress. What in the field should change so that we can hope to do better than what has been done before? Having considered the question, I propose the following answer: The single biggest obstacle to real progress in the philosophy of mathematics is a lack of reflection on which questions should be addressed with what methods. And this is particularly apparent in the role that formal tools have in the discipline as it is today. Formal tools encompass paradigmatically formal, artificial languages, formal logic expressed with such languages, and mathematical proofs about such languages. These tools were developed during the rise of logic over 100 years ago, and they are ubiquitous in the philosophy of mathematics today. But even though formal tools have been used with great success in other parts of inquiry, in the philosophy of mathematics they have done a lot of harm, besides quite a bit of good. In my contribution to this volume, I would thus like to think a bit about what role formal tools should have in the philosophy of mathematics. I will argue that they should have merely a secondary role, unless one holds certain substantial views in the philosophy of mathematics, ones that very few people hold. The role of formal tools is thus tied to substantial questions in the philosophy of mathematics.

Much work done under the heading of "philosophy of mathematics" consists of proofs, precise mathematical proofs. Many talks at philosophy of mathematics conferences present new proofs, and often a proof is taken to be the surest sign of progress in the field. But this should be a bit puzzling. Proof is the method to establish results in mathematics. But it is rather unusual that the method to achieve results in the philosophy of X, some discipline, is the same as the method for achieving results in X. For example, physics achieves results via experimentation, amongst other methods. But the philosophy of physics does not. It might learn about the results of physics, achieved amongst others via experimentation. But it itself does not proceed via experimentation to do whatever it hopes to do. In general the method of the philosophy of X is distinct from the method of X, and this might be particularly compelling when X has a very distinct method, as does mathematics with that of precise proof. So, why should philosophers of mathematics hope to achieve results with the same method with which the discipline they hope to understand achieves its results?

The common justification for the role of proof in the philosophy of mathematics is this: metamathematics! Mathematical languages can be formulated with mathematical precision, and the notion of a proof can be characterized precisely as a consequence. Given this we can carry out proofs about proofs, including proofs about what can't be proven. And this justifies the central role of proofs and formal tools in the philosophy of mathematics, or so the story goes. This story is partly right, but also full of holes and sources of error. In this chapter, we will have a closer look at what role formal tools should have in the philosophy of mathematics. We will discuss both the use of formal languages and the use proof about them in metamathematics.

The role of formal tools depends crucially on what the answer is to two other questions: first, the question of the status of axioms in mathematics, and, second, the question whether or not all of mathematics is philosophically equal. We will discuss these two questions in more detail below. Thus these three questions are closely related:

1. What is the proper role of formal tools?
2. What is the status of axioms?
3. Is all of mathematics philosophically equal?

In the end, I will argue that much damage has been done in the philosophy of mathematics by relying on formal tools. Their role in the philosophy of mathematics is limited, unless one holds a certain position with respect to the other two questions. These answers to the other two questions are not widely accepted, but I do accept them. Even though I hold that the role of formal tools is limited, I think most philosophers should think that it is even more limited.

Much of the discussion in the following will have a special emphasis on formal tools in the philosophy of arithmetic, but this is merely our example. The general issues carry over to other parts of mathematics as well, and the conclusion I hope to reach is general.

2   Formal tools

Mathematics is first and foremost an activity carried out in ordinary natural language with the use of symbols. Mathematical conversations are carried out in natural language, mathematical papers are written in natural language, but with the additional use of symbols. When we wonder about the role of formal tools, what is not at issue is the role of symbols. To make this clear, let us distinguish three different uses of, for example, what is pronounced "four" in mathematics. First there is the *natural language expression* "four". It is part of ordinary English, and it occurs in ordinary English sentences as

(1) You shouldn't drink more than four beers.

Secondly, there is the symbolic use of "4". It will occur in simple mathematics, as when an elementary school teacher writes on a blackboard

(2) What is 4+2?

And it also occurs in serious arithmetic, of course. How is the *symbolic use* of "4" related to the English word "four"? It is tempting to think that it is merely a symbolic abbreviation of English word. After all, for every number word in English there is a symbolic one that is pronounced just the same way. And symbolic uses of number words occur in ordinary written English, as in

(3) Over their life, the average Bavarian drinks 15432 beers.

Finally, there is the *formal use* of "4", as it occurs in the artificial languages of formal logic.[1] Such languages are introduced and defined by logicians, and various expressions in it are given a meaning. Even if symbolic uses of number words are merely abbreviations of natural language expressions, there is a real question as to how formal uses of number words relate to the other ones.

This division of three different uses carries over to a variety of other expressions as well. For example, "for all" and symbolic and formal uses of "∀". As a symbol it can merely be an abbreviation of a natural language expression. But as a formal expression, there is an issue how it relates to the other two uses.

When we are wondering what role formal tools should have in the philosophy of mathematics, we are not wondering what role symbols should have, but what role in particular artificial languages should have. Such artificial languages are perfectly precise and thus subject to mathematical proofs about them. Much technical work in the philosophy of mathematics is closely related to mathematical proofs about such artificial languages. The question will be what role such proofs should have. But first, lets make clearer what is and what isn't at issue.

3   Two projects in the philosophy of mathematics

When we ask what role formal tools should have in the philosophy of mathematics it of course depends on what one means by the philosophy of mathematics. The philosophy of mathematics is a diverse discipline and in a sense formal tools clearly play a central role in some parts. So, there is an uncontroversial case in favor of formal tools, and there is a controversial case about their status. Let us distinguish two large-scale projects, both usually carried out under the umbrella of the philosophy of mathematics. One of them is *foundational*. The foundational project is directly concerned with issues tied to formal mathematical theories. It tries to find out, say, which axioms of a formal theory are needed to prove what theorems. Or, whether a certain formal theory can be represented in another. Or, whether a certain sentence is independent of a certain set of axioms. Much great work is done here, and all of it is essentially tied to using formal languages, formal logic, and formal tools more generally. This is the uncontroversial part. Next, there is the *large-scale philosophical* project. This project concerns the large-scale philosophical questions related to such notions as truth, knowledge, fact of the matter, as they apply to mathematics, and questions about the large-scale relationship between mathematics and other parts of inquiry, and so on. These are the questions that philosophy is traditionally concerned with, and here they are directed at mathematics. These questions are not directly about any formal languages, and so here there is a real issue whether formal tools play or should play a role in making progress in answering them. This is our first topic, and this is the more controversial part. This question is related to, but not exactly the same as, what role the foundational project should have in the large-scale philosophical project. The foundational project does not, of course, need to have any role in the large-scale philosophical one to be a good and legitimate project. There are other interesting things in the world than large-scale philosophical questions. But many people who engage in the foundational project have the ambition for their work to be of significance for the large-scale philosophical project as well. And there might well be a role for formal tools in the large-scale philosophical project that goes beyond the foundational project. What is at issue for us here is the role of formal tools in the large-scale philosophical project, which is related to, but not quite the same, as the role of the foundational project in the large-scale philosophical project. Many philosophers throughout the history of the philosophy of mathematics have given formal tools a central role in the project of finding answers to the large-scale philosophical questions. A clear example is Hilbert, who thought that the main missing part for the defense of a certain large-scale philosophical picture was a certain technical result: a proof of the consistency of a certain formal axiom system by finitistic means. Once that had been done, Hilbert thought, a certain large-scale picture of mathematics would be vindicated.[2] We will see more about this below.

Next, we should also distinguish a *substantial role* from an *insubstantial role*. Formal tools might be useful as tools of representation. We might use them to represent something or other about our mathematical activity or language, and such a representation might be useful in various ways. There is no question that formal tools can be useful this way, but this is insubstantial. But they might also be used to do something that can't easily be done without them, they might really make a substantial difference. They might play a crucial role in achieving results, not just representing them. Whether formal tools can have such a substantial role is what is at issue.

Finally, we should distinguish a *revisionary* from a *descriptive* role of formal tools. Someone might hold that formal tools should play a central role in the philosophy of mathematics since mathematics as it is presently done should be replaced with a discipline that relies exclusively on formal languages. But this revisionary approach will simply be sidelined here. What is at issue is to understand mathematics from a philosophical point of view, as it actually is. We are concerned with the descriptive project, not the revisionary one, in part because it is hard to see how the revisionary could be made plausible, given how well mathematics is doing. Revision in any discipline should be triggered by some problem in that discipline. Mathematics as it is today does not seem to need a large-scale revision.

Whether formal tools should have a substantial role in the descriptive, large-scale philosophical project is what is at issue. We will divide the discussion by considering various options where one might think formal tools might have such a role in the large-scale philosophical project.

4 Mathematical activity

Mathematics is, amongst other things, an activity. It is something people do, within a certain community, with results, winners and losers, and all that. To understand this activity, one might employ the social sciences. And the social sciences might rely on formal methods in their study of the social aspect of mathematics. Of course, such formal methods would not likely involve the formal tools we talked about above, namely artificial languages. They will more likely be formal methods from statistics to, for example, show that problems posed by mathematicians with tenure are given more weight than ones posed by those without tenure. But I take it that reliance on results from the social sciences that are achieved with the help of formal methods, like the ones from statistics, are only insubstantial for showing the role of formal tools in the large-scale philosophy of mathematics. Even if these tools are substantial parts of the social sciences, it remains to be seen why the results of the social sciences are a substantial part of the philosophy of mathematics. It is a big step from the sociology of any science to the philosophy of that science, and it seems no different in the philosophy of mathematics. In any case, though, this is not what anyone has in mind when they want to defend the role of formal tools in the philosophy of mathematics. What is more crucial for us is the role of the formal tools of formal logic, the use of artificial languages in the philosophy of mathematics.

What the activity of mathematics is like is a largely empirical question. It is to be studied like other questions about the activity of individuals or groups of individuals. This simple point seems to be somewhat under-appreciated, though. Some positions that are hotly debated in the philosophy of mathematics are nothing more than questions about mathematical activity, and thus largely empirical questions. However, in the philosophical debate they are hardly treated as such. Take, for example, the debate about fictionalism. Most philosophical defenses of fictionalism are not in the ballpark of what kind of considerations are required to defend it. At most, they can be taken to argue that fictionalism is coherent or that it is a route to nominalism or that it is compatible with this, that, or the other philosophical position. But to argue that it is true one has to argue that actual utterances involve a pretense, or that mathematical activity in fact involves a certain stance by its participants, or the like. These are empirical questions, ones that can be properly addressed in psychology or the social sciences. But defenses of fictionalism hardly even consider these.

I will suggest below that several questions that are traditionally considered part of the philosophy of mathematics are in fact solidly empirical questions. That doesn't mean that they should not be considered part of the philosophy of mathematics any more, but it should reflect on the methods with which they should be addressed.

5 Mathematical language

Traditionally, the role of formal tools comes through the analysis of mathematical language. The language of first-order arithmetic is taken to represent, or analyze, the language used in mathematical practice. And important features of the language of mathematical practice can be read off, and are made explicit by, the language of first-order arithmetic. This move, from the language used in mathematics to the language of first-order arithmetic, is so common that it is hardly made explicit. In the early days of our discipline, that is, after Frege, this was more explicitly discussed, but the transition has mostly moved to the background.

This is where the real damage is done in the use of formal tools in the philosophy of mathematics. Consider, for example, the language in which arithmetic is ordinarily carried out, and the language of first-order Peano Arithmetic, $\mathcal{L}_{PA}$. $\mathcal{L}_{PA}$ here is an artificial language in standard first-order logic, with constant symbols, function symbols, quantifiers and variables, and so on.[3] $\mathcal{L}_{PA}$ does about as badly as it could as a way to analyze the language of arithmetic. $\mathcal{L}_{PA}$ gets almost everything wrong about the language in which arithmetic is ordinarily carried out. I will explain this in this section. But still, I will argue in section 6 that $\mathcal{L}_{PA}$ is supremely useful and good, even though it gets almost everything about the language of arithmetic wrong. What it is good at, and why it doesn't matter for this that it gets so much wrong, will be discussed after we see what it gets wrong.

5.1 Syntax

When we take $\mathcal{L}_{PA}$, the language of first order Peano Arithmetic, as an analysis of mathematical language, there is clearly a lot that it gets badly wrong. For example, in natural language, quantified noun phrases can be arguments of predicates, like transitive verbs:

(4) John kicked a man.

which has a syntactic argument structure like this

(5) kick(John, a man)

But in first-order logic only terms can be arguments of predicates, and quantifiers aren't terms. So, instead of the quantifier phrase "a man" we have to make a variable the argument, and put the quantifier somewhere else:

(6) $\exists x(kick(John, x) \wedge man(x))$

Similarly, the noun phrase (NP)–verb phrase (VP) structure of natural language sentences gets badly misrepresented in first-order languages. First-order languages do not split up sentences into NPs and VPs in the same way in which natural languages do. They both have a predicate–argument structure in general, but what is a predicate and what is an argument is quite different. More sophisticated formal theories of natural language have moved on, away from first-order languages, to ones more capable of representing NP–VP structure, predicate–argument structure, and so on.

This failing of first-order languages very much applies to first-order arithmetical theories. Consider a mathematical statements in natural language, for example,

(7) Every number is less than some number.

It has a a clear argument structure, splits up into a (quantified) noun phrase and a verb phrase, and so on. But the first-order statement in $\mathcal{L}_{PA}$

(8) $\forall x \exists y (x < y)$

completely misrepresents these aspects of the English sentence. Other formal languages do much better at getting that aspect of the syntax of English right, but such languages are in many other ways more complicated, and less suitable for what $\mathcal{L}_{PA}$ is good for. Even though $\mathcal{L}_{PA}$ does badly when it comes to the syntax of the language of mathematics, $\mathcal{L}_{PA}$ nonetheless is obviously a great formal tool. It gets something right, but not that.

To say this is not to dismiss formal language in general, in particular when it comes to understanding the syntax of natural languages. There is no question that there are formal languages that capture various syntactic aspects of natural languages well. But $\mathcal{L}_{PA}$ is not one of them. And, as we will see below, $\mathcal{L}_{PA}$ is really good for something, and those languages that capture the syntax of natural languages better aren't as good at that as $\mathcal{L}_{PA}$. The lesson so far simply is as follows: don't hope to learn about the syntax of ordinary talk about numbers from the syntax of $\mathcal{L}_{PA}$. The syntax of ordinary talk about numbers is studied in syntax, the sub-discipline of linguistics, and is then well or badly captured in various formal languages. But to find out what the syntax of ordinary talk about numbers is we have to look at linguistics. And to find out how well a certain formal language captures it, we have to first look at linguistics, and then compare what it tells us with the formal languages.

5.2 Number words

$\mathcal{L}_{PA}$ treats number words as either constants (in the case of "0") or complex terms (in the case of all other number words). All terms in first-order logic pick out an object in any model, and so number words are treated as expressions that pick out objects, in other words as referring expressions. That number words are referring expressions is widely agreed upon among philosophers of mathematics. That is, that number words aim to refer is agreed upon by philosophers of mathematics. Whether or not they succeed in referring to anything is very controversial, at least in literal uses. But whether or not number words in natural language are referring expressions is far from clear. As Frege had already observed in his ground-breaking *Grundlagen* (1884), there is a puzzle about the uses of number words in natural language, which we can call *Frege's other Puzzle*,[4] unrelated to *Frege's Puzzle*, which is about belief ascriptions and identity statements.[5] In natural language, number words can occur like adjectives, modifying nouns:

(9) Jupiter has four moons.

but apparently they can also occur like names or singular terms:

(10) The number of moons of Jupiter is four.

But, in general, the grammatical category of a singular term is different from that of an adjective. In general, a singular term cannot occur in the position of an adjective, and the other way round. How come number words can do this?

Although almost all philosophers of mathematics take number words to be referring expressions, like proper names, linguists usually focus on number words as they are used in completely different ways, for example, as parts of quantified noun phrases. And both groups have good reasons for their point of view. There is a real puzzle here.

I have argued in Hofweber (2005a, 2007b) that number words in natural language are not referring expressions. They are rather determiners that appear syntactically in the position of a singular terms for a variety of different reasons. Furthermore, number words in arithmetic are merely symbolic abbreviations of natural language number words, and thus they, too, are not referring expressions. In addition, arithmetic statements are literally true, although no reference is even attempted in them. Since there are different reasons why number words appear syntactically as singular terms, there are different cases to consider. The reason why number words appear as singular terms in examples like (10) is quite different from the reason why they appear as singular terms in arithmetical statements. I will briefly give those reasons in the following.

In Hofweber (2007b), I argued that the difference between (9) and (10) is one analogous to the difference between

(11) Otávio knows judo.

and

(12) Its Otávio who knows judo.

Both communicate the same information, but they do so with a different emphasis or focus. In addition, the focus effect is the result of the syntactic structure, not the result of special intonation as in

(13) OTÁVIO knows judo.

Similarly, the difference between (9) and (10) is one of a focus effect that is the result of the syntactic structure of (10). I have also argued that explanation for why there is a focus effect in (10) involves that the number word "four" in it is still a determiner, as it is in (9), but one that was displaced from its usual syntactic position. If this is correct, then "four" is not a referring expression in (10).

The more important case for the philosophy of mathematics is, however, a completely different one. I here disagree with Frege and neo-Fregeans, who take great inspiration from examples like (10) for the philosophy of arithmetic. The important case for arithmetic is the occurrence of number words in arithmetical statements, like

(14) $2 + 2 = 4$

In Hofweber (2005a), I have argued, on largely empirical grounds, that even there number words are not referential expressions, although such statements are literally true. One key to this argument is to notice that (14) when read out aloud is commonly read in two different kinds of ways, one corresponding to a plural reading, the other to a singular one:

(15) Two and two are four.
(16) Two and two is four.

and similarly using "plus", or "make". In Hofweber (2005a), I argued that (i) in the plural reading number words are "bare determiners", determiners without their noun argument present, and that (ii) the singular reading is the result of "cognitive type coercion", coercing the syntactic form of the plural statement into a different one for cognitive reason, thereby leaving the semantic function of the relevant expressions untouched. I won't repeat the details, but if it is correct, then even in arithmetical equations like (14) number words are still determiners, and not referential expressions. In addition, it follows that arithmetical statements are literally true, and their truth is independent of what there is, including how many things there are.[6]

My point for the rest of this chapter doesn't depend on that you believe any of this about the language of arithmetic. It is rather that there is a real issue whether or not $\mathcal{L}_{PA}$ correctly captures the semantic function of number words as picking out objects. And that this issue is one in the study of the language of mathematics, as it is actually carried out. This issue is not at all easily settled, and the rather simplistic pictures of natural language that were popular during the time when first-order languages ruled the study of natural language are long overthrown. What is important is to keep in mind that we need to settle this issue, in linguistics and the philosophy of language, and then, once we have settled it, we can see whether or not $\mathcal{L}_{PA}$ represents the semantic function of number words correctly. The hard work here is done somewhere else, in the study of natural language, or whatever the language of mathematics is. But unfortunately many philosophers are too quick to conclude from the facts that $\mathcal{L}_{PA}$ clearly gets something right (we will see shortly what) that it gets many other things right as well, including what the semantic function of number words is. This is a great source of error in the philosophy of mathematics. The success of $\mathcal{L}_{PA}$ as the default formal language of arithmetic has more than once led to unjustified conclusions about features of number words in natural language from those in $\mathcal{L}_{PA}$. But what $\mathcal{L}_{PA}$ gets right does not indicate one way or another whether it gets the semantics of number words right. Before we can see what it gets right, lets look at one more thing that it gets wrong.

5.3 Quantifiers

$\mathcal{L}_{PA}$ is a first order language, and as such it comes with a certain picture of quantification. On the one hand, quantification is given a certain syntactic treatment. Quantifiers are never directly arguments of predicates, for examples, contrary to ordinary English. We have already noted above that this, uncontroversially, misrepresents ordinary mathematical talk in its syntactic aspect. On the other hand, there is the semantics of quantifiers. Quantifiers in first-order logic are, standardly, given a model theoretic semantics, where they range over a domain of objects that are the objects that the terms in the language denote. This semantics of the quantifiers does fairly uncontroversially correspond to a reading that quantifiers have in ordinary natural languages, one where we make claims about whatever is out there in reality, our domain of quantification. But, I hold, this is not the only use that quantifiers have in natural language, and moreover, ordinary uses of quantification over natural numbers is not in accordance with this "domain" use of quantifiers. This can be argued for on the basis of considerations about natural language semantics and the particular uses of quantifiers over natural numbers, and it coheres with the picture of number words outlined above. I make my case for these claims in Hofweber (2005a, 2005b). Whether or not this is correct is not central in the following. But we should agree that whether or not $\mathcal{L}_{PA}$, as a first-order language, correctly captures certain aspects of ordinary quantification over numbers is an issue that is to be settled by first finding out how ordinary quantification over number works, then how it is represented in $\mathcal{L}_{PA}$, and finally to see whether $\mathcal{L}_{PA}$ represents it correctly. Quantification in natural language is much more complicated along a variety of dimensions than quantification in first or higher-order languages. It won't be too easy to figure this out, and we shouldn't draw any conclusions from how quantification works in $\mathcal{L}_{PA}$ to how it works in ordinary talk about numbers.

I hold that $\mathcal{L}_{PA}$ gets basically everything wrong about the syntax and semantics of talk about natural numbers. This, I take it, is uncontroversial for the question of the syntax, but controversial for the question of the semantics. But whether or not this is correct, the question what $\mathcal{L}_{PA}$ gets right and what it gets wrong has to be assessed via looking at the semantics of ordinary uses of quantifiers over natural numbers, and the semantic function of number words, and then to compare it to how these expressions are taken to be, semantically, in $\mathcal{L}_{PA}$. One should not draw any direct conclusions from the way they are represented in $\mathcal{L}_{PA}$ to how they are in ordinary talk about numbers.

Even though I think $\mathcal{L}_{PA}$ gets all this wrong, I do think that $\mathcal{L}_{PA}$ is great, and supremely useful, despite getting all this wrong. This is because $\mathcal{L}_{PA}$ gets something important right, and getting that right is independent of getting the syntax and semantics of number talk wrong. Here is what $\mathcal{L}_{PA}$ gets right.

6 Inferential relations

What we get right with the use of $\mathcal{L}_{PA}$ is capturing the inferential relations among the sentence of arithmetic. When we talk about numbers we use ordinary natural language. The sentences in our natural language stand in certain inferential relations to each other. Certain ones imply other ones, and there are discernible patterns of implication among them. We can properly capture those patterns of implication with the use of formal languages. First, we specify an artificial language, and some technical notion of consequence among sentences in this language. Then, we systematically assign sentences of our natural language that we use to talk about natural numbers to sentence in this artificial language such that inferential relations are exactly mirrored. Whenever an English sentence about natural numbers implies another one, in the ordinary sense of implication, then the corresponding sentence in the formal language implies, in the technical sense, the other corresponding sentence, and the other way round. This is what $\mathcal{L}_{PA}$ is good at. It manages to mirror the inferential relations among our ordinary sentences about natural numbers. And given the mathematical precision of the artificial language, it makes mathematical proofs about inferential relationships possible. Let me spell this out a little more.

Let E be a fragment of English, in particular the fragment consisting of the sentences employed in arithmetical reasoning. I take it that there is a notion of logical consequence that applies to sentences of E, call it "implies$_E$", and I take this notion as primitive for present purposes. Let L be an artificial language, say $\mathcal{L}_{PA}$. I assume that there is a defined notion of consequence that applies to the sentences of L, call it "implies$_L$". Let Φ be a systematic assignment of sentence of E onto sentences of L. We then say that Φ is *inferentially adequate* iff: for all sentences α and β of E, α implies$_E$ β iff $\Phi(\alpha)$ implies$_L$ $\Phi(\beta)$. Thus, an assignment of sentences in a fragment of a natural language to an artificial language is inferentially adequate just in case it exactly mirrors the inferential relations among the groups of sentences. We will also call an artificial language inferentially adequate if the "intended" or "standard" assignment is inferentially adequate. And it turns out that if we restrict ourself to certain simple sentences of English (ones without "finitely many" in it, or the like), then there is a inferentially adequate assignment of these sentences into $\mathcal{L}_{PA}$. How the assignment goes is not precisely specified, although I presume it could be. We do this on the fly, but in a systematic way. In general, we do it by rephrasing our English sentences so that they very closely mirror sentences of $\mathcal{L}_{PA}$, and assign them such sentences this way. It would be a mistake, though, to think that we simply assign English sentences ones in the formal language with the same truth conditions. Such sentence don't have truth conditions, by themselves. They are merely true in certain models, but this is secondary. Note that whether or not an assignment is inferentially adequate is not subject to mathematical proof since it relates to questions about the primitive notion of implication, which is not fully precise. But given presumed inferential adequacy, mathematical proofs can be carried out about what follows from what.[7]

An assignment of sentences of a formal language to those in a natural language can be inferentially adequate without getting the syntax of the sentences in the natural language right, and without getting the semantic features right. All that is required is that whatever features are gotten wrong are gotten wrong systematically, so that all errors cancel themselves out in the end. This is what makes $\mathcal{L}_{PA}$ so great. It is inferentially adequate for a very important fragment of ordinary talk about natural numbers. Of course, the fact that $\mathcal{L}_{PA}$ is inferentially adequate might be taken as evidence that it does get the syntax and semantics right as well. After all, wouldn't it be the easiest explanation why it gets the inferential relations right that it gets everything else right as well? There is something to be said for this line, and I think that it gets the inferential relations right is *some* evidence that it gets the rest right. But the other evidence, that it gets the syntax and semantics wrong, is much stronger. I take the case of the syntax to be obvious. The syntax of natural language even when restricted to simple talk about numbers is not the same as the syntax of $\mathcal{L}_{PA}$. The evidence for this is conclusive. Nonetheless, $\mathcal{L}_{PA}$ gets the inferential relations right. And we can see quite easily how that can be. It captures enough of the inferentially relevant parts of that language in a way that inferential relations can be mirrored, while the syntax and semantics is different.

Since $\mathcal{L}_{PA}$ is a precise formal language, and the relevant notion of consequence is a precise notion, this gives rise to the possibility of precise proofs about what follows from what. And given inferential adequacy, this will imply that the corresponding inferential relations hold between the corresponding English sentences. Here formal tools can do real work. And they can do all this work even if, as I claimed above, $\mathcal{L}_{PA}$ gets basically everything wrong about the syntax and semantics of the language of arithmetic. And to carry out proofs about inferential relationships, formal tools do not play a secondary role, they are essential. Without precision in the notion of consequence and the language to which it applies no such proof would be possible. Here, there is hope for a positive and substantial role for formal tools.

Our question about the role of formal tools in the large-scale philosophy of mathematics is the following: what role do proofs about inferential relations have in this endeavor? Why do proofs about inferential relations matter for the large-scale philosophical questions? To answer this question, we will have to look at two other questions: the question about the status of axioms, and the question about whether all of mathematics is philosophically equal. I will argue in the following section that if one holds a certain position on these two other questions, then formal tools can have such a substantial role. However, most philosophers don't have that view about the other two questions. (I hold this view, but I won't be able to defend it here.)

7 The status of axioms

Axioms are special mathematical statements. They are just like other mathematical statements in many ways: they are in the same language, involving the same mathematical expressions, and so on. But they are also, somehow, special, and distinct from other mathematical statements in the very same part of mathematics. In what sense they are distinct from other statements in the same part of mathematics is a major dividing line for how to understand that mathematical discipline. We can distinguish the *descriptive* from the *constitutive* conception of (particular) axioms. On the descriptive conception, axioms are special, but they are mostly special for us. Among all the true statements in that part of mathematics, axioms are ones that are especially compelling to us, and that are especially useful for deriving consequences. They might be a nice way to put together a group of statements that allow us to derive a lot of other ones, and that are each fairly simple, in particular for us. But there is otherwise no special connection between what is true in that part of mathematics and what is an axiom. Axioms are just some especially compelling and systematizing true statements. On the constitutive conception of axioms, on the other hand, there is such a connection. Here the axioms, somehow, determine what is the case in this domain. The axioms are the basis for all truth in the domain. All truth flows from the axioms. On the constitutive conception of axioms for a particular part of mathematics, axioms are not just especially compelling true statements, they are the source of all truth in that domain. Both conceptions of axioms hold that axioms are true statements, but they differ in their assessment of priority. For descriptive axioms the mathematical facts are prior to the axioms, whereas on the constitutive conception of axioms it is the axioms that are prior to the facts. This difference seems intuitive, but it is hard to spell it out more precisely. In fact, whether or not it can be spelled out properly is an open question, but I will count on it being clear enough for the following.

Whether axioms can ever be constitutive is problematic. How could it be that all truth flows from the axioms in the sense, for example, that what is true in that domain of mathematics is just what is implied by the axioms? After all, the axioms imply themselves, and so how could they guarantee their own truth? But we can get a sense how this could be by seeing an analogy with fiction (although constitutive axioms might not require an association with fictionalism in the philosophy of mathematics). What is true in a fiction flows from (in part at least) what is implied by the story. The story implies itself, and so it guarantees its own truth. Not literal truth, of course, but truth in the relevant sense. Even though the difference between these two conceptions of axioms is hard to make precise, and even though it isn't clear whether axioms can ever be constitutive, the difference is compelling enough to appear in many places in the philosophy of mathematics.[8] Let's elaborate on this with some examples.

It is plausible that the Peano Axioms are not constitutive for arithmetic. By the Peano Axioms, I simply mean the particular arithmetical statements that are these axioms, not the formal axiom system of PA. The Peano Axioms are thus ordinary mathematical statements, not ones in an artificial language. One reason for thinking that the Peano Axioms are not constitutive for arithmetic is the independence of some Π_1^0 statements $\forall x \Phi(x)$ where every instance $\Phi(n)$ can be proven from the Peano Axioms. Even though the universally quantified statement can't be proven from the axioms, it nonetheless is true, and so truth goes beyond what can be proven from the axioms. This argument of course assumes that constitutive axioms only determine facts that can be proven from them, which might not be so if we understand what facts axioms determine in a different way, for example, by bringing in second order logic. I would be hesitant to do so, but this is certainly debatable. All I am claiming is that it is plausible that the Peano Axioms are not constitutive for arithmetic.

A further reason against the Peano Axioms being constitutive for arithmetic requires a close reflection on what makes any axioms constitutive, which we won't be able to give here, but which we can nonetheless appreciate in outline. Being constitutive for a certain domain of truths is not the same as implying all these truths. Many subsets of all the truths will imply all the truths, but that does not make them constitutive axioms of the domain. More is required for the axioms to be constitutive. The constitutive axioms are those that play some role in that domain of mathematics that, explicitly or implicitly, are involved in its conception. And this seems to require that the axioms can't be an afterthought, something discovered long after the domain has been established as an active part of mathematics. Peano's Axioms seem more like a discovery after the fact, rather than something involved at the beginning. But again, all I am saying is that it is plausible that the Peano Axioms are not constitutive for arithmetic.

One tempting example of constitutive axioms are algebraic axioms, for example, the axioms of group theory. What it is to be a group is fully determined by these axioms since the notion of a group is defined by these axioms. However, the results of *group theory*, the mathematical discipline, do not have the group axioms as constitutive axioms. That there are any groups at all is not settled by the group axioms, nor do the group axioms alone imply almost any of the results of group theory. What is needed in addition is the rest of mathematics, or at least the relevant parts of it. So, when one wants to show that there is an infinite cyclic group one does this by pointing out that the integers under addition have all the required features. But that they do have these features is not settled by the group axioms. Whether algebraic axioms, axioms like the ones of group theory, are good examples of constitutive axioms for a mathematical discipline, or merely useful definitions, doesn't have to be settled here. Algebraic axioms are often brought up as examples of constitutive ones, and if they help to illustrate the difference, even if they in the end don't fall on the relevant side of the difference, then that should be enough for us now.

If there is a part of mathematics that has constitutive axioms, then formal tools will have a greater role in the large-scale philosophical project than if this is never true. If truth in a part of mathematics is determined by what follows from the axioms, since the axioms are constitutive for that part, then investigating what does and doesn't follow from the axioms is tied to investigating where the facts are and where they end. To find out, via a proof of an inferentially adequate formal language, that a certain statement is independent of the constitutive axioms, would shed quite a bit of light on the factual status of that claim. It would show that the statement is outside of the realm of the facts.

But if the axioms are descriptive, then nothing should be taken to follow from the fact that a certain statement is independent of the axioms. What is true in that domain is not in any special way tied to the axioms, and so no conclusion should be drawn from the statement having or not having a certain relation to the axioms. The status of the axioms is crucial for the consequences of certain results about what follows from what.[9]

8 An epistemic role?

Might formal tools not have an epistemic role even if axioms are descriptive? For example, couldn't they help us to acquire knowledge about what is or isn't true in a certain domain? No, not in the relevant way. A proof that something follows from the axioms when they are descriptive shows that this statement is true, just like the axioms, but this we could have established by just proving the statement from the axioms directly. Formal tools might aid in such proofs, but this is not the substantial role we were looking for. But what about a proof of the independence of a statement from the axioms? This also does not have as great an epistemic significance as sometimes claimed, if the axioms are descriptive. Nothing could be concluded about there being no fact of the matter with regard to that independent statement, nor can it be concluded that we can't know the answer. The axioms are not our sole source of evidence when they are descriptive. They are merely one way for us to write up what compels us in a systematic way.

No assumption can be made that the axioms are an exhaustive source of evidence. The epistemic role of formal tools so understood is nothing to write home about. If the axioms are constitutive, then things are different, in the way discussed above.

9 Pluralism

The overall role of formal tools is further tied to another question, which is itself a central question in the large-scale philosophy of mathematics. It is the question whether or not all of mathematics is philosophically equal. This question has an innocent reading, where the answer is clearly: no. Some parts of mathematics are simple, known with more certainty, and so on. There clearly are differences among different parts of mathematics that extend to their philosophical assessment. But the question is intended in a grander scheme: Among the large-scale philosophical pictures that we philosophers have of mathematics, does one apply to all of mathematics, or do different ones apply to different parts of mathematics? Let's call those who think that all of mathematics is equal with respect to the large-scale philosophical picture *monists*, and those who think that it is not *pluralists*. Most philosophers, I think it is fair to say, are monists. Usually, fictionalists are fictionalists about all of mathematics, and platonists or structuralists are platonists or structuralists about all of mathematics. A philosophy of mathematics is often conceived as a philosophical story of mathematics as a whole, intended to fit all the parts equally. But there are some prominent exceptions. Notably, Hilbert thought that some part of mathematics, the finitist part, was in large-scale philosophical ways different from the rest of mathematics. Similarly, contemporary predicativists, like Solomon Feferman,[10] hold that predicative mathematics is in important philosophical ways different from other non-predicative parts of mathematics. Those who are pluralists thus hold that at least two parts of mathematics are different in important philosophical ways, and they thus have to say (i) how they are different and (ii) how they nonetheless come together into one discipline: mathematics. The pluralists mentioned so far are *dualists*, they hold that from a large-scale philosophical point of view there is one crucial difference which divides mathematics into two parts: the finitist versus the rest, or the predicative versus the rest. But in principle, pluralists could hold that there are a variety of different large-scale philosophical accounts of different parts of mathematics.

One way to be a pluralist is to hold that for some parts of mathematics axioms are descriptive, while for others they are constitutive. This marks a large-scale difference between these two parts of mathematics, and it is one of the ways in which formal tools can have an even greater role. The issue is the following: If there is a part of mathematics that has constitutive axioms, then the question of the consistency of these axioms becomes of central importance. To see this, consider what happens when there are inconsistent axioms in a part of mathematics where the axioms are descriptive, and thus not constitutive. Here, the axioms are merely a simple way for us to systematize the truths in that part of mathematics. They might be epistemically compelling, especially powerful in allowing us to derive all kinds of important results from them, and so on. But if they are inconsistent, then this should not be taken to be the end of that part of mathematics. It just shows that we did badly in finding basic principles from which the statements that are true in that part follow. And it might show that what we take to be compelling in this field isn't even all true.[11] If the axioms are inconsistent, then we just have to find better ones. The domain of mathematical truths that the axioms where supposed to capture is unaffected by this. Our conception of that domain might be in trouble, but the domain is not. Inconsistent axioms reflect badly on us, not what they are about.

If the axioms are constitutive, then things are different. The axioms are constitutive for what is true in that domain. If the axioms themselves are inconsistent, then the domain itself is in deep trouble. Constitutive axioms can't be separated from the domain they are about in the way that descriptive axioms can be. If constitutive axioms are inconsistent, then the domain that they were supposed to be constitutive for goes down the drain. Without consistency no meaningful domain has been constituted by the axioms. The question of the consistency of descriptive axioms is of interest, but the question of the consistency of constitutive axioms is a matter of survival.

Given that we hold that formal tools like $\mathcal{L}_{PA}$ get the inferential relations of ordinary mathematical sentences right, it follows that the consistency of an ordinary mathematical theory can properly be captured by a statement about artificial languages, like $\mathcal{L}_{PA}$. A set of axioms, in ordinary mathematical language, is consistent just in case not everything follows from it. This is a claim about what follows from what, and it thus is properly captured with a claim about what follows from what in a formal language. And this claim, again, can be stated with mathematical precision since it is a claim about a precise language and a precise notion of consequence. And since it can be stated with mathematical precision, it in effect becomes a mathematical statement itself. It can be stated in the language of arithmetic, and it can be stated in the language of PA, as the formal consistency statement of PA, CON_{PA}.[12] This is central for our main issue here, for the following reason.

Suppose, we have a part of mathematics with constitutive axioms. Then, the consistency of these axioms is of large-scale philosophical importance since without it there is no meaningful domain at all in this discipline. No facts are determined by the axioms since no facts can be inconsistent. If the axioms are constitutive, then it is central that they are consistent. But the consistency claim of the axioms is, in effect, itself a mathematical claim. To show that the axioms are consistent is to show that a certain mathematical claim is true. But what about the part of mathematics in which this is supposed to be shown? Does it have constitutive axioms as well? If it does, then the proof of the consistency of the first axioms would depend on the consistency of the second ones in a way that is too close for comfort. If the constitutive axioms of the part of mathematics in which the claim of consistency is stated are inconsistent, then that very claim of consistency loses all content. It would be a mathematical claim in an ill-defined domain. Thus, if these axioms are inconsistent then this would not only put the proof of consistency in question (which it would whether or not the axioms are constitutive), but it would put the whole statement of consistency itself in question. It is a mathematical claim after all, and it would be a mathematical claim in a part of mathematics where the axioms are supposed to be constitutive, but are inconsistent. The statement of the consistency of the axioms itself would be a statement in a domain that has not been properly established, and thus the statement of consistency itself would be most dubious. It would be not just dubious with respect to its truth value, but with respect to whether a coherent statement has been made at all. The status of the axioms of arithmetic (and finite set theory, and similar parts of mathematics) thus has a special standing among all other mathematical disciplines, philosophically. Only if these axioms are not constitutive can we deal with constitutive axioms in other parts of mathematics without things turning out badly. If the axioms of arithmetic are constitutive, then the claim of their consistency is only fully meaningful if they are consistent. This does not give us any ground to stand on.

But what if this is true: suppose the axioms of the parts of mathematics where consistency claims live, arithmetic and the like, are not constitutive. Instead, there is an objective domain of mathematical facts, somehow, and it is not constitutively tied to any axioms. Suppose further that some other part of mathematics does have constitutive axioms. Then the claim of the consistency of these axioms is well-grounded since it is a claim in a domain of mathematics where we have established facts. And a proof in arithmetic of the consistency of the axioms of the other part of mathematics has large-scale philosophical implications. It would show that there is a meaningful domain of mathematical facts in this other part. And the proof that this is so would not itself be subject to the further question if the domain of mathematical facts relied on to prove it is itself legitimate. If such a scenario obtains, then formal tools will have a large-scale philosophical role, even though only a limited one. They would be required to show that a certain domain of mathematics is well-grounded and meaningful. Inconsistent constitutive axioms destroy their domain. A proof of their consistency establishes it as a legitimate part of mathematics.

It turns out that the view about arithmetic referred to above, and defended in Hofweber (2005a) very much fits into this picture. Arithmetic axioms are not constitutive, and there is an objective domain of arithmetical facts. However, the reason why this is so does not carry over to other, higher, parts of mathematics. Whatever the large-scale philosophical story is about them, it will have to be fundamentally different.

For formal tools to ever have this greater role it must be true that

1. Arithmetic, or a similar part of mathematics, is about a domain of mathematical facts, but does not have constitutive axioms.
2. Some other part of mathematics does have constitutive axioms.

This would give a role to the statement of consistency, which can be seen as an arithmetical statement and so part of the domain of facts. The role of formal tools is even greater if this is true:

3. You can prove in arithmetic that (at least a sufficiently large part of) the axioms of the other part are consistent.

I don't know how many philosophers believe all these to be the case, but those who do can maintain that formal tools have a greater role than those who deny it. I believe this is indeed true. The view of arithmetic defended in Hofweber (2005a) has the required features, in particular there is a domain of arithmetical facts that is in no way constitutively tied to any axioms. Which parts of mathematics have constitutive axioms is a difficult question, but I would put money on that some do. However, I can't hope to give a defensible candidate here, nor can I hope to give a defensible account of what constitutive axioms are more precisely.

The third requirement, a consistency proof in arithmetic, might seem to be the most dubious. Hasn't Gödel shown that this is hopeless? Yes, and no. What matters is an informal proof of consistency, and in it we are certainly entitled to rely on the consistency of, for example, PA. So, all that is necessary is to give a relative consistency proof in PA of a certain set of axioms, which we can then take to sufficiently establish the consistency of the axioms. A relative consistency proof can often be given for axioms that appear to be substantially stronger than arithmetic. A series of results along these lines have been established in "reductive" proof theory, using the technique of proof theoretic reduction.[13] But even if it is not possible to give a relative consistency proof of all of the axioms, it might be possible to give such a proof for a sufficiently large part of the axioms, a part large enough to be taken to establish a core of the domain of mathematics in question. Some of the constitutive axioms might be taken to form the core of the domain, while others might be constitutive for the particular way to develop the core, but also optional in that the core might be developed along other lines. If this is true, then a consistency proof of the core would still be of large-scale importance since it at least established the core. Thus, consistency proofs and proof theoretic reductions of subsystems of a certain set of axioms are also of large-scale importance on this line, although it remains to be seen what counts as the core and what is merely a development of the core.

10 Conclusion

Formal tools are easily a source of error, and in fact have been such a source. When it comes to the understanding of mathematical activity and mathematical language, formal tools merely have a secondary, representational role. They are at best used to represent results established by other means. This is particularly vivid and neglected when it comes to understanding the semantic function of number words, where in the philosophy of mathematics they are almost uniformly and without much discussion understood as referring expressions, in analogy with their treatment in formal theories of arithmetic. And all this even though there is much evidence that in ordinary discourse, including in mathematical discourse, they have a different semantic function. Whatever their function is in the end, this has to be established by empirical means, in the study of mind and language. Formal tools can then be used to represent the result, but they shouldn't inspire us to expect one result or another.

However, there are two questions that determine how big the role of formal tools is in the large-scale philosophy of mathematics. First, the question about the status of axioms, second the question about pluralism. In particular, if one holds that axioms are sometimes constitutive, but they are not for arithmetic, then formal tools will have a much larger, but still limited, role. The real questions will still have to be addressed with other means. Whether the axioms are constitutive is not a question that is to be settled with formal tools, nor is the question if all of mathematics is in relevant ways the same. These will be the more central and harder questions. And similarly, for many other of the central questions in the philosophy of mathematics, formal tools will play little or no role.

Even though our main example in the discussion of these claims was arithmetic, the same, I maintain, carries over to other parts of mathematics, like set theory. My claims that $\mathcal{L}_{PA}$ gets basically everything wrong about the language of arithmetic (except inferential relations) does, of course, not carry over to the language of ordinary set theory and $\mathcal{L}_{ZF}$.[14] Even here there are real issues about how much $\mathcal{L}_{ZF}$ gets right about the language of set theory, as it is ordinarily carried out, informally with the use of symbols. Whether or not expressions for sets are referential is an issue that is less obvious than one might think. Here, real work needs to be done in the study of mathematical language.

So, for the future, I hope that the field of the philosophy of mathematics will have a clearer sense of with what methods which questions should be addressed. In particular, I hope that we will have a clearer sense of which questions should be addressed with formal tools and why. Some of the questions we are dealing with are empirical questions, some are technical questions, and some are philosophical questions. Depending on where a particular question falls will have a great effect on how we can hope to answer it. Once it is clear what kind of question it is, we can see with what method it should be addressed. To see what kind of question we are dealing with is often hard. Sometimes real progress can be made by showing that a particular question can be understood as a technical question. The history of logic has given us a few such cases, but the success of these few cases has lead to treating many other questions as being too closely tied to the formal tools that were successfully used in those cases. If we keep these issues apart, then things will be better. We won't be anywhere near done, but things will be better.[15]

Notes

1. Although the language of Peano Arithmetic does not contain "4", it is commonly introduced as a new term with the definition "4 = S(S(S(S(0))))".
2. See Hilbert (1925). The philosophical significance of Hilbert's use of formal tools and their development in contemporary "reductive proof theory" is discussed in Hofweber (2000).
3. I will use "PA" for the axiom system, and $\mathcal{L}_{PA}$ for the language in which these axioms are given. $\mathcal{L}_{PA}$ is a standard first-order language, that is, no free logic, no ϵ terms, and so on. These modifications soften some of the blows mentioned below, but don't avoid the general issue that the success of $\mathcal{L}_{PA}$ is independent of getting all these other things right or wrong. It will not matter for my points below just what precise the basic vocabulary of $\mathcal{L}_{PA}$ is taken to. We can include "<" or not, and either just have a constant for 0, or constants for both 0 and 1, or constants for all numbers. My main points merely rely on $\mathcal{L}_{PA}$ being a standard first-order language.
4. See Hofweber (2005a).
5. See Salmon (1986).
6. A different picture of number words is presented in Øystein Linnebo's contribution to this volume.
7. I am neglecting in this section that strictly speaking I don't believe that the formal notion of consequence used in $\mathcal{L}_{PA}$ is adequate for the informal notion of consequence. This has nothing to do with the shortcomings of first-order logic, but with another issue in the philosophy of logic which right now would just sidetrack us. See Hofweber (2007a, 2008). I am also putting aside those who hold that there is no clear enough set of facts about what implies what in terms of ordinary implication among English sentences. See, for example, Resnik (1997, 161 ff.) for such a stance toward facts about what follows from what.
8. The difference between descriptive and constitutive conceptions of axioms is related to Hellman's (2003) distinction between "assertory" and "algebraic" conceptions of mathematics, which is discussed in Mary Leng's article in this volume.
9. I am indebted to an anonymous referee for several helpful suggestions on this section.

10. See, for example, the essays in Feferman (1998).
11. I am assuming here that "axioms" in this sense, in which they can turn out to be inconsistent, don't have to be true. I am not taking it to be an option that mathematics itself might be inconsistent. For inconsistent mathematics, the same issue will arise for axioms to be non-trivial.
12. I am skipping here the issue of which one of various candidates should be seen to properly capture the consistency statement. Not because there isn't something important here, but just because that is not our main issue. I am simply assuming that it can be properly captured. See Feferman (1960).
13. See Feferman (1988) for a survey of such results.
14. $\mathcal{L}_{ZF}$ is the language of formal ZF set theory. Some parts of set theory are carried out in this language directly, and in those no mismatch can occur. But others are carried out in ordinary language, with the use of symbols, and here there is a real question about what features are correctly represented in $\mathcal{L}_{ZF}$.
15. Thanks to Øystein Linnebo, Mike Resnik, and an anonymous referee for many helpful comments on an earlier draft. A predecessor of this paper was presented at Stanford at a conference in honor of Solomon Feferman, on the occasion of his retirement. As one of his students, I would like to once more express my gratitude for his teaching, guidance, and support.

References

Feferman, S. (1960). Arithmetization of metamathematics in a general setting. *Fundamenta Mathematicae*, 49:35–92.
Feferman, S. (1988). Hilbert's program relativized. *Journal of Symbolic Logic*, 53: 364–384.
Feferman, S. (1998). *In the Light of Logic*. Oxford University Press, Oxford, UK.
Frege, G. (1884). *Die Grundlagen der Arithmetik: eine logisch mathematische Untersuchung über den Begriff der Zahl*. W. Koebner.
Hellman, G. (2003). Does category theory provide a framework for mathematical structuralism? *Philosophia Mathematica*, 11(2):129–157.
Hilbert, D. (1925). Über das Unendliche. *Mathematische Annalen*, 95:161–190.
Hofweber, T. (2000). Proof-theoretic reduction as a philosopher's tool. *Erkenntnis*, 53:127–146.
Hofweber, T. (2005a). Number determiners, numbers, and arithmetic. *The Philosophical Review*, 114(2):179–225.
Hofweber, T. (2005b). A puzzle about ontology. *Noûs*, 39:256–283.
Hofweber, T. (2007a). The ideal of deductive logic. Unpublished manuscript.
Hofweber, T. (2007b). Innocent statements and their metaphysically loaded counterparts. *Philosophers' Imprint*, 7(1):1–32.
Hofweber, T. (2008). Validity, paradox, and the ideal of deductive logic. In Beall, J. (ed.), *Revenge of the Liar: New Essays on the Paradox*. Oxford University Press, Oxford, UK.
Resnik, M. D. (1997). *Mathematics as a Science of Patterns*. Oxford University Press.
Salmon, N. (1986). *Frege's Puzzle*. MIT Press, Cambridge, MA.

10
The Individuation of the Natural Numbers

Øystein Linnebo

1 Introduction

It is sometimes suggested that criteria of identity should play a central role in an account of our most basic ways of referring to objects. The view is nicely illustrated by an example due to Quine (1950). Suppose you are standing on the bank of a river, watching the water that floats by. What is required for you to refer to the river, as opposed to a particular segment of it, or the totality of its water, or the current temporal part of this water? According to Quine, you must at least implicitly be operating with a criterion of identity that informs you when two sightings of water count as sightings of the same referent. For unless you have at least an implicit grasp of what is required for your intended referent to be identical with another object with which you are directly presented, you would not succeeded in singling out a unique object for reference.

This view goes back at least to Frege. In his *Foundations of Arithmetic*, Frege first argues that the natural numbers are abstract objects. Then Frege (1953, §62) asks how these objects "are given to us". Unlike ordinary concrete objects, we cannot have any "ideas or intuitions" of the natural numbers. How then do we manage to refer to natural numbers? Frege answers as follows.

> If we are to use the symbol *a* to signify an object, we must have a criterion for deciding in all cases whether *b* is the same as *a*, even if it is not always in our power to apply this criterion. (*ibid.*)

This passage and surrounding ones show that Frege took criteria of identity to play a very important role in an account of reference to the natural numbers.

The view that criteria of identity play a central role in our most basic forms of reference is an attractive one. As Quine's example nicely brings out, the idea that reference to an object crucially involves an ability to distinguish

220

the actual referent from other candidate referents enjoys great pre-theoretic plausibility. Another attraction of the view is its great generality. Since the notion of a criterion of identity is applicable to all kinds of objects, this approach to the problem of reference is applicable not just to concrete objects but also to abstract ones. This is a major advantage over competing approaches. For the language of mathematics abounds with apparent cases of reference to mathematical objects. And this language seems to succeed in expressing all kinds of truths. This phenomenon calls for an explanation, not a dismissal. But many approaches to the problem of reference are unable to accommodate reference to abstract objects. For instance, if the relation of reference requires some form of causal interaction, there can be no reference to abstract objects.[1] By contrast, since criteria of identity are found in the abstract realm as well as in the concrete, the approach in question extends naturally to abstract objects.

The aim of this chapter is to investigate how an approach to the problem of reference which gives pride of place to criteria of identity can be applied to the natural numbers. Other than the informal considerations adduced above, I will not attempt any direct defense of this kind of approach. My hope is rather that my investigation will produce an account of reference to the natural numbers which is attractive enough to provide indirect support for this approach.

The chapter is organized as follows. I begin by clarifying the notions of individuation and criterion of identity, which play a central role in my investigation. Then I explain two competing criteria of identity which have been argued to play a central role in reference to the natural numbers. One criterion regards the natural numbers as *cardinal numbers*, individuated by the cardinalities of the collections that they number. This account is favored by classic logicists such as Frege and Russell and by their followers.[2] The other criterion regards the natural numbers as *ordinal numbers*, individuated by their positions in the natural number sequence. This account is favoured by many constructivists and non-eliminative structuralists.[3] Next I outline some arguments to the effect that the ordinal account provides a better fit with our practices of thinking about and referring to natural numbers. I end by developing the ordinal account in more detail and discussing its implications concerning the metaphysical status of the natural numbers.

A broader lesson arising from this chapter is that the philosophy of mathematics ought to pay more attention to psychology and the philosophy of language than is currently done.[4]

2 Individuation and criteria of identity

My clarification of the notions of individuation and criterion of identity is organized around three important distinctions.

The first distinction is between two senses of the word "individuation"—one semantic, the other metaphysical. In the *semantic* sense of the word, to individuate an object is to single it out for reference in language or in thought. The problem discussed above of how this singling out is effected is thus a problem concerning semantic individuation. By contrast, in the *metaphysical* sense of the word, the individuation of objects has to do with "what grounds their identity and distinctness."[5] Sets are often used to illustrate the intended notion of "grounding." The identity or distinctness of sets is said to be "grounded" in accordance with the principle of extensionality, which says that two sets are identical iff they have precisely the same elements:[6]

$$\text{SET}(x) \wedge \text{SET}(y) \to [x = y \leftrightarrow \forall u(u \in x \leftrightarrow u \in y)] \tag{Ext}$$

The metaphysical and semantic senses of individuation are quite different notions, neither of which appears to be reducible to or fully explicable in terms of the other. Since I am not convinced that sufficient sense can be made of the notion of "grounding of identity" on which the metaphysical notion of individuation is based, my focus will be on the semantic notion of individuation. I find this notion clearer and more amenable to systematic investigation. This choice of focus means that our investigation will be a broadly empirical one. For the study of how an object is singled out for reference in language or thought, we will obviously have to draw on empirical linguistics and psychology.

Next, what is the relation between the semantic notion of individuation and the notion of a *criterion of identity*? According to the approach to the problem of reference with which I will be concerned, there is a very close relation. It is by means of criteria of identity that semantic individuation is effected. Singling out an object for reference involves being able to distinguish this object from other possible referents with which one is directly presented.[7]

The final distinction is between two types of criteria of identity. A *one-level criterion of identity* says that two objects of some sort F are identical iff they stand in some relation R_F:

$$Fx \wedge Fy \to [x = y \leftrightarrow R_F(x, y)] \tag{1L}$$

Criteria of this form operate at just one level in the sense that the condition for two objects to be identical is given by a relation on these objects themselves. An example is the set-theoretic principle of extensionality.

A *two-level criterion of identity* relates the identity of objects of one sort to some condition on entities of another sort. The former sort of objects are typically given as functions of items of the latter sort, in which case the criterion takes the following form:

$$f(\alpha) = f(\beta) \leftrightarrow \alpha \approx \beta \tag{2L}$$

where the variables α and β range over the latter sort of item and $\approx$ is an equivalence relation on such items.[8] An example is Frege's famous criterion of identity for directions:

$$d(l_1) = d(l_2) \leftrightarrow l_1 \parallel l_2 \qquad \text{(2L-Dir)}$$

where the variables l_1 and l_2 range over lines or other directed items. An analogous two-level criterion relates the identity of geometrical shapes to the congruence of things or figures having the shapes in question. Some terminology will be useful. Let's refer to the items over which the variables α and β range as *presentations*. The idea is that these are items that present the objects with whose identity we are concerned. Let's refer to the equivalence relation $\approx$ as a *unity relation*.

My decision to focus on the semantic notion of individuation makes it natural to focus on two-level criteria. For two-level criteria of identity are much more useful than one-level criteria when we are studying how objects are singled out for reference. A one-level criterion provides little assistance in the task of singling out objects for reference. In order to apply a one-level criterion, one must already be capable of referring to objects of the sort in question. By contrast, a two-level criterion promises a way of singling out an object of one sort in terms of an item of another and less problematic sort. For instance, when Frege (1953, §62) investigated how directions and other abstract objects "are given to us", although "we cannot have any ideas or intuitions of them", he proposed that we relate the identity of two directions to the parallelism of the two lines in terms of which these directions are presented. This would be explanatory progress since reference to lines is less puzzling than reference to directions.

3   The individuation of the natural numbers

How are the natural numbers individuated? That is, what is our most basic way of singling out a natural number for reference in language or in thought? The views found in the literature naturally fall into two types as follows: those that take the natural numbers to be individuated as *cardinal numbers*, and those that take them to be individuated as *ordinal numbers*.

According to the former type of view, the natural numbers are individuated by the cardinalities of the concepts or the collections that they number. For instance, our most basic way of thinking of the number 5 is as the cardinal measure of quintuply instantiated concepts or five-membered collections. This view naturally corresponds to a two-level criterion of identity. According to Frege, a number is presented by means of a concept, which has the number in question, and two concepts F and G determine the same number just in case the Fs and the Gs can be put in one-to-one correspondence. Let $F \approx G$ abbreviate the standard second-order formalization of this requirement. Frege's claim is then that the natural numbers are subject to the following criterion of identity, which has become known as *Hume's Principle*:

$$\#F = \#G \leftrightarrow F \approx G \qquad \text{(HP)}$$

Similar views have been defended by Russell and the neo-Fregeans Bob Hale and Crispin Wright.[9]

The view that the natural numbers are finite cardinals individuated by (HP) has some attractive features. Many philosophers find it attractive that the view builds the application of the natural numbers as measures of cardinality directly into their identity conditions; for instance, the number three is individuated as the number that counts all triples. But the most impressive feature is no doubt a technical result known as *Frege's Theorem*. Consider the theory that consists of pure second-order logic and (HP) as a sole non-logical axiom. Frege's Theorem says that this theory and some natural definitions suffice to derive all the familiar axioms of second-order Dedekind-Peano Arithmetic and thus all of ordinary arithmetic.[10]

According to the competing view, the natural numbers are individuated by their ordinal properties, that is, by their position in the natural number sequence. For instance, our most basic way of thinking of the number 5 is as the fifth element of this sequence. This view too corresponds naturally to a two-level criterion of identity. A natural number is presented by means of a "counter" or a numeral, which occupies a unique position in a sequence. Two such "counters" or numerals determine the same number just in case they occupy the same position in their respective orderings. For instance, the decimal numeral "5" is equivalent to the Roman numeral "V" because both occupy the fifth position in their respective orderings. If we symbolize this latter relation by $\sim$, this can be expressed as the following criterion:

$$N(\bar{m}) = N(\bar{n}) \leftrightarrow \bar{m} \sim \bar{n} \qquad \text{(2L-}\mathbb{N}\text{)}$$

This criterion of identity will be developed in more detail in Section 5.

Although I cannot defend the view here, I believe that both the cardinal and the ordinal conceptions are legitimate. That is, I believe either conception could successfully be used by a community in order to single out objects. The question I will investigate here is rather which of the conceptions provides the most plausible analysis of our most basic form of reference to the natural numbers. In the next two sections, I argue that the answer is the ordinal conception.

However, before embarking on this argument I would like to pause briefly to reflect on the question we are discussing. Does our question presuppose that there really are natural numbers? Since the notion of individuation with which we are concerned is the semantic one, the answer is *no*. Our question is concerned with the features of our conception of natural numbers which are responsible for singling them out for reference. And these features can be investigated regardless of whether there are objects of which this conception is (even approximately) true. There is thus no need to presuppose the existence of natural numbers. Nevertheless, for ease of expression I will often write as if this presupposition is made.

Next, does our question presuppose that there are robust and general facts of the matter about what is responsible for our ability to single out individual natural numbers for reference? Might there not exist different but equally legitimate conceptions of the natural numbers?[11] There might, for instance, be different conceptions of the natural numbers in cultures with no formal education, among educated lay people, and among professional mathematicians. I don't find this scenario particularly threatening. Firstly, even if human referential practices were too heterogeneous to admit of a single "pure" analysis, it would still be of significant interest to articulate and explore the various options. For instance, the neo-Fregean program has tended to assume without argument that any abstractionist approach to the natural numbers must be based on Hume's Principle. The ordinal-based alternative mentioned above and further developed below shows this assumption to be false.

Secondly, even if human referential practices were highly heterogeneous, our question might still allow of an interesting answer when suitably relativized to a community. (If so, my first move would be to relativize the question to educated lay people from contemporary Western culture.) More thoroughgoing skeptics may worry that even relativized to a community, our question may not admit of a unique answer. I admit that this is a possibility and will accordingly not assume that this skeptical worry is unfounded. My response to this worry is more pragmatic. Let's see how much sense can be made of the question and, in light of this, to what extent it admits of a unique answer.

4 Against the cardinal conception

I grant that the cardinal conception provides one possible way of thinking and talking about the natural numbers. But I deny that this is how we actually single out the natural numbers for reference in our most basic arithmetical thought and reasoning. I will now present some objections to the cardinal conception as an account of our actual arithmetical practice. The objections will be presented roughly in the order of increasing strength.

It should be noted that my concern in this section is quite different from that of some leading advocates of the cardinal conception. For instance, Hale and Wright are not concerned that their analysis should match our actual arithmetical practice. Their goal is rather to establish a possible route to *a priori* knowledge of an abstract realm of natural numbers.[12] This is clearly a legitimate and interesting goal. But in my view the more pressing question is to what extent actual arithmetical practice provides such knowledge. In particular, how and to what extent do ordinary educated lay people achieve such knowledge?[13]

4.1 The objection from special numbers

This objection seeks to show that, if the cardinal conception had been correct, then certain special numbers would have been obvious and unproblematic in a way that they are not.

One such number is zero. It takes very little sophistication to know that certain concepts have no instances. For example, the concept of being a lump of gold in my pocket has no instances. So, if our basic conception of natural numbers had the form #F, then zero should have been a very obvious number. By contrast, the view that our most basic conception of the natural numbers is as ordinals predicts that zero should be just as non-obvious and problematic as the negative numbers; for every sequence of numerals has a first but no zeroth element. As it turns out, it was only at a very late stage in the history of mathematics that zero was admitted into mathematics as a number in good standing.[14] This suggests that our most basic conception of the natural numbers is ordinal-based rather than cardinal-based.

How convincing is this objection? I believe it has some force against the view that our most basic conception of a natural number has the form #F where F is a concept. But the cardinal-based approach can perhaps be modified so as to block the objection. One may, for instance, take the most basic conception of a natural number to have the form #xx where xx is a plurality of objects. Since there are no empty pluralities, this would block the easy route to the number zero. However, this response would be a significant deviation from the traditional Frege–Russell view. And more seriously, it would undermine Frege's impressive bootstrapping argument for the existence of infinitely many natural numbers. For this argument makes essential use of the number zero.[15]

The objection from special numbers can also be developed for various infinite cardinals. If our basic conception of a natural number had been of the form #F, then infinite cardinals should have been much more obvious and natural than they in fact were. For instance, not much sophistication is needed to grasp the concept of self-identity. And with a sufficient grasp of arithmetic comes a grasp of the concept of being a natural number. The cardinal conception therefore predicts that the numbers that apply to these two concepts should have been fairly obvious. But the history of mathematics shows that it took the great creative genius of Cantor to accept and explore the idea of infinite cardinal numbers. And when he did so, he met with wide-spread incomprehension and opposition. This historical evidence is better explained by the view that our basic conception is an ordinal one based on numerals. For where there are some very simple concepts with infinitely many instances, it is far from obvious that there are sequences of numerals of the sort that are required to present infinite ordinals. A natural response to this objection would be to modify (HP) so as to assign numbers only to finite concepts.[16] Let FIN(*F*) be some formalization of the claim that the concept *F* has only finitely many instances. Then let *Finite Hume* be the following principle:

$$\text{FIN}(F) \wedge \text{FIN}(G) \to (\#F = \#G \leftrightarrow F \approx G) \qquad \text{(FHP)}$$

This modification clearly blocks the present argument. However, this way of blocking the argument is problematic. For as will be argued in Section 5.1, it is implausible that people with ordinary arithmetical competence should grasp, however implicitly, the concept of finitude.

4.2 The objection from the philosophy of language

This objection begins with the claim that in natural language, expressions of the form "the number of *F*s" are definite descriptions rather than genuine singular terms. This claim enjoys strong linguistic evidence. Certainly the surface structure of the expression indicates that it is a definite description rather than a genuine singular term. Moreover, expressions of this form are easily seen to be non-rigid. For instance, the number of bicycles in my possession is 4; so had I bought another, the number of bicycles in my possession would have been 5. By contrast, numerals are genuine singular terms and rigid designators.

The objection continues by claiming that our most basic conception of an entire category of objects cannot be based entirely on definite descriptions but must also involve some more direct form of reference. For in order to understand a definite description, one must be capable of some more direct way of referring to the objects in question. Consider, for instance, the sentence "The tallest person in this room is male". In order to understand this sentence one needs an ability to identify people which is prior to and independent of what is provided by definite descriptions. More generally, if *F* is a sortal and Φ a predicate defined on *F*s, then in order to understand "the *F* is Φ" one needs some more direct way of referring to *F*s than is provided by definite descriptions.[17]

The relevant contrast between numerals and descriptions of the form "the number of *F*s" has to do with their internal semantic articulation. My claim is that the descriptions cannot serve as a basic mode of reference to numbers because they have an internal semantic articulation, which presupposes some more basic form of reference to numbers. By contrast, numerals are semantically simple expressions with no internal semantic articulation. This does not mean that their reference is primitive and unanalyzable; indeed I have argued that their reference admits of an analysis in terms of criteria of identity. But this is an account of what the reference of a semantically simple expression consists in, not an account of how the semantic value of a complex expression (such as a definite description) is determined by the semantic values of its simple constituents.[18]

4.3   The objection from lack of directness

If there are five apples on a table, we can think of the number 5 as the number of apples on the table. Or, following Frege, we can think of 5 as the number of (cardinal) numbers less than or equal to 4. But neither or these ways of thinking of the number 5 feels particularly direct or explicit. Indeed, many people with basic arithmetical competence won't find it immediately obvious that the number of numbers between 0 and 4 (inclusive) is 5 as opposed to 4.

Rather, the only perfectly direct and explicit way of specifying a number seems to be by means of some standard numeral in a system of numerals with which we are familiar. Since the numerals are classified in accordance with their ordinal properties, this suggests that the ordinal conception of the natural numbers is more basic than the cardinal one.[19]

It may be objected that, since we don't have transparent access to all features of our thought, we cannot take at face value the kind of phenomenological evidence that I have just adduced about what kinds of reference *feels* most direct. This is a perfectly legitimate concern. But although the above considerations are particularly powerful when presented from a first-person point of view, there is no need to present them in that way. I conjecture that the results would be confirmed by a more objective, third-personal investigation of cognitive processing of arithmetical claims, for instance in terms of reaction times.[20]

4.4   Alleged advantages of the cardinal conception

I end this section by briefly considering two alleged advantages of the cardinal conception.

One alleged advantage is that the applications of the natural numbers are built directly into their identity conditions. This is just an instance of a more general requirement sometimes known as *Frege's constraint*.[21] But should this constraint be respected? We obviously need *some* account of how mathematics is applied. But why should the account have to build the applications of mathematical objects directly into their identity conditions?[22] Besides, even if Frege's constraint could be defended, this would not obviously favor the cardinal conception. For the natural numbers lend themselves to ordinal as well as cardinal applications, and Frege's constraint does not settle which of these applications should be built into the identity conditions of numbers.

Another alleged advantage of the cardinal conception is that it allows for Frege's famous "bootstrapping argument" for the principle that every number has an immediate successor. Mathematically this argument is extremely elegant and interesting. But as an account of people's actual arithmetical reasoning or competence it is implausible. The argument was developed only in the 1880s and is complicated enough to require even trained mathematicians to engage in some serious thought. So, this is unlikely to be the source of ordinary people's conviction that every number has an immediate successor.[23]

5 Developing the ordinal conception

I now develop the ordinal conception of natural numbers in more detail.[24]

5.1 Refining the criterion of identity

Arithmetic teaches us that the natural numbers are *notation independent* in the sense that they can be denoted by different systems of numerals. In fact, an awareness of this notation independence is implicit already in basic arithmetical competence. Even people with very rudimentary knowledge of arithmetic know that the natural numbers can be denoted by ordinary decimal numerals, by their counterparts in written and spoken English and other natural languages, and by sequences of strokes. Many people also know alternative systems of numerals such as the Roman numerals and the numerals of position systems with bases other than ten, such as binary and hexadecimal numerals. To accommodate the notation independence of the natural numbers, we need an account of acceptable numerals and a condition for two numerals to denote the same number.

I will take a numeral to be any concrete object that can stand in a suitable ordering. On this very liberal view, a numeral need not even be a syntactical object in any traditional sense. For instance, if a pre-historic shepherd counts his sheep by matching them with cuts in a stick, then these cuts count as numerals. The purpose of a numeral is just to mark a place in an ordering. But of course, one and the same object can inhabit different positions in different orderings. The syntactical string "III" can, for instance, mean either 3 or 111 depending on the ordering in which it is placed. We should therefore make explicit the ordering in which a numeral is placed. So I will take a numeral to be an ordered pair $\langle u, R \rangle$, where u is the numeral proper and R is some ordering in which u occupies a position.

How should the ordering R be understood? One question is whether R should be understood in an extensional manner (say as a set or list) or in intensional manner (say as an algorithm or procedure). I see two reasons to favor the latter option. Firstly, in order to be learnable and effectively useable, the ordering must be computable. This means that the ordering must be understood as some sort of procedure rather than as an infinite collection of ordered pairs. Secondly, there are only finitely many numeral tokens of any learnable and effectively useable numeral system. This means that the ordering must be defined not only on actual numerals but in a way that extends to possible further numerals as well. This is best accounted for by considering orderings in an intensional sense.

Another question is about the order type of the relation R. A minimal requirement is that R be a discrete linear ordering with an initial object. For we want it to be the case that no matter how far we have counted, there is a unique next numeral—provided there are further numerals at all. The question is whether we have any reason to go beyond this minimal requirement.

We could, for instance, add the further requirement that R be of order type ω.[25] But I don't think that would be a good idea. The notion of being of order type ω is conceptually quite demanding. This notion was not clearly grasped until rather late in the history of mathematics, at the earliest in the late sixteenth century when mathematical induction was first explicitly articulated as one of the core principles of arithmetic. In fact, the problems that many people have with the idea of mathematical induction suggests that even an implicit grasp of the notion goes beyond what is required for basic arithmetical competence. This makes it doubtful that the notion should play a central role in people's most basic conception of the natural numbers.

What about the weaker requirement that R be without an end point? This requirement too seems undesirable. For it seems plausible to allow numeral systems based on finite orderings to denote natural numbers. (This may actually be the case on one construal of the Roman numerals.)

Finally, what about the requirement that R be well-founded or that its order type not exceed ω? These requirements too seem too demanding to be implicit in basic arithmetical competence. The fact that all traditional numeral systems are well-founded and of order type at most ω is explained better and much more simply by observing that these are the kinds of orderings with which we are familiar from the recursive formation rules of natural language.

I conclude that no requirement needs to be imposed on the order type of R other than the minimal one that R be a discrete linear ordering with an initial object. On this view, the natural numbers are clearly distinguished from non-standard numbers only when mathematical induction is introduced as a basic arithmetical principle.

I turn now to the equivalence relation that must hold between two numerals $\langle u, R \rangle$ and $\langle u', R' \rangle$ for them to determine the same number. This equivalence relation must clearly be a matter of the two objects u and u' occupying analogous positions in their respective orderings. More formally, $\langle u, R \rangle$ and $\langle u', R' \rangle$ are equivalent just in case there is a relation C which is an order-preserving correlation of initial segments of R and R' such that $C(u, u')$. I write $\langle u, R \rangle \sim \langle u', R' \rangle$ to symbolize that the two ordered pairs are equivalent in this sense.[26]

A slight refinement of our previous two-level criterion of identity for natural numbers can now be expressed as follows:

$$N\langle u, R\rangle = N\langle u', R'\rangle \leftrightarrow \langle u, R\rangle \sim \langle u', R'\rangle \qquad \text{(2L-N')}$$

where the variables u and u' range over concrete objects, and R and R' range over relations of the sort characterized above.[27] Let "NUM(x)" be a predicate that holds of all and only the objects that can be presented in this way.

5.2 Justifying the axioms of Dedekind–Peano Arithmetic

I now show how this conception of the natural numbers, supplemented with some natural and conceptually very simple definitions, allows us to justify all the axioms of Dedekind–Peano Arithmetic (PA). I proceed in three stages. First, I introduce three relations on numerals which correspond to the basic arithmetical relations of succession, addition, and multiplication. Then, I show that these relations on numerals induce corresponding relations on the natural numbers, thus allowing us to define the non-logical primitives of the language of PA. Finally, I show how we can justify the axioms of PA on this basis.

We can define a successor relation $S^{\#}$ on numerals by letting $S^{\#}(\langle u, R\rangle, \langle u', R'\rangle)$ just in case u' is the R'-successor of some v such that $\langle u, R\rangle \sim \langle v, R'\rangle$. We now wish to define relations on numerals $A^{\#}$ and $M^{\#}$, which correspond to the arithmetical relations of addition and multiplication on natural numbers. Let $I^{\#}(x)$ formalize the claim that the numeral x is first in its ordering. Following the practice of ordinary counting, we think of such initial numerals as representing the number 1. Then, the two desired relations are defined by means of the following recursion equations (where the free variables are implicitly understood as ranging over numerals):

$$I^{\#}(y) \rightarrow [A^{\#}(x, y, z) \leftrightarrow S^{\#}(x, z)] \qquad \text{(A1)}$$

$$S^{\#}(y, y') \wedge A^{\#}(x, y, z) \rightarrow [A^{\#}(x, y', z') \leftrightarrow S^{\#}(z, z')] \qquad \text{(A2)}$$

$$I^{\#}(y) \rightarrow [M^{\#}(x, y, z) \leftrightarrow x \sim z] \qquad \text{(M1)}$$

$$S^{\#}(y, y') \wedge M^{\#}(x, y, z) \rightarrow [M^{\#}(x, y', z') \leftrightarrow A^{\#}(z, x, z')] \qquad \text{(M2)}$$

It is easily verified that the three relations $S^{\#}$, $A^{\#}$, and $M^{\#}$ do not distinguish between numerals that are equivalent under $\sim$. These relations therefore induce corresponding relations on the natural numbers themselves, defined by

$$S(N(x), N(y)) \leftrightarrow S^{\#}(x, y) \qquad \text{(Def-S)}$$

$$A(N(x), N(y), N(z)) \leftrightarrow A^{\#}(x, y, z) \qquad \text{(Def-A)}$$

$$M(N(x), N(y), N(z)) \leftrightarrow M^{\#}(x, y, z) \qquad \text{(Def-M)}$$

232   *Øystein Linnebo*

Following the practice of ordinary counting, we let 1 be the first number. For instance, 1 may be presented as $N\langle"1", D\rangle$, where D is the standard ordering of the decimal numerals.

We can now verify that the various axioms of PA hold. Some are quite straightforward.

$$\text{Num}(1) \tag{PA1}$$

$$\text{Num}(x) \rightarrow \neg S(x, 1) \tag{PA2}$$

$$S(x, y) \wedge S(x', y) \rightarrow x = x' \tag{PA3}$$

$$S(x, y) \wedge S(x, y') \rightarrow y = y' \tag{PA4}$$

$$A(x, 1, z) \leftrightarrow S(x, z) \tag{PA5}$$

$$S(y, y') \wedge A(x, y, z) \rightarrow [A(x, y', z') \leftrightarrow S(z, z')] \tag{PA6}$$

$$M(x, 1, z) \leftrightarrow x = z \tag{PA7}$$

$$S(y, y') \wedge M(x, y, z) \rightarrow [M(x, y', z') \leftrightarrow A(z, x, z')] \tag{PA8}$$

(PA1) is trivial. For (PA2), let x be any number. Then there is a numeral $\langle u, R \rangle$ such that $x = N(\langle u, R \rangle)$. If $S(x, 1)$, then u would precede the initial object of the relevant ordering, which is impossible. (PA3) follows from the observation that any two numerals both of which immediately precede a third numeral are equivalent under $\sim$. (PA4) follows from the observation that any two numerals that immediately succeed a third are equivalent. (PA5)–(PA8) follow readily from (A1), (A2), (M1), and (M2).

It is less obvious how to justify the next axiom:

$$\forall x(\text{Num}(x) \rightarrow \exists y S(x, y)) \tag{PA9}$$

It would clearly suffice if we could assume that every numeral bears $S^\#$ to some further numeral. But this assumption is doubtful, at least as concerns actual numeral tokens. Consider instead the weaker modal claim that necessarily, for any numeral, there could be some further numeral to which it bears $S^\#$:

$$\Box \forall u \, \forall R \, \Diamond \, \exists u' \, \exists R' \, S^\#(\langle u, R \rangle, \langle u', R' \rangle) \tag{1}$$

This weaker claim is extremely plausible. For assume we are given a numeral $\langle u, R \rangle$. Then it is possible that there is an object u' distinct from u and all of its R-predecessors. Let R' be the result of adding the pair $\langle u, u' \rangle$ to the initial segment of R ending with u. Then $\langle u', R' \rangle$ is as desired. (PA9) follows from (1) and the claim that numbers exist necessarily if at all.[28]

Finally, we need to specify some condition of finitude with which to restrict the numbers such that we get all and only the natural numbers.

I claim that this condition is simply that mathematical induction be valid of the natural numbers. That is, x is a natural number (in symbols: $\mathbb{N}(x)$) just in case the following open-ended schema holds:

$$\phi(1) \wedge \forall u \forall v[\phi(u) \wedge S(u, v) \rightarrow \phi(v)] \rightarrow \phi(x)$$

This allows us to justify the final axiom of PA, namely the induction scheme:

$$\phi(1) \wedge \forall u \forall v[\phi(u) \wedge S(u, v) \rightarrow \phi(v)] \rightarrow \forall x(\mathbb{N}(x) \rightarrow \phi(x)) \qquad \text{(PA10)}$$

Prior to the adoption of this scheme, it is doubtful that people had uniquely singled out the natural number structure as opposed to the objects of some non-standard model of arithmetic.

6 The metaphysical status of natural numbers

Does the account of reference to the natural numbers that I have outlined tell us anything about their metaphysical status? I now show that the account opens for the possibility of a "reductionist" interpretation of arithmetical discourse. But I argue that this interpretation should not be seen as a vindication of nominalism but rather as a reductionist account of what reference to natural numbers consists in.

It is useful to begin by reminding ourselves of an important feature of most non-mathematical predications. Consider for example the question whether a physical body has the property of being a solid sphere. This question cannot be answered on the basis of any one of the body's proper parts; rather, information will be needed about the entire body. And there is nothing unusual about this example. Whether a body has some property generally depends on all or many of its parts. So a body has most of its properties in an irreducible way, that is, in a way that isn't reflected in any property of any one of its proper parts. This means that physical bodies play an ineliminable role in making predications true: for the truth of such predications cannot be reduced to a matter of how the body is presented.

The natural numbers are very different in this regard. Consider the question whether a natural number n has some arithmetical property, say the property of being even. Unlike the case of the roundness of a physical body, any numeral $\langle u, R \rangle$ by means of which n is presented suffices to determine an answer to the question, based on whether or not the numeral proper u occurs in an even-numbered position in the ordering R. There is no need to examine other presentations of n (or the number itself, whatever exactly that would involve). In fact, this observation can be seen to generalize: for any arithmetical property P, the question whether n possesses P can be reduced to a question about any numeral by means of which n is presented.

A natural number is in this way "impoverished" compared to the numerals that present it, as all of its properties are already implicitly contained in each of these numerals. This opens for the possibility of a form of reductionism about natural numbers as questions about such objects can be reduced to questions about their presentations. It then becomes, in the apt metaphor of Wright (1992), pp. 181–182, hard to resist the idea that the natural numbers are mere "shadows of syntax".

Given this reductionism, does it still make sense to say that numerals refer to natural numbers? I believe this is best understood as the question whether it still makes sense to ascribe semantic values to numerals. There is strong *prima facie* reason to do so. When we analyze English and the language of arithmetic, singular terms such as "5" and "1001" seem to function just like terms such as "Alice" and "Bob". The default assumption is therefore that all these terms function in similar ways. Since singular terms such as "Alice" and "Bob" clearly have semantic values (namely, the physical bodies that they refer to), this provides *prima facie* reason to think that arithmetical singular terms such as "5" and "1001" have semantic values as well.

It may be objected that this *prima facie* reason is overridden by our discovery that questions about natural numbers can be reduced to questions about the associated numerals. Since this reduction shows that it suffices to talk about the numerals themselves, there is no need to ascribe any sort of semantic values to numerals. But this objection is much too quick. For the reducibility of questions about numbers to questions about numerals allows of two different kinds of explanation, only one of which allows the objection to go through. To see this, we need to distinguish between *semantics* and what is sometimes called *meta-semantics*.[29] Semantics ascribes semantic values to expressions and studies how the semantic value of a complex expression depends on the semantic values of its various simple constituents. But the relation between a linguistic expression and its semantic value is never a primitive one but one that obtains in virtue of some other and more basic facts. Compare the relation of ownership, which also isn't a primitive one. When I bear the ownership relation to my bank account, this isn't a primitive fact but one that obtains in virtue of some other and more basic facts. Meta-semantics is the study of what it is in virtue of which expressions have semantic values.

The distinction between semantics and meta-semantics points to two different ways of understanding the reducibility of questions about numbers to questions about numerals. The first way, which is the one presupposed in the above objection, is a form of *semantic reductionism*, according to which numerals don't have semantic values but serve some alternative semantic purpose. This would yield a nominalist interpretation of the language of arithmetic. The second way of understanding the reducibility is as a form of *meta-semantic reductionism*, namely, a reductive analysis of the relation that obtains between a numeral and its semantic value. On this second analysis,

it is perfectly true to say that the numerals have numbers as their semantic values. The point is rather that this truth admits of a reductionist analysis.
I defend the second account elsewhere.[30] If correct, what consequences will this have for the question of mathematical platonism? If by "mathematical platonism" we mean simply the view that there are true sentences some of whose semantic values are abstract, then my view is obviously a platonist one.[31] But given how lightweight the relevant semantic values are, this may be more of a reason to sharpen one's definition of "mathematical platonism" than for platonists to declare victory.[32]

Notes

1. Indeed, according to the standard definition, an object is *abstract* just in case it is non-spatiotemporal and does not stand in causal relations.
2. See for instance Frege (1953), Russell (1919), Wright (1983), and Hale and Wright (2001).
3. The account is defended by Dedekind (1996) and, more recently, by various non-eliminative mathematical structuralists, such as Parsons (1990), Resnik (1997), and Shapiro (1997).
4. Thomas Hofweber's contribution to this volume draws a similar lesson (although our attempts to heed this lesson yield quite different results).
5. See e.g. Lowe (2003).
6. Note that this "grounding" is guaranteed to be informative only in the presence of the axiom of Foundation.
7. See Lowe (2003) for a comparison of the metaphysical notion of individuation with a metaphysical notion of a criterion of identity.
8. An approach based on two-level criteria of identity is found in Williamson (1990), chapter 9.
9. See also Roy Cook's contribution to this volume. I here gloss over some important differences which are irrelevant to our present concerns. In particular, Frege and Russell wanted to reduce (HP) to what they took to be more basic class-theoretic notions. For details see Frege (1953), Russell (1919), Wright (1983), and Hale and Wright (2001).
10. This was observed in Parsons (1965) and discussed at length in Wright (1983). For a nice proof, see Boolos (1990).
11. See also Alexander Paseau's contribution to this volume, where it is argued that our conception of the natural numbers is compatible with a reduction of arithmetic to set theory.
12. See e.g. Hale and Wright (2000).
13. This alternative goal is defended in Heck (2000) and Linnebo (2004), p. 168.
14. Fibonacci is often credited with the introduction into the European tradition of an explicit numeral for zero in the early thirteenth century. But apparently even he did not regard zero as a proper number. According to the three-volume history of mathematics Kline (1972), "By 1500 or so, zero was accepted as a number" (p. 251).
15. See Shapiro and Weir (2000) for a discussion.
16. See Heck (1997), who at pp. 590–591 provides a definition (due to Frege) of what it is for a concept to be finite, which does not assume an antecedent conception of natural number.

17. For a more developed argument of this sort, see Evans (1982), esp. Section 4.4.
18. In the terminology of Stalnaker (1997) (which will be explained in Section 6), the former account belongs to *meta-semantics*, whereas the latter belongs to *semantics proper*.
19. A view of this sort is defended in Kripke (1992).
20. Indeed, most cognitive psychologists appear to think that our capacity for exact representations of numbers (other than very small ones) is based on our understanding of some system of numerals.
21. See Wright (2000) for a discussion and partial defense of this constraint.
22. See also Parsons (2008), Section 14 for criticism of Frege's constraint.
23. See Linnebo (2004), pp. 168–169.
24. For a broadly similar account, see Parsons (1971), Section 3.
25. Mathematical structuralists are often attracted to this requirement. See the works cited in footnote 3.
26. Note that ∼ is guaranteed to be one-to-one only on finite numerals. This means that the "numbers" corresponding to any infinite numerals cannot always be identified with ordinal numbers in the contemporary sense. Some of these "numbers" will be unintended and pathological objects which will be ruled out as soon as one's conception of natural number is sophisticated enough to include the principle of induction, as described below.
27. (2L-ℕ′) can for instance be formulated in a two-sorted language with one sort of variables and function-terms for numbers and another sort for non-numbers. The variables u and u' are then non-numerical, whereas the function-terms involving the functor N are numerical. This blocks any attempt to derive from (2L-ℕ′) a version of the Burali–Forti paradox, as would otherwise be possible. The requirement that the numerals proper be concrete objects is not an *ad hoc* restriction to avoid paradox but is independently motivated by the role of numerals as presentations of numbers. For these presentations have to be epistemically and semantically more "accessible" than the numbers themselves. However, the question will nonetheless arise whether it is possible to remove the requirement that the numerals proper be concrete. I outline one way in which this can consistently be done in Linnebo (2009), where I develop a modal theory of abstraction understood as a "dynamic" process. (Thanks to Stewart Shapiro for questions that forced me to get clearer on the threat of a Burali–Forti paradox.)
28. The proof uses "the Brouwerian axiom" $p \rightarrow \Box \Diamond p$. Another strategy is to settle for a modal theory of arithmetic based on the principle that necessarily for any number there could be a larger number. One can then give a very natural interpretation of ordinary Peano Arithmetic in this modal theory by interpreting the ordinary arithmetical quantifiers $\forall n$ and $\exists n$ as respectively $\Box \forall n$ and $\Diamond \exists n$. See my Linnebo (unpublished manuscript) for details.
29. See e.g. Stalnaker (1997).
30. See e.g. Linnebo (2009). I hope to return to this issue in future work.
31. See Agustín Rayo's contribution to this volume for another discussion of lightweight forms of platonism.
32. I am grateful to Roy Cook, Frode Kjosavik, Peter Koellner, Richard Pettigrew, Stewart Shapiro, and the audience at the New Waves conference at the University of Miami for comments and discussion. This paper was written during a period of AHRC-funded research leave (grant AH/E003753/1). I gratefully acknowledge their support.

References

Boolos, G. (1990). The Standard of Equality of Numbers. In Boolos, G. (ed.), *Meaning and Method: Essays in Honor of Hilary Putnam*. Harvard University Press, Cambridge, MA. (Reprinted in Boolos, 1998.)

Boolos, G. (1998). *Logic, Logic, and Logic*. Harvard University Press, Cambridge, MA.

Dedekind, R. (1996). Was Sind und Was Sollen die Zahlen? In Ewald, W. (ed.), *From Kant to Hilbert: A Source Book in the Foundations of Mathematics*, volume II, pp. 790–833. Oxford University Press, Oxford.

Evans, G. (1982). *Varieties of Reference*. Oxford University Press, Oxford.

Frege, G. (1953). *Foundations of Arithmetic*. Blackwell, Oxford. Transl. by J. L. Austin.

Hale, B. and Wright, C. (2000). Implicit Definition and the A Priori. In Boghossian, P. and Peacocke, C. (eds), *New Essays on the A Priori*. Oxford University Press, Oxford. (Reprinted in Hale and Wright, 2001.)

Hale, B. and Wright, C. (2001). *Reason's Proper Study*. Clarendon, Oxford.

Heck, R. G. (1997). Finitude and Hume's Principle. *Journal of Philosophical Logic*, 26:589–617.

Heck, R. G. (2000). Cardinality, Counting, and Equinumerosity. *Notre Dame Journal of Formal Logic*, 41(3):187–209.

Kline, M. (1972). *Mathematical Thought from Ancient to Modern Times*. Oxford University Press, Oxford.

Kripke, S. Logicism, Wittgenstein, and *de re* Beliefs about Numbers. Unpublished Whitehead Lectures, delivered at Harvard in May 1992.

Linnebo, Ø. (2009). Bad Company Tamed. *Synthese*, 170(3). DOI 10.1007/sll229-007-9265-7.

Linnebo, Ø. The Potential Hierarchy of Sets. Unpublished manuscript.

Linnebo, Ø. (2004). Predicative Fragments of Frege Arithmetic. *Bulletin of Symbolic Logic*, 10(2):153–174.

Linnebo, Ø. (2009). Frege's Context Principle and Reference to Natural Numbers. In Lindstrom, S. (ed.), *Logicism, Intuitionism, and Formalism: What Has Become of Them?*, volume 341 of *Synthese Library*, pp. 47–68. Springer.

Lowe, E. (2003). Individuation. In Loux, M. and Zimmerman, D. (eds), *Oxford Handbook of Metaphysics*, pp. 75–95. Oxford University Press, Oxford.

Parsons, C. (1965). Frege's Theory of Number. In Black, M. (ed.), *Philosophy in America*. Cornell University Press. (Reprinted in Parsons, 1983.)

Parsons, C. (1971). Ontology and Mathematics. *Philosophical Review*, 80:151–176. (Reprinted in Parsons, 1983.)

Parsons, C. (1983). *Mathematics in Philosophy*. Cornell University Press, Ithaca, NY.

Parsons, C. (1990). The Structuralist View of Mathematical Objects. *Synthese*, 84: 303–346.

Parsons, C. (2008). *Mathematical Thought and Its Objects*. Cambridge University Press, Cambridge.

Quine, W. (1950). Identity, Ostension, and Hypostasis. *Journal of Philosophy*, 47: 621–633. (Reprinted in Quine, 1953.)

Quine, W. (1953). *From a Logical Point of View*. Harvard University Press, Cambridge, MA.

Resnik, M. (1997). *Mathematics as a Science of Patterns*. Oxford University Press, Oxford.

Russell, B. (1919). *Introduction to Mathematical Philosophy*. Allen & Unwin, London.

Shapiro, S. (1997). *Philosophy of Mathematics: Structure and Ontology.* Oxford University Press, Oxford.

Shapiro, S. and Weir, A. (2000). 'Neo-Logicist' Logic is Not Epistemically Innocent. *Philosophia Mathematica*, 8(2):160–189.

Stalnaker, R. (1997). Reference and Necessity. In Hale, B. and Wright, C. (eds), *Blackwell Companion to the Philosophy of Language*, pp. 534–554. Blackwell, Oxford.

Williamson, T. (1990). *Identity and Discrimination.* Blackwell, Oxford.

Wright, C. (1983). *Frege's Conception of Numbers as Objects.* Aberdeen University Press, Aberdeen.

Wright, C. (1992). *Truth and Objectivity.* Harvard University Press, Cambridge, MA.

Wright, C. (2000). Neo-Fregean Foundations for Analysis: Some Reflections. *Notre Dame Journal of Formal Logic*, 41(4):317–334.

11
Toward a Trivialist Account of Mathematics

*Agustín Rayo**

The aim of this chapter is to defend mathematical trivialism—the view that the truths of pure mathematics have trivial truth-conditions and the falsities of pure mathematics have trivial falsity-conditions.

I doubt there can be an easy argument for trivialism, for two reasons. The first is that the debate got off to a bad start. Discussions in the philosophy of mathematics tend to presuppose a certain conception of the conceptual landscape that makes little room for trivialism. I think this conception is mistaken, and that once it is set aside, trivialism can be seen to be a plausible position. But old habits die hard. The second reason is that trivialists face an important challenge. They need to explain what the point of mathematical knowledge could be if mathematics deals with trivialities. I think there is a good answer to the challenge, and that the resulting picture of mathematical knowledge is independently attractive. But it is a picture that is not easy to set up, and is unlikely to seem compelling at first.

When Otávio and Øystein invited me to participate in this volume, I was faced with a choice. I could write a paper outlining the entire case for trivialism in very broad strokes, or I could focus on some particular portion of the argument, discussing it in detail at the expense of the rest of the material. In the end, I decided to do the former. The result is a paper that is somewhat impressionistic. Although many of the omissions are addressed in other work, you will find that crucial moves are made with insufficient discussion and that arguments are often sketchy. Please bear with me. In return, I will try to give you a sense of the "big picture" underlying the combination of views I defend.

1 Against conventional wisdom

Platonists and nominalists disagree about *ontology*: Platonism is the view that there are mathematical objects; nominalism is the view that there aren't any.

Committalists and noncommittalists, in contrast, disagree about the *truth-conditions* of mathematical statements: committalism is the view that everyday mathematical assertions carry commitment to mathematical objects; noncommittalism is the view that they don't. These two distinctions yield a fourfold partition of logical space, but the two most popular positions are Platonism+committalism and nominalism+noncommittalism. *Error-theories* (i.e. nominalism+committalism) and *irrelevance-theories* (i.e. Platonism+noncommittalism) are both consistent, but they are not as well represented in the literature. (A notable exception is the error-theory espoused in Field (1980).)

Conventional wisdom has it that each of the two most popular positions has an advantage over the other: nominalism+noncommittalism does a better job of accounting for mathematical knowledge since it doesn't need to explain how we could have knowledge of the abstract realm; but Platonism+committalism does a better job accounting for mathematical discourse since it doesn't need to postulate a non-standard semantics (or a less-than-straightforward connection between what is communicated by an assertion and the truth-conditions of the sentence asserted).

It seems to me that this is an unhelpful picture of the terrain. It focuses on the wrong distinction when it comes to epistemology, and it relies on a questionable conception of the way language works when it comes to mathematical discourse. I shall discuss each of these points in turn.

Say that a sentence has *trivial* truth-conditions if any scenario in which the truth-conditions fail to be satisfied would be unintelligible. (More on the relevant notion of intelligibility below.) We have no trouble making sense of a scenario in which there are no elephants, so we should take "there are elephants" to have *non*-trivial truth-conditions. But (most of us) are unable to make sense of a scenario in which something fails to be self-identical. So we should take a logical truth like "$\forall x(x = x)$" to have *trivial* truth-conditions. (Some dialetheists would disagree.)

Trivialism is the view that true sentences of pure mathematics have trivial truth-conditions (and that false sentences of pure mathematics have trivial falsity-conditions). According to the trivialist, *nothing* is required of the world in order for the truth-conditions of a mathematical truth to be satisfied: there is no intelligible possibility that the world would need to steer clear of in order to cooperate with the demands of mathematical truth. This means, in particular, that there is no need to go to the world to check whether any requirements have been met in order to determine whether a given mathematical truth is true. So once one gets clear about the sentence's truth-conditions—clear enough to know that they are trivial—one has done all that needs to be done to establish the sentence's truth. (Keep in mind that getting clear about the truth-conditions of a given mathematical sentence can be highly non-trivial. So determining whether the sentence is true is not, in general, a trivial affair.)

For the trivialist, our knowledge of pure mathematics can be understood on the model of our knowledge of pure logic. (More on this below.) The non-trivialist, however, owes us an account of what is required of the world in order for the truth-conditions of a given mathematical truth to be satisfied, and an explanation of how we might be in a position to check whether the relevant requirement has been met. So when it comes to the task of accounting for mathematical knowledge, the trivialist has an advantage over the non-trivialist. But it is important to note—and this is where conventional wisdom proves unhelpful—that the distinction between trivialism and non-trivialism *cuts across* the distinction between nominalism+noncommittalism and Platonism+committalism. In particular, one can be a nominalist and a noncommittalist without being a trivialist, and one can be a Platonist and a committalist while being a trivialist.

Suppose, for example, that one is a nominalist and embraces a version of noncommittalism whereby the truth-conditions of a sentence of pure mathematics are given by its *universal Ramseyfication*. (If ϕ is an arithmetical sentence, its universal Ramseyfication is the universal closure of $\ulcorner(\mathscr{A} \to \phi)^*\urcorner$, where $\mathscr{A}$ is the conjunction of a suitable list of axioms and ψ^* is the result of uniformly substituting variables for mathematical vocabulary in ψ.) Then as long as one is able to make sense of a finite world, one will take oneself to be a non-trivialist. To see this, consider an arbitrary arithmetical falsehood, $\mathscr{F}$. Since $\mathscr{A}$ can only be true if there are infinitely many objects, the universal closure of $\ulcorner(\mathscr{A} \to \mathscr{F})^*\urcorner$ can only be false if there are infinitely many objects. So the falsity-conditions of $\mathscr{F}$ will fail to be satisfied if the world is finite, and are therefore non-trivial.[1] Of course, a non-trivialist non-committalist might have a story to tell about how it is that the truth-conditions of mathematical truths (and the falsity conditions of mathematical falsehoods) can be known to be satisfied, even though they are non-trivial. The Universal Ramseyfier, for example, might have a story to tell about why it is that we're entitled to the assumption that there are infinitely many things. The point is that, unlike the trivialist, she needs a story to tell—non committalism does not, by itself, deliver epistemological innocence.

We have seen that one can be a nominalist and a noncommittalist without being a trivialist. But one can also be Platonist and a committalist while being a trivialist. Traditional Platonists think that even though numbers exist, it is intelligible that they do not exist. *Subtle* Platonists maintain, in contrast, that for the number of the Fs to be *n just is* for there to be *n* Fs. Accordingly, the view that there are no numbers is not just false, but unintelligible. ("Suppose there are no numbers. For the number of Fs to be 0 *just is* for there to be no Fs. So the number 0 must exist after all!"). This means that even if one is a committalist one should think that *nothing* is required of the world in order for the truth-conditions of "there are numbers" to be satisfied. For there is no intelligible possibility that the world needs to steer clear of. So there is room for a committalist who is also a subtle Platonist to be a trivialist. (Each of the following can be interpreted as defending a version of subtle Platonism: Frege (1884), Parsons (1983), Wright (1983), Stalnaker (1996), and Linnebo's contribution to this volume.)

Moral: if one is concerned with mathematical knowledge, the most interesting place to look is not the contrast between nominalism+noncommittalism and Platonism+committalism. It is the contrast between trivialism and non-trivialism.

I have argued that conventional wisdom delivers a potentially misleading picture of the epistemological terrain. I would now like to explain why I think it delivers a misguided picture of the linguistic terrain. According to conventional wisdom, Platonism+committalism does a better job than nominalism+noncommittalism when it comes to accounting for mathematical discourse. For consider a mathematical sentence such as "2 is prime". Proponents of Platonism+committalism can take "2" to refer to a particular object—the number 2—and claim that the sentence is true just in case that object has the property expressed by "is prime". But proponents of nominalism+noncommittalism think there are no numbers. So they lack a straightforward way of specifying truth-conditions for "2 is prime". They must either claim that the logical structure of mathematical statements shouldn't be taken at face value, or claim that the information conveyed by mathematical assertions is very different from what the sentences asserted literally say.

The problem with this way of approaching the issue is that it is based on a questionable picture of the workings of language: the idea that there is a certain kind of correspondence between the structure of language and the structure of the world. More specifically, what is presupposed is the following: (1) there is a particular carving of the world into objects which is more apt, in some metaphysical sense, than any potential rival—a carving that is in accord with the world's true "metaphysical structure"; (2) to each legitimate singular term there must correspond an object carved out by the world's metaphysical structure; and (3) satisfaction of the truth-conditions of an atomic sentence requires that the objects corresponding to singular terms in the sentence bear the property expressed by the sentence's predicate. (Should one be a deflationist about properties, and claim that for an object to have the property of Fness *just is* for the object to be F? Or, should one admit properties as separate items in one's ontology? Different versions of the view will address the issue in different ways.)

This conception of language is a close cousin of the "picture theory" that Wittgenstein defended in the *Tractatus*.[2] And it seems to me that it ought to be rejected for just the reason Wittgenstein rejected the picture theory in his later writings. Namely, if one looks at the way language is actually used, one sees that usage is not beholden to the constraint that an atomic sentence can only be true if its logical structure is in suitable correspondence with the structure of the world.

Assertions are tools for communication. Suppose you are organizing a dinner party and are thinking about seating arrangements. I say "There will be an odd number of people at the table". In doing so, my objective is not to represent the structure of reality as somehow corresponding to the logical structure of the sentence I uttered. In particular, I do not mean to commit myself to a non-trivial ontological thesis about numbers, and go on to represent numbers as bearing a certain relation to people and the table. All I want to do is help you discriminate amongst the possibilities at hand. Suppose, for example, that you are trying to decide between using the round table and using the rectangular table. Then the point of my assertion will be fully satisfied if I get you to opt for the former, and get you to understand why this is the right decision—a rectangular table will make for awkward seating. If I also happen to succeed in limning the structure of reality with my assertion, by choosing a sentence with the right logical structure, that is no part of what I set out to do.

If *assertions* of sentences involving mathematical vocabulary are not intended to limn the structure of the world, what could be the motivation for thinking that the truth-conditions of the sentences themselves play this role? As far as I can tell, it is nothing over and above the idea that the logical structure of atomic sentences should correspond to the structure of the world. Remove this idea and there is no motivation left.

It is true that our language is compositional. But the role of compositionality is to allow for the production of large numbers of sentences from a restricted lexicon. To claim that compositionality plays the *additional* role of allowing for the representation of the structure of reality is to set forth a doctrine that is not supported by our linguistic usage. It is to start out with a preconception of the way language ought to work, and impose it on our linguistic theorizing from the outside—from beyond what is motivated by the project of making sense of our linguistic practice. (For further discussion of Tractarian conceptions of language, see Heil (2003).)

We can now see where conventional wisdom goes wrong. The idea was supposed to be that Platonism+committalism is able to take the logical structure of a sentence like "2 is prime" at face value because it has a matching ontology to offer: the reference of "2" is taken to exist and is taken to have the property expressed by "is prime". Nominalism+noncommittalism, in contrast, lacks the needed ontology, so it must choose between treating "2 is prime" as saying something literally false and meddling with its logical structure. But once one abandons the doctrine that the logical structure of an atomic sentence must correspond to the structure of reality, there is room for a distinction between the *semantic values* of expressions occurring in a sentence—a piece of theoretical machinery used to explain how the meanings of complex expressions depend on the meanings of its parts—and the objects that must exist in order for the sentence's truth-conditions to be satisfied. In particular, a friend of nominalism+noncommittalism can take the logical structure of "2 is prime" at face value, and assign semantic values to "2" and "is prime" in the course of developing a compositional semantic theory, while resisting the conclusion that satisfaction of the sentence's *truth-conditions* requires that the semantic value of "2" exist and instantiate the semantic value of "is prime". This is a tricky point, so I shall dwell on it further.

Consider a sentence like "roses are red *and* violets are blue". Standard semantic theories assign a semantic value to "and"—a certain kind of *function*—but it would be a mistake to go from this to the conclusion that the sentence carries commitment to functions: that part of what is required of the world in order for the sentence's truth-conditions to be satisfied is that there be functions. To take the additional step would be to misjudge the role of semantic values in our semantic theorizing. The point of assigning a semantic value to "and" is that we want our semantics to be *compositional*—we want a systematic way of determining the semantic properties of sentences of the form $\ulcorner\phi$ and $\psi\urcorner$ on the basis of the semantic properties of ϕ and ψ—not to get the result that sentences involving "and" count the semantic value amongst their ontological commitments.

Most philosophers take a similar attitude toward the semantic values of predicates. Consider a sentence like "Susan runs". Standard semantic theories assign a semantic value to "runs"—in the simplest case, an extension—but most of us would want to resist the conclusion that "Susan runs" carries commitment to extensions: that part of what is required of the world in order for the sentence's truth-conditions to be satisfied is that there be extensions. Again, the point of assigning a semantic value to "runs" is that we want our semantics to be *compositional*—we want a systematic way of determining the semantic properties of expressions of the form $\ulcorner t$ runs$\urcorner$ on the basis of the semantic properties of t—not to get the result that sentences involving "runs" count the semantic value amongst their ontological commitments.

When it comes to singular terms, however, it is common for philosophers to think of semantic values as playing a broader role. Philosophers often presuppose that the semantic value of a singular term t does more than just deliver a systematic way of determining the semantic properties of sentences of the form $\ulcorner t$ Fs$\urcorner$ on the basis of the semantic properties of F. There is the additional requirement that the semantic value of t be counted amongst the ontological commitments of $\ulcorner t$ Fs$\urcorner$. How is this expanded role for the semantic values of singular terms supposed to be motivated? As far as I can tell, it is nothing over and above the idea that the logical structure of atomic sentences should correspond to the structure of the world. Remove this idea and there is no motivation left.

This is not to say, of course, that a sentence of the form ⌜t Fs⌝ should *always* remain uncommitted to the semantic value of t. In many cases— when t is "Susan" and F is "runs", for example—commitment to the semantic value of t is the right result. The point I wish to make is that commitment to the semantic value of t shouldn't be regarded as *automatic*. When appropriate, it can be secured by assigning the right semantic values to t and F, and specifying the right rule for extracting truth-conditions from the semantic values of sentences. But one shouldn't assume that a sentence of the form ⌜t Fs⌝ must carry commitment to a certain object merely on the grounds that that object has been assigned as t's semantic value. As in the case of lexical items falling under different syntactic categories, the reason for assigning semantic values to singular terms is to allow for compositionality, not to secure a correspondence between the logical structure of our sentences and the structure of the world.

In a semantic theory where the role of semantic values is exhausted by considerations of compositionality—together with the principle that the semantic value of a sentence must determine truth-conditions for the sentence, relative to suitable contextual parameters—it is possible to develop a semantics for mathematical discourse that runs contrary to conventional wisdom. One can take the logical structure of mathematical sentence at face value and still get the conclusion that all that is required of the world in order for the truth-conditions of "the number of the planets is eight" to be satisfied is that there be eight planets, and the conclusion that any true sentence in the language of pure mathematics gets assigned trivial truth-conditions and any false sentence in the language of pure mathematics gets assigned trivial falsity-conditions. (See my "On Specifying Truth-Conditions" for details. For a related discussion, see Hofweber's contribution to this volume.)

I hear some complaints:

1. *Objection:* You claim that one can give a semantic theory according to which mathematical sentences carry no objectionable commitments. But the semantic theory itself will carry all sorts of commitments (functions, extensions, etc.). So we're stuck with the problematic commitments anyway! How is this any help to a friend of nominalism?

   Response: Suppose we've established the view that sentences of the object-language carry no commitment to abstract objects. Then we've established that there can be a type of quantifier that has the same syntax and inferential patterns as the committalist's quantifier but carries no commitment to abstract objects. A friend of nominalism+noncommittalism will claim that she uses such quantifiers when she uses the object-language—and, therefore, that she incurs no problematic commitments in ordinary mathematical discourse. But she will wish to make an additional claim: she will claim that she also uses the special quantifier when she does *semantics*. So she can consistently claim that problematic commitments are incurred neither when she uses the object-language to do mathematics nor when she uses the metalanguage to do semantics. (Whether she can use this move to convince the unconvinced is a different matter—again, see *OSTC* for details.)

2. *Objection:* Suppose I buy your semantics, and you convince me that all that is required of the world in order for the truth-conditions of "the number of the planets is eight" to be satisfied is that there be eight planets. Why should I conclude from this that "the number of the planets is 8" carries no problematic commitments? After all, everyone should believe that "the number of the planets is eight" is true just in case the number of the planets is eight. Isn't this enough for the conclusion that "the number of the planets is eight" carries commitment to numbers?

   Response: In asking about the ontological commitments of a sentence, there is a potential ambiguity. On one reading of the question, one wants to know which of the objects that are carved out by the world's metaphysical structure need to exist in order for the sentence's truth-conditions to be satisfied. If this is how you think about the matter, you should say that the truth-conditions of "the number of the planets is eight" can be stated in "more fundamental" and "less fundamental" terms. The more fundamental statement—the one that tells us which of the objects carved out by the world's metaphysical structure must exist in order for the sentence to be true—is the one delivered by the proposed semantics: that there be eight planets. A less fundamental statement—one that makes no effort to limn metaphysical structure—is that the number of the planets be eight. (I am not myself able to understand what people mean by "metaphysical structure" or "metaphysically fundamental", but Cameron (forthcoming) and Williams (typescript) have developed the proposal in this direction.) On a different reading of the question—the reading that I prefer—one is unconcerned with metaphysical structure. In asking about a sentence's ontological commitments, all one wants is an informative statement of how the world must be in order for the sentence's truth-conditions to be satisfied. On this second reading, the proposed semantics can be used to argue that even though one can accurately specify the truth-conditions of "the number of the planets is eight" by saying "that the number of the planets be eight", it would be just as accurate to say "that there be eight planets". Neither of these statements counts as "more fundamental" than the other: for the number of the planets to be eight *just is* for there to be eight planets. And it is in this sense that "there are eight planets" can be said to carry no problematic commitments.

Moral: The claim that only committalists are in a position to take the logical structure of mathematical sentences at face value is based on a questionable conception of language. When the problematic assumptions are dropped, there is no obstacle to taking logical structure at face value while being a non-committalist—or, indeed, a trivialist.

★ ★ ★

Where does all of this leave us? First and foremost, I would like to urge you to consider becoming a trivialist. In doing so, you would put yourself in a position to give a satisfying answer to a question that has been a source of endless woe for epistemologists of mathematics: how do we know that the world satisfies the requirements that would need to be satisfied in order for the truths of pure mathematics to be true and its falsehoods to be false? While the non-trivialist is searching for a way to justify non-trivial claims to the effect that the realm of abstract objects has a certain property, or that the world is infinite, you will confidently proclaim: "The relevant requirements can be known to be satisfied because *nothing* is required of the world in order for a truth of pure mathematics to be true, and *nothing* is required of the world in order for a falsity of pure mathematics to be false."

If you were beholden to conventional wisdom, you would worry that the price to be paid for trivialism is semantic awkwardness—you would fear that by becoming a trivialist you would have had to choose between meddling with the logical form of mathematical statements and claiming that they are used to convey information which is very different from what they literally say. But now you know that such fears would be doubly mistaken. You know, first of all, that as long as one is a *subtle* Platonist one can be a trivialist even if one is also a committalist (and there was never any worry about committalists being led into semantic awkwardness). You know, moreover, that the claim that *non*committalists are faced with semantic awkwardness is based on a questionable conception of language. So you know that even if one were unhappy about subtle Platonism, one could be a trivialist by becoming a noncommittalist and abandoning the idea that truth can only be achieved through a correspondence between the structure of language and the structure of reality.

The previous paragraph identifies two ways of being a trivialist without plunging into semantic awkwardness: subtle Platonism and post-Tractarian noncommittalism. It is worth emphasizing that these views more closely related than one might think. Suppose you are a subtle Platonist. You believe that for the number of the Fs to be *n just is* for there to be *n* Fs. So when the committalist claims that satisfaction of the truth-conditions of "the number of the planets is 8" requires of the world that the planets be numbered by the number 8, and the noncommittalist counters that all that is required is that there be eight planets, you will see them as stating the very same requirement—for the committalist's requirement to be met *just is* for the noncommittalist's requirement to be met.

2 Knowing trivial truths

Suppose trivialism is right and the truths of pure mathematics have trivial truth-conditions. What could the point of mathematical knowledge be? The purpose of this section is to answer that question. But it will take some time to set things up—the crucial discussion won't take place until Section 2.3.

2.1 Intelligibility

Let a *story* be a set of sentences in some language we understand. I shall assume that stories are read *de re*: that every name used by the story is used to say of the name's actual bearer how it is according to the story, and that every predicate used by the story is used to attribute the property actually expressed by the predicate to characters in the story. Accordingly, in order for a story that says "Hesperus is covered with water" to be true it must be the case that Venus itself is covered with H_2O. (I shall ignore names that are actually empty, such as "Sherlock Holmes", and predicates that are actually empty, such as "... is composed of phlogiston" or "... is a unicorn".)

Sometimes we describe a story as unintelligible on the grounds that it is too complicated for us to understand. That is not the notion of unintelligibility I have in mind here. As I understand the term, a story is *unintelligible* for a subject if her best effort to make sense of a scenario in which the story is true would yield something she regards as *incoherent*. (Intelligibility can then be defined as non-unintelligibility.) Let me give you some examples of what I have in mind.

Consider a story that says "A fortnight elapsed in only 13 days". My best effort to make sense of a scenario in which this story is true ends in incoherence. "Fortnight" *means* "period of 14 days". So a scenario verifying "A fortnight elapsed in only 13 days" would have to be a scenario in which a period of 14 days lasts only 13 days, which is something I regard as incoherent. (Of course, it would be easy enough to make sense of a scenario in which *language* is used in such a way that the expression "a fortnight elapsed in only 13 days" is true. But that won't help with the question of whether the original story is intelligible, in the relevant sense.)

The preceding example might tempt you to think that only "conceptually inconsistent" stories count as unintelligible. But consider a story that says "Hesperus is not Phosphorus"—presumably an example of a "conceptually consistent" statement. My best effort to make sense of a scenario in which this story is true ends in incoherence. For a scenario in which the story is true would have to be a scenario in which Hesperus itself (i.e. Venus) fails to be identical with Phosphorus itself (i.e. Venus), and the non-self-identity of Venus is something I regard as incoherent. Another example is

as follows: consider a story that says "there is a lake with water but no H_2O".
For something to contain water *just is* for it to contain H_2O. So a scenario
verifying "there is a lake with water but no H_2O" would have to be a scenario
in which something that is filled with H_2O fails to be filled with H_2O—which
I regard as incoherent. (As before, it would be easy enough to make sense of
a scenario in which *language* is used in such a way that the expression "Hesperus is not Phosphorus" or "there is a lake with water but no H_2O" is true.
But that is irrelevant to the issue at hand.)

Objection: Given that you insist on a *de re* reading of stories, I can see why
you want to treat "Hesperus is not Phosphorus" and "there is a lake with
water but no H_2O" as unintelligible. But consider a scenario in which
the first celestial body to be visible in the evenings is not the last celestial body to disappear in the morning. I grant you that "Hesperus is not
Phosphorus" is not literally true in this scenario. But surely there is some
derived sense of "verify" such that "Hesperus is not Phosphorus" is verified by this scenario. So there is a certain sense in which "Hesperus is not
Phosphorus" is intelligible after all.

Reply: In order to claim that this alternate notion of intelligibility is
well-defined when we go beyond toy examples like "Hesperus is Phosphorus", one needs a substantial assumption: the assumption that every
name and every predicate has a "narrow content" or "primary intension"
which can be used to determine which scenarios will count as verifying
a given sentence in the derived sense of "verify". I myself am skeptical
of this assumption since I find no evidence for it in linguistic practice.
(As far as I can tell, all that is required by our actual linguistic usage is
the ability to determine which of a highly restricted set of contextually
salient possibilities would count the sentence asserted as expressing a true
proposition—see my "Vague Representation" for details.) But nothing in
this chapter hinges on rejecting the assumption. If you think the alternate characterization of intelligibility is legitimate, that's fine. Just keep
in mind that it's not the one that will be relevant in this chapter.

Moral: As it is understood here, the intelligibility of a story (for a subject) is
a highly non-*a priori* matter. For what one finds unintelligible depends on
whether one believes that Hesperus is Phosphorus, or that water is H_2O. And
knowledge of such truths is far from *a priori*.

2.2 Intelligibility and identity

I would like to suggest that there is a close connection between intelligibility
and *identity*.

Statements of the form "$a = b$" are identity statements. But they are only a
special case. Consider the following sentences:

SIBLING
To be a sibling *just is* to share a parent.
[In symbols: "Sibling$(x) \equiv_x \exists y \exists z (\text{Parent}(z,x) \land \text{Parent}(z,y) \land x \neq y)$"]

HEAT
To be hot *just is* to have high mean kinetic energy.
[In symbols: "Hot$(x) \equiv_x$ High-Mean-Kinetic-Energy(x)"]

WATER
To be composed of water *just is* to be composed of H_2O.
[In symbols: "Composed-of-water$(x) \equiv_x$ Composed-of-$H_2O(x)$"]

In these three sentences, the expression "just is" (or its formalization "$\equiv_x$")
is functioning as an identity-predicate of sorts. To accept "$F(x) \equiv_x G(x)$" is not
simply to accept that all and only the Fs are Gs. If you accept SIBLING, for
example, you believe that there is *no difference* between being a sibling and
sharing a parent with someone; you believe that if someone is a sibling it
is *thereby* the case that she shares a parent with someone. (Compare: If you
accept "Hesperus is Phosphorus", you believe that someone who travels to
Hesperus has *thereby* traveled to Phosphorus.)

One might be tempted to describe SIBLING, HEAT and WATER as expressing identities amongst *properties* (e.g. "the property of being a sibling = the
property of sharing a parent".) I have no qualms with this description, as
long as property-talk is understood in a suitably deflationary way. But I will
avoid property-talk here because it is potentially misleading. It might be
taken to suggest that one should only assert SIBLING if one is prepared to
countenance a traditional Platonism about properties—the view that even
though it is intelligible that there be no properties, we are lucky enough to
have them. The truth of SIBLING, as I understand it, is totally independent
of such a view. If one wishes to characterize the difference between "$\equiv_x$"
and the standard first-order identity predicate "=", the safe thing to say is
that whereas "=" takes a singular-term in each of its argument-places, "$\equiv_x$"
takes a first-order predicate in each of its argument places. I shall therefore
refer to sentences of the form $\ulcorner a = b \urcorner$ (where *a* and *b* are singular terms)
as *first-order* identity statements, and sentences of the form $\ulcorner \phi(x) \equiv_x \psi(x) \urcorner$
(where $\phi(x)$ and $\psi(x)$ are first-order predicates) as *second-order* identity
statements.

Sometimes one is in a position to endorse something in the vicinity of a
second-order identity statement even though one has only partial information. Suppose you know that the chemical composition of water includes
oxygen but don't know what else is involved. You can still say:

Part of what it is to be composed of water is to contain oxygen.
[In symbols: "Composed-of-water$(x) \ll_x$ Contains-Oxygen(x)."]

I shall call this as a *semi-identity statement*. Think of it as a more idiomatic a way of saying:

> To be composed of water *just is* (to contain oxygen and to be composed of water).

(Please note that it is no part of the view that "$F(x) \ll_x G(x)$" entails that something is an F "in virtue" of being a G, or that being a G is "more fundamental" than being an F, or that being G is part of the "essence" of an F.)

As in the case of second-order identity statements, it is tempting to think of semi-identity statements in terms of properties (e.g. "the property of being water has the property of containing oxygen as a part".) Again, I have no objection to this sort of description, as long as property-talk is taken in a suitably deflationary spirit. But I will avoid it here because of its potential to mislead.

Second-order identity statements can be dispensed with in the presence of semi-identity statements. WATER, for example, is equivalent to the conjunction of "part of what it is to be water is H_2O" and "part of what it is to be H_2O is to be water". And, in general, "$F(x) \equiv_x G(x)$" is equivalent to the conjunction of "$F(x) \ll_x G(x)$" and "$G(x) \ll_x F(x)$". Note, moreover, that the content of a first-order identity-statement "$a = b$" can be expressed by way of the second-order identity-statement "$x = a \equiv_x x = b$"—to be *a just is* to be *b*. (More precisely: "$a = b$" is equivalent to the conjunction of "$x = a \equiv_x x = b$" and "$\exists x(a = x)$".) This means that semi-identity statements can be used to do the work of both first- and second-order identity statements.

A few paragraphs back I hinted at a close connection between intelligibility and identity. I can now tell you what I take the connection to be. To wit: the sole source of unintelligibility for a subject is inconsistency with the semi identities she accepts. (More precisely: a story will be unintelligible for a subject just in case she is in a position to derive, using inferences she takes to be logically valid, something she regards as incoherent from the result of adding the story to the set of semi-identities she accepts.)

Three observations: (1) In suggesting a connection between identity and intelligibility, I do not mean to suggest that one of these notions is "prior" to the other. The claim is that the two notions are connected, and that one can better understand them by understanding the connection. (2) So far we have focused our attention on intelligibility for a subject. What about intelligibility *simpliciter*? If you think there is an objective fact of the matter about which semi-identities are true, you can go on to say that a story is intelligible *simpliciter* just in case it is logically consistent with the set of true semi-identities. (3) So far we have focused on the intelligibility of a story. What about the intelligibility of a *scenario*? The usual way of picking out a scenario is by setting forth a story (or some other kind of representation);

in this special case, one can say that a scenario is intelligible just in case the story used to pick it out is intelligible. (For a more detailed discussion of intelligibility, see my "An Account of Possibility".)

The connection between identity and intelligibility can help us get a grip on the elusive notion of a sentence's truth-conditions. To see this, note that one can think of a sentence's truth-conditions as a requirement imposed on the world—the requirement that the world be a certain way. Knowing whether a scenario we take to be intelligible must fail to obtain in order for the requirement to be met is valuable because it gives us an understanding of how the world would need to be in order for the requirement to be satisfied. But knowing whether a scenario we take to be *un*intelligible must fail to obtain in order for the requirement to be met is not very helpful. For when one is unable to make sense of a scenario, the claim that it must fail to obtain gives one no understanding of how the world would need to be in order for the requirement to be satisfied. The lesson is that one can model a sentence's truth-conditions as a partition of the space of scenarios one takes to be intelligible. Accordingly, the connection between semi-identity and intelligibility yields a connection between semi-identity and a sentence's truth-conditions: since the scenarios one regards as intelligible will depend on the semi-identities one accepts, and since a sentence's truth-conditions can be modeled as a set of intelligible scenarios, one's beliefs about the range of possible truth-conditions will depend on the semi-identities one accepts.

A first consequence of this observation is that a subject is committed to seeing any sentence she takes to follow from semi-identity statements she accepts as having trivial truth-conditions. If, for example, you think that to be composed of water *just is* to be composed of H_2O, then you are committed to thinking that nothing is required of the world in order for the truth-conditions of "if a lake contains water, it contains H_2O" to be satisfied. And if you think that to be Hesperus *just is* to be Phosphorus, you are committed to thinking that nothing is required of the world in order for the truth-conditions of "if Hesperus is a planet, Phosphorus is a planet" to be satisfied.

It is worth considering some additional consequences of the connection between semi-identity and intelligibility. The semi-identity operator "$\ll$" can bind more than one variable. For instance,

Sisters$(x, y) \ll_{x,y} \exists z(\text{Parent}(z, x) \wedge \text{Parent}(z, y))$
[*Read:* part of what it is for x and y to be sisters is for x and y to share a parent.]

But it can also bind no variables at all

a wedding takes place $\ll$ someone gets married
[*Read:* part of what it is for a wedding to take place is for someone to get married.]

The same is true of second-order identity statements. For instance,

a wedding takes place $\equiv$ someone gets married
[*Read:* for a wedding to take place *just is* for someone to get married.]

In light of the connection between semi-identity statements and truth-conditions, this yields the result that a subject who accepts "for a wedding to take place *just is* for someone to get married" is committed to the view that "a wedding takes place" and "someone gets married" have the same truth-conditions.

Now let T be a sentence which is known to have trivial truth-conditions. One can use $\ulcorner p \equiv T \urcorner$ to capture the thought that p has trivial truth-conditions. For instance,

$\forall x(x = x) \equiv T$
[*Read:* that everything is identical is trivially the case.]

Second-level identity is an equivalence relation. So if one accepts $\ulcorner p \equiv T \urcorner$ and accepts $\ulcorner q \equiv T \urcorner$ one is committed to accepting $\ulcorner p \equiv q \urcorner$ and $\ulcorner q \equiv p \urcorner$. If one takes seriously the idea that a logical truth has trivial truth-conditions, then part of what one does when one recognizes p as a logical truth is to accept $\ulcorner p \equiv T \urcorner$. So one is committed to accepting $\ulcorner p \equiv q \urcorner$ for any p and q which one recognizes as logical truths. (When I say "logical truth" I mean "truth of a free logic" since we want to avoid the result that $\ulcorner \exists x(x = c) \urcorner$ has trivial truth-conditions whenever c is a proper name.)

The same goes for mathematics according to the trivialist. Suppose you think that nothing is required of the world in order for the truth-conditions of "there are numbers" to be satisfied—equivalently

there are numbers $\equiv T$
[*Read:* that there be numbers is trivially the case.]

You have thereby committed yourself to

there are numbers $\equiv \forall x(x = x)$
[*Read:* for there to be numbers *just is* for everything to be self-identical.]

For if it is trivially the case that there are numbers and it is trivially the case that everything is self identical, then there is no more to the world's satisfying the one than there is to the world's satisfying the other: if the world is such that there are numbers, it is *thereby* such that everything is self-identical, and if it is such that everything is self-identical, it is *thereby* such that there are numbers. So "there are numbers" and "$\forall x(x = x)$" have the same truth-conditions. More generally, every true sentence of pure mathematics has the same truth-conditions as any other.

Moral: Suppose you buy into the idea that a story is intelligible for a subject just in case the subject takes the story to be consistent with the set of semi-identities she accepts. Then you should also buy into the idea that a subject's views about the range of possible truth-conditions—and therefore her views about which sentences share their truth-conditions—will be shaped by the identity statements she accepts.

2.3 Knowledge

The discussion in the preceding sections suggests that our cognitive attitudes toward semi-identity statements and ordinary statements play different roles: by accepting semi-identity statements, we fix the limits of what we take to be intelligible; by accepting ordinary statements, we partition the space of intelligible scenarios into regions that are treated as candidates for truth and regions that are ruled out as false. In this section, I will sketch an epistemological picture based on this idea.

In rough outline, the picture is as follows. In an effort to satisfy our goals, we develop strategies for interacting with the world. Fruitful strategies allow us to control what the world is like and predict how it will evolve under specified circumstances. They also allow us to direct our research in ways that lead to the development of further fruitful strategies. In order to articulate the strategies we adopt, we do three things at once: firstly, we develop a *language* within which to formulate theoretical questions; secondly we set forth *theoretical claims* addressing some of these questions; and finally, we endorse a family of *semi-identity statements*. The third task is connected to the other two because the semi-identity statements we endorse help determine which theoretical questions are worth investigating and which are not. To a certain extent, the endeavor is a holistic one. It is sometimes possible to vary the semi-identity statements one endorses without significantly affecting the success of one's methods of inquiry, provided one makes compensating adjustments in the theoretical claims one accepts. (For a more detailed discussion, see my "An Account of Possibility".)

Consider, for example, our acceptance of "part of what it is to be water is to contain hydrogen and oxygen". It affects our overall theorizing by ruling out certain questions as pointless while allowing others as fruitful. (See Block and Stalnaker (1999) for further details.) For example, we would regard it as wrong-headed to try to understand *why* every portion of water contains oxygen and hydrogen. ("Water *just is* H_2O!") But we see it as a worthwhile endeavor to try to understand *why* liquid water is colorless. Now suppose we vary the water-related semi-identities we accept. Say we reject "part of what it is to be water is to contain hydrogen and oxygen" and accept instead "part of what it is to be water is to be a colorless liquid". Then we will see a different range of questions as worth pursuing. It will now seem pointless to ask why liquid water is colorless ("that's just part of what it *is* to be water!") but interesting to ask why every portion of water contains oxygen and hydrogen. The change would also affect the way certain theoretical claims are formulated. For instance, we would need to reformulate the principle that water at sea-level freezes at 0 degrees celsius. Nevertheless, someone sufficiently committed to the alternate semi-identity might be able make enough adjustments elsewhere in the system to secure a successful articulation of her methods of inquiry.

Is there an objective fact of the matter about which choice of semi-identity statements is correct? My own view is that there is not. All one can say is that different choices are more or less amenable to the development of a successful articulation of one's methods of inquiry, given the way the world is. The empirical facts make it the case that acceptance of "part of what it is to be water is to contain hydrogen and oxygen" leads to particularly fruitful theorizing, and one has strong reasons to accept it on this basis. But there is no more to be said on its behalf. And when it comes to other semi-identity statements, the empirical pressures are even milder. For example, whether it is a good idea to accept "part of what it is to be a reptile is to have a certain lineage" rather than "part of what it is to be a reptile is to have a certain phenotype" might to a large extent depend on the purposes at hand. I would like to emphasize, however, that nothing in this chapter will turn on taking a pragmatic attitude toward semi-identity statements. If you think there is an objective fact of the matter about which semi-identities are correct, that's fine for present purposes.

Let us now turn our attention to logical truth. Suppose you learn that it fails to be the case that $\neg p$. Then, if you are a friend of classical logic, you will be in a position to conclude that p. But not just that: you will think that your understanding of why it fails to be the case that p is *already* an understanding of why it is the case that p. There is no need to add an explanation of why the transition is valid since there is nothing to be explained. For it to fail to be the case that $\neg p$ *just is* for it to be the case that p. There is no intelligible scenario in which the transition fails, whose absence would need to be accounted for. Asking "I can see that if it is not the case that $\neg p$ it will be the case that p, but *why* is this so?" is as wrong-headed as asking "I can see that water is H_2O, but *why* is this so?".

The point generalizes to more complex logical truths. When one treats a sentence ϕ as *logically* true, one does more than simply treat it as true. One is, in effect, accepting the higher-order identity statement $\ulcorner \phi \equiv \mathsf{T} \urcorner$ (which is equivalent to accepting $\ulcorner \psi \equiv \theta \urcorner$ when ϕ is of the form $\ulcorner \psi \leftrightarrow \theta \urcorner$, and equivalent to accepting $\ulcorner \psi \ll \theta \urcorner$ when ϕ is of the form $\ulcorner \psi \rightarrow \theta \urcorner$). This means that the result of treating $\ulcorner \psi \rightarrow \theta \urcorner$ as a logical truth is not just that one will take oneself to be justified in accepting θ whenever one feels justified in accepting ψ. One will think that one's understanding of why ψ's truth-conditions are satisfied is *already* an understanding of why θ's truth-conditions are satisfied.

As before, there is no need to add an explanation of why the *transition* from ψ to θ is valid: that θ's truth-conditions be satisfied is *part of what it is* for ψ's truth-conditions to be satisfied. Of course, when $\ulcorner \psi \to \theta \urcorner$ is sufficiently complex, coming to recognize it as a logical truth may be a highly non-trivial process. And throughout that process one might be justified in asking oneself *why* $\ulcorner \psi \to \theta \urcorner$ is a logical truth. But one's query can be addressed by setting forth a sufficiently illuminating proof. And once one has understood such a proof, one will see that there is no intelligible scenario in which the transition from ϕ to θ fails, and therefore that one's understanding of why ψ's truth-conditions are satisfied is *already* an understanding of why θ's truth-conditions are satisfied.

We are now in a position to answer an important question about logical knowledge: if the truths of pure logic have trivial truth-conditions, what could the point of knowing a logical truth be? The answer is that in learning a logical truth one increases one's ability to distinguish between intelligible and unintelligible scenarios, and therefore one's ability to use old information in new ways. (See Stalnaker (1984) chapter 5, and Stalnaker (1999) chapters 13 and 14.) An example will help illustrate the point.

Suppose that there are seventeen apples, and that you have counted them. This gives you a certain range of abilities. You are able to determine whether you got short-changed at the market, or whether there are enough apples for your recipe. You are also able to answer questions of the form "How many apples?" One might represent that you have such a range of abilities by claiming that you know that there are seventeen apples [in symbols: $\exists_{17}!x(\text{Apple}(x))$]. Now suppose that there are twenty-nine pears, and you have also counted them. This, again, gives you a distinctive range of abilities, a fact that might be represented by claiming that you know that there are twenty-nine pears [in symbols: $\exists_{29}!x(\text{Pear}(x))$]. Perhaps you are able to combine these two cognitive accomplishments in the service of a single task. You might, for instance, be in a position to determine whether there are more apples than pears. But other tasks might elude you. Say you know that every relevant piece of fruit is an apple or a pair, and that no piece of fruit is both an apple and a pair. Then you have all the information you need to answer questions of the form "How many pieces of fruit?". But you may still not be in a position to use the information at you disposal for that particular task, at least not immediately. What is missing is knowledge of a logical truth:

$$\exists_{17}!x(\text{Apple}(x)) \land \exists_{29}!x(\text{Pear}(x)) \land \neg\exists x(\text{Apple}(x) \land \text{Pear}(x)) \to$$

$$\exists_{46}!x(\text{Apple}(x) \lor \text{Pear}(x))$$

In performing the relevant computation, do you acquire novel information about the world? It is tempting to say that you do since you will learn there are forty six pieces of fruit. But I think the right thing to say is that you don't. For you already knew that every piece of fruit is an apple or a pair (but not both) and that there are seventeen apples and twenty nine pears, and *part of what it is* for that to be the case is that there be forty six pieces of fruit. In carrying out the computation, your cognitive accomplishment consists not in the acquisition of new information, but in the ability to deploy old information in new ways. Before you carry out the computation you are unsure about whether a scenario in which there are, say, *thirty*-six pieces of fruit could be genuinely intelligible while respecting the information you already had about apples and pears. What the computation reveals is that it is not. You have increased your ability to distinguish between intelligible and unintelligible scenarios, and this gives you the ability to see how to answer questions of the form "How many pieces of fruit?" in light of the information you had at your disposal all along. (I have greatly benefitted from discussion with Adam Elga on these topics.)

When one embraces a logical system one adopts a framework for settling questions of intelligibility. In deciding which logic to accept, one must therefore strike a delicate balance. If one's logic is too strong, it will commit one to treating as unintelligible scenarios that might have been useful in making sense of the world. By weakening one's logic one opens the door to a larger range of intelligible scenarios, all of them candidates for truth. In discriminating amongst them, one will have to explain why one favors the ones one favors. The relevant explanations may sometimes lead to fruitful theorizing about the world. But they may also prove burdensome. Consider a friend of intuitionistic logic, who denies that for it to fail to be the case that ¬*p just is* for it to be the case that *p*. She thinks it might be worthwhile to ask why it is the case that *p* even if you fully understand why it is not the case that ¬*p*. In the best case scenario, making room for an answer will lead to fruitful theorizing. But things may not go that well. One might come to see the newfound conceptual space between a sentence and its double negation as a pointless distraction, demanding explanations in places where there is nothing fruitful to be said. (For a particularly insightful discussion of intuitionistic logic, see Wright (2001).)

There can sometimes be empirical pressure of a more or less direct kind in favor of a particular semi-identity statement—think of "part of what it is to be water is to contain hydrogen and oxygen". But when it comes to choosing a logical system, the decision to accept the relevant semi-identities is likely to be driven by considerations of a more general nature. We want a framework for settling questions of intelligibility that is flexible enough to allow for interesting questions to be posed but constrained enough to make our insights transferable to a large range of contexts.

Could one be an *objectivist* about logic, and claim that there is an objective fact of the matter about which logical system is correct? Perhaps the thought is that the world has one true "logical structure", and that a logical system is correct to the extent that it does justice to the logical structure of the world. (How do we know which logical system is objectively correct? Maybe we get evidence of correctness when a logical system delivers a useful framework for settling questions of intelligibility.) Nothing I have said is in tension with making such additional claims. But, as far as I can tell, there they would be unmotivated without a prior commitment to the objectivist standpoint.

Let us finally turn to the case of mathematics. If you are a non-trivialist, you think there is a world of difference between logic and mathematics. Whereas the truths of pure logic have trivial truth-conditions (and its falsities have trivial falsity-conditions), there are intelligible scenarios that fail to satisfy the truth-conditions of some mathematical truth (or the falsity-conditions of some mathematical falsehood). Accordingly, one needs some sort of entitlement to the view that the rogue scenarios fail to obtain before one can claim to know that the relevant mathematical truths are true (or that the relevant mathematical falsehoods are false). For the trivialist, in contrast, there is no deep difference between logic and mathematics. As in the case of logic, mathematical truths have trivial truth-conditions (and mathematical falsities have trivial falsity-conditions). The difference is simply that the language of mathematics enjoys expressive resources that the language of logic lacks.

These enhanced expressive resources are important in two ways. First, they allow us to articulate requirements on the world that cannot be articulated in the language of pure logic, or that can only be articulated with significant awkwardness. By using the sentence "$\#_x\text{Apple}(x) = \#_x\text{Pear}(x)$", for example, one can express the thought that there be just as many apples than pairs— something that cannot be done within the language of first-order logic in any straightforward sense. (Can the trivialist generalize this point, and give a recipe that specifies ontologically innocent truth-conditions for arbitrary mathematical sentences? This is not as easy as one might think, but see my *OSTC*.)

Second, the enhanced expressive resources of mathematics improve our ability to sort out the intelligible from the unintelligible. Consider the rather unlovely logical truth that I mentioned a few paragraphs back:

$$\exists_{17}!x(\text{Apple}(x)) \wedge \exists_{29}!x(\text{Pear}(x)) \wedge \neg\exists x(\text{Apple}(x) \wedge \text{Pear}(x)) \rightarrow$$

$$\exists_{46}!x(\text{Apple}(x) \vee \text{Pear}(x))$$

By accepting this sentence, one acquires the ability to rule out as unintelligible a scenario in which there are seventeen apples, twenty-nine pears and anything other than forty-six apple-or-pairs. But in accepting the (far simpler) mathematical sentence "$17+29 = 46$", one acquires a more general ability—the ability to rule out as unintelligible a scenario in which there are seventeen Fs, twenty-nine Gs and anything other than forty-six F-or-Gs (provided no Fs are Gs). And, of course, an improved ability to sort out the intelligible from the unintelligible is important because it gives us an improved ability to transfer insights from one context to another. To pick a simple example, knowledge of the basic facts of multiplication puts you in a position to use the insight gained from counting the rows and the insight gained from counting the columns for the purposes of answering questions of the form "How many tiles?". And, of course, this is only the beginning.

Moral: Even if the trivialist believes that the truths of pure mathematics have trivial truth-conditions, she is able to explain why mathematical knowledge is worthwhile.

Notes

*For their many helpful comments, I would like to thank Ross Cameron, Roy Cook, Matti Eklund, Caspar Hare, John Heil, Øystein Linnebo, Alejandro Pérez Carballo, Brad Skow and an anonymous reviewer for this volume.

1. Other versions of non-trivialist non committalism include Hodes (1984, 1990), Fine (2002) II.5, Rayo (2002), and Yablo (2002), as well as Bueno and Leng's contributions to this volume. The proposal in McGee (1993) may or may not be interpreted as noncommittalist, but it is certainly non-trivialist. An example of trivialist noncommittalism is Hofweber (2005). Assessing modal versions of noncommittalism can be tricky—see my "On Specifying Truth-conditions" for discussion.
2. Here I have in mind a traditionalist interpretation of the *Tractatus*, as in Hacker (1986) and Pears (1987). See, however, Goldfarb (1997).

References

Block, N., and R. Stalnaker (1999) "Conceptual Analysis, Dualism, and the Explanatory Gap," *Philosophical Review* 108, 1–46.
Boolos, G., ed. (1990) *Meaning and Method: Essays in Honor of Hilary Putnam,* Camridge University Press, Cambridge.
Bottani, A., M. Carrara, and P. Giaretta, eds. (2002) *Individuals, Essence, and Identity: Themes of Analytic Metaphysics,* Kluwer Academic, Dordrecht and Boston.
Cameron, R. (forthcoming) "Quantification, Naturalness and Ontology." In Hazlett (ed.) (forthcoming).
Field, H. (1980) *Science Without Numbers,* Basil Blackwell and Princeton University Press, Oxford and Princeton.
Fine, K. (2002) *The Limits of Abstraction,* Oxford University Press, Oxford.
Frege, G. (1884) *Die Grundlagen der Arithmetik.* English Translation by J. L. Austin, *The Foundations of Arithmetic,* Northwestern University Press, Evanston, IL, 1980.
Goldfarb, W. (1997) "Metaphysics and Nonsense: On Cora Diamond's *The Realistic Spirit,*" *Journal of Philosophical Research* 22, 57–73.
Hacker, P. (1986) *Insight and Illusion: Themes in the Philosophy of Wittgenstein,* Oxford University Press, Oxford.
Hazlett, A., ed. (forthcoming) *New Waves in Metaphysics,* Palgrave-Macmillan.
Heil, J. (2003) *From an Ontological Point of View,* Clarendon Press, Oxford.

Hodes, H. T. (1984) "Logicism and the Ontological Commitments of Arithmetic," *Journal of Philosophy* 81:3, 123–149.

Hodes, H. T. (1990) "Ontological Commitments: Thick and Thin." In Boolos (ed.) [1990], 347–407.

Hofweber, T. (2005) "Number Determiners, Numbers and Arithmetic," *The Philosophical Review* 114, 179–225.

McGee, V. (1993) "A Semantic Conception of Truth?" *Philosophical Topics* 21, 83–111.

Morton, A., and S. Stich, eds. (1996) *Benacerraf and His Critics*, Basil Blackwell, Oxford.

Parsons, C. (1983) *Mathematics in Philosophy*, Cornell University Press, Ithaca, NY.

Pears, D. (1987) *The False Prison: A Study of the Development of Wittgenstein's Philosophy*, volume 1, Oxford University Press, Oxford.

Rayo, A. (2002) "Frege's Unofficial Arithmetic," *The Journal of Symbolic Logic* 67, 1623–1638.

Rayo, A. (2008a) "On Specifying Truth-Conditions," *The Philosophical Review* 117, 385–443.

Rayo, A. (2008b) "Vague Representation," *Mind* 117, 329–373.

Stalnaker, R. C. (1984) *Inquiry*, MIT Press, Cambridge, MA.

Stalnaker, R. C. (1996) "On What Possible Worlds Could Not Be." In Morton and Stich (eds.) [1996], 103–119. Reprinted in Stalnaker [2003], 40–54.

Stalnaker, R. C. (1999) *Context and Content*, Oxford University Press, Oxford.

Stalnaker, R. C. (2003) *Ways a World Might Be: Metaphysical and Anti-Metaphysical Essays*, Clarendon Press, Oxford.

Williams, J. R. G. (typescript) "Fundamental and Derivative Truths."

Wittgenstein, L. (1922) *Tractatus Logico-Philosophicus*, Routledge and Kegan Paul, London. Translation by C.K. Ogden. Published as "Logisch-Philosophische Abhandlung", in *Annalen der Naturphilosophische* Vol. XIV, 3/4, 1921, pp. 184–262.

Wright, C. (1983) *Frege's Conception of Numbers as Objects*, Aberdeen University Press, Aberdeen.

Wright, C. (2001) "On Being in a Quandary: Relativism, Vagueness, Logical Revisionism," *Mind* 110, 45–98.

Yablo, S. (2002) "Abstract Objects: A Case Study," *Nous* 36, supp. 1, 255–286. Originally appeared in Bottani et al. (2002).

Part V

From Philosophical Logic to the Philosophy of Mathematics

12
On Formal and Informal Provability

Hannes Leitgeb

This article is a philosophical study of mathematical proof and provability. In contrast with the prevailing tradition in philosophy of mathematics, we will not so much focus on "proof" in the sense of proof theory but rather on "proof" in its original intuitive meaning in mathematical practice, that is, understood as "a sequence of thoughts convincing a sound mind" as Gödel (1953, p. 341) expressed it. Call provability in the former sense *formal provability* and provability in the latter sense *informal provability*. So our aim is to investigate informal provability, both conceptually and extensionally. However, our main method of doing so will be, on the one hand, to demarcate informal provability from formal provability, and on the other hand, to study informal provability by formal means. Moreover, the whole investigation will be carried out in a somewhat restricted setting: our primary focus will be on informal provability in pure mathematics rather than in applied mathematics, and within pure mathematics we will concentrate just on informal provability in the more mundane areas of mathematics, such as number theory and analysis, rather than in the more foundational areas.[1] Furthermore, we will only deal with informal proofs as far as their justificatory role in mathematics is concerned, disregarding other roles that proofs can have (see Auslander (2008) on other roles; see Detlefsen (2008) for a general discussion on the significance of proofs in mathematics and of a variety of different methods of proof).

There is not a lot of literature on informal provability and how it relates to formal provability, apart from some classical remarks made by Gödel which we are going to track in Sections 3–5 of this article. Most famously, Myhill (1960) deals with informal provability when he speaks of "absolute provability". But this is really just a difference of terms: when we say "informal provability", we mean "provability that is not relativized to a *formal system*", whereas Myhill emphasizes the non-relativity aspect of "provability that is *not relativized* to a formal system". Myhill defends the view that absolute or informal provability is a primitive notion which is not conceptually reducible to (combinations of) concepts from syntax, model theory, or psychology, and that by the Incompleteness Theorems "there exist, for any [provably][2] correct formal system containing the arithmetic of natural numbers, correct inferences which cannot be carried out in that system" (1960, p. 462). In many respects, this article is going to follow his lead.

In the field of epistemic arithmetic, an absolute or "intuitive" provability operator is added to the language of arithmetic in order to study epistemic properties of arithmetical statements: see Shapiro (1985a) and Reinhardt (1986), as well as Flagg (1985), and Horsten (1998); we will make use of this approach and of some of its findings in Sections 4 and 5 of this chapter when we investigate the logic of informal provability and how informal provability relates to truth.

Finally, Gödel's (1951) dichotomy of either the mind surpassing the powers of any Turing machine or some statements in mathematics being absolutely undecidable has triggered new interest recently; see Feferman (2006a) and Koellner (2006). We will have to say more about this in Section 5 of this article.

There are of course various philosophical accounts of mathematics in which informal aspects of proof are highlighted; for instance, in various ways, our views will be close to Brouwer's (except for his Intuitionism); Lakatos' (1976) seminal study on proof analysis in his *Proofs and Refutations* is another paradigm case example. More recently, Rav (1999) gives a take on mathematical provability from the viewpoint of the "everyday" mathematician that is very close in spirit to ours, and so do Suppes (2005) from the viewpoint of neurophilosophy, Bundy et al. (2005) from the angle of automated reasoning, Tieszen (1992) from a phenomenological point of view, and Mayo-Wilson (unpublished draft) from a Wittgensteinian perspective on philosophy of mathematics. Furthermore, for a detailed account of the discovery aspects of geometrical proof, see Giaquinto (2007). Unfortunately, neither of these books or papers, nor this very article itself, amounts to more than a mere advertisement for a future theory of informal or absolute provability; it seems that we are still lacking some theoretical resources from philosophy of mathematics, logic, epistemology, and cognitive science that would be necessary to actually develop such a theory. So this is yet another set of *Prolegomena*; hopefully it is one that triggers further research.

Here is the plan of the chapter: Section 2 will make the conceptual distinction between formal and informal provability both clearer and sharper. In Section 3, we will present a Gödelian perspective on informal provability which derives from some of Gödel's considerations on intuition; indeed the whole project is very much inspired by some of Gödel's views on the topic. Section 4 considers the logic of informal provability operators and predicates. Finally, in Section 5, we are asking (but not quite answering) the eternal question: Are there true but informally unprovable statements? But before we turn to these topics in more detail, a few preliminary remarks are in order, which is the subject of Section 1.

1 Preliminary remarks: Provable, proof, and proving

It would be in the tradition of classic analytic philosophy to start this chapter with an explicit definition of "(informally) provable", as well as of related terms such as "(informal) proof" and "(informal) proving". Given the ways in which some areas of analytic philosophy have advanced, however, it should not come as a big surprise that we will not be able to do so, even though it would be nice if we could.

Take *knowledge* as an example: Recently, in epistemology, the view seems to prevail that this notion is regarded best as conceptually primitive.[3] Of course, it would be a sad story, if this was the end of it; rather an informative theory in which "knows that" figures as undefined but meaningful expression has to be developed and that is exactly what philosophers try to do. We suggest to approach "informally provable", "informal proof", and "informal proving" in a similar spirit. Moreover, the proposal of aligning these latter notions with knowledge and knowability is not a mere coincidence: After all, the primary purpose of a mathematical proof is to justify a mathematical statement and indeed to give grounds for knowing that statement to be true. Indeed, if a mathematical statement is provable in principle, then it is knowable in principle. This is not to say that the converse is trivially the case, too, as there might be ways of knowing a mathematical statement on empirical grounds or on the basis of expert testimony or in some other way. In this chapter, we will solely deal with the only *mathematical* manner in which a mathematical statement can become knowledge, that is, by mathematical proof. Although we will not be in the position to put forward any definition of "(informal) proof" or "(informal) proving" – whereas "(informally) provable" may be defined quite trivially in terms of either of them – we will nevertheless try to say something informative about the concepts they express.

As a starting point, the following scant and preliminary remarks on how provability, proof, and proving relate to each other shall suffice:

We regard "(informally) provable" as a modal expression with some sort of normative force, where we leave open whether the modality in question is epistemic, logical, metaphysical, deontic, or something else (or, which is most likely, of some mixed type). An analysis of "it is (informally) provable that A" in terms of "it is possible that it is proven that A"[4] would not seem to be on the right track, since mathematical provability seems much more feasible and easily comprehensible than (epistemic, logical, ...) possibility simpliciter, and hence no progress would be made. However, understanding "x is (informally) provable" as "there is an (informal) proof of x" looks much more promising, since existence in itself is a logical and indeed quite unproblematic concept, at least compared with possibility. So naturally we are led to discuss informal prov*ability* and informal *proof* at the same time.

Whereas published mathematical proofs – as of 2009 – might be identified with certain traces of chalk on blackboards, printed texts in mathematics journals, or electronic patterns in some technical devices, we regard mathematical proofs *per se* as abstract entities which are independent of any material instantiation. It is tempting therefore to consider proofs as sequences of propositions, and maybe this is a useful way of putting it at this point, but the details might be much messier: For example, what if a member of such an abstract sequence were not of a declarative sort (as expressed in natural language by a descriptive sentence) but rather of an imperative sort (as expressed by a command)? So, maybe the notion of proposition that is to be employed here has to be a broad one. Or, what if there were components of proofs in this abstract sense of the term which were both non-propositional and non-conceptual and which thus could not be components of propositions at all?[5] Then considering proofs as sequences of propositions would simply be too restrictive.

Apart from (informal) provability and (informal) proof, we will also deal with (informal) proving, that is, with the process of going through or following or verifying a proof:[6] this may be taken to be a mental process token in a concretely existing mathematician's mind or brain, or, alternatively and more generally, as an abstract type of mental process which might not get realized in any human being ever; we opt for the latter option. The term "mental" should not be misunderstood as entailing that the informal concept of proving is "merely" psychological: just as the concept of knowledge is partly normative, even though knowledge is a mental state type, the concept of proving may be partly normative – and it is – though informal proving may still be a mental process type. Furthermore, "mental" should be understood quite broadly and subject to idealization here.[7] For example, the "mental agent" in question might well be a social community of mathematical minds rather than one mathematical mind in isolation, or it might be an ideal proof agent, which is in itself an abstract entity of which we say that it "carries out" a proof in some abstract sense of "carrying out". In either case, the psychological or sociological aspects of the informal concept of proving are by far not exhaustive of it, just as the concept of knowledge is not exhausted by its psychological or sociological aspects. In analogy with our definition in terms of (informal) proofs, "*x* is (informally) provable" can also be understood as "there is an (informal) proving process that leads to *x* as its final step", which is why provability, proof, and proving are a natural triad of concepts to consider jointly. We will do so below, and we will take the liberty of jumping from the one to the other frequently without much further comment. The advantage of also dealing with proof processes lies in the greater flexibility of the term "process" and the structural surplus of processes vis à vis the static entities that are denoted by "proof": It is clear that there can be both diagnostic (declarative) and executive (imperative) steps in a proof process; it is also obvious that a mental process presupposes mental representations on which it operates, some of which might even be of a non-conceptual nature, and this might be so even when some of these mental representations do not show up anymore in any propositional summary of the overall process.[8] Perhaps an informal proof ought to be regarded as nothing but a recipe or a guide or a set of instructions of how to generate an informal proving process: the latter would do the actual job of proving a statement whereas the former would only be a convenient tool to make the essentials of the former intersubjectively accessible. One should not mistake the recipe for the cooking.

In any case, we will understand both "(informal) proof" and "(informal) proving" sufficiently inclusive so that on the side of provability we will only deal with *in principle* provability, so that all practical and temporal issues as "it is provable but we do not have enough space to write down the proof", "it is provable, but we will not live long enough to prove this", and "this may be provable, but at present we cannot prove it" become irrelevant. In other words, the concept of provability we are concerned with is meant to transcend all boundaries of a merely pragmatic kind. At the same time, we also want to avoid straightforward trivializations of the form "God surveys all mathematical truths, so for him mathematical truth and mathematical provability coincide (extensionally)": we simply would not want to classify this godly type of "surveying" as proving in the required sense of the word. The really interesting question is whether there is some stable and theoretically prolific concept of provability that covers some middle ground between the two extremes, that is, which is sufficiently, but not excessively, idealized. We think that the standard concept of provability that mathematicians use – or a reasonable extrapolation thereof – is exactly of this kind, and as the consideration in the next section will show, it is indeed a concept of informal provability that is not tied to any formalized system.

2 Formal versus informal provability: A conceptual distinction

Fortunately, there is no need to explain what we mean in general by "formally provable/formal proof (in a recursively axiomatized system T)" – every logic textbook can do the job. In contrast, "informally provable/informal proof/informal proving"[9] in the sense of standard mathematics are much less clear notions. We will try to take first steps towards an analysis of the latter concepts by contrasting them with the former ones. When we turn to this comparison between formal and informal provability, we will only be interested in what mathematicians *mean* by "provability" (or "proof" or "proving"), that is, only conceptual issues. Serious empirical studies would be needed to scrutinize – confirm or disconfirm – some of the empirical claims about how mathematicians understand these terms. Unfortunately, at this point, it is only the meagre observations and interpretations of a single philosopher–mathematician that will have to serve as an empirical basis.

Here follows a table of the contrasting features that we want to highlight:

Formal Provability	Informal Provability
formal syntax	not syntactically determined
logical level: 1st order, 2nd order, ...	not logically determined
terms, formulas	interpreted terms, interpreted formulas
logical rules, syntactically encoded	truth preservation, evident steps
logical axioms	no logical axioms
mathematical axioms	lack of axioms,
	axioms as partial definitions,
	deductive "gaps" vs.
	true and evident foundational axioms

In order to determine a recursively axiomatized system T, and hence its corresponding formal provability predicate, one must determine first the syntax of the formal language in question, which is achieved by fixing the vocabulary and the recursive syntactic rules of the language. Compare this to provability in mathematical practice: it is certainly not part of the meaning of "proof" that the language in which a proof is carried out is formalized in the sense just explained; in fact, mathematicians are generally not reflecting on the languages of their mathematical articles, monographs, textbooks, and lectures at all, and *a fortiori* they do not require that proofs be stated in a formal language. This does not mean, of course, that actual proofs do not contain any "formal" expressions. What mathematicians do is to extend natural language by some (maybe new) distinguished symbols and to use this language in much the same way as lawyers extend natural language by legal terms for their own purposes. The language(s) of mathematics are thus best viewed as "organic" entities which grow and change in time, both with the natural languages they extend and with respect to their specific mathematical resources. One might also invoke Tarski's insights into their different semantic properties in order to see the crucial differences between formal languages on the one hand and natural languages – whether extended or not – on the other. Accordingly, this yields the first crucial conceptual difference between formal and informal provability.

When the syntax of a recursively axiomatized system T is laid down, also the logical level on which the system is going to operate has to be decided: Shall one use a first-order language? Will there be second-order quantifiers? Modal operators? Which logical signs will be primitive? In contrast, the logical level of an everyday mathematical theorem or proof is indeterminate. Or is it? Is it not the case that we can reconstruct large parts of today's mathematics in, say, first-order set theory, and doesn't that show that standard mathematics must be construed in first-order logic? Not at all. It only shows that large fragments of mathematics can be *reconstructed* this way, which might tell us more about the method of reconstruction than about what gets reconstructed. After all, there might be more than just one reconstruction: for example, should we formalize a concrete instance of complete induction over natural numbers in a "real world" proof in number theory or in real analysis as an instantiation of the first-order scheme of induction or as an instance of the second-order axiom of induction? No part or aspect of the actual mathematical proof seems to necessitate an answer, and the manner in which mathematicians communicate, check, and learn from such a proof does not do so either.[10] Secondly, while every successful reconstruction of an actual mathematical theory is bound to preserve some of the theory's properties, it is unclear whether the logical level on which the theory is located belongs to its preserved features, or indeed whether there is anything of that sort that can be preserved at all. The logical structure of a reconstructing formal system is, at best, underdetermined by the reconstructed fragment of mathematics, but it is more likely that there is simply no fact of the matter at all whether that fragment of mathematics is first-order, second-order, third-order, or whatever else.

With the language and logical level of a formal system in place, we can take a closer look at what this system is a system of, that is, the terms and formulas it determines. It belongs to the meaning of "formal provability" that these terms and formulas are viewed as mere sequences of symbols which are well-formed according to an explicitly stated system of syntactic rules. However, it is essential to the terms and formulas which are used in concrete mathematical proofs that they come with an interpretation. Try out the following: give a mathematician a statement in number theory or analysis and then ask her to forget about the meaning of the statement and nevertheless to prove it! Or, ask her if it is possible to prove a mathematical statement without that statement being true. The impossibilities involved are of a conceptual nature: it is built into the mathematician's understanding of "proof" that what is proven is meaningful and indeed true. "False mathematical theorem proven" would make a nice newspaper headline perhaps, but the only reaction by mathematicians would be as follows: obviously, this cannot be a proof (but only a failed proof attempt). None of this applies to proofs in the formal sense of the word. This does not mean that mathematicians *never* take any steps in a proof on the basis of purely syntactic considerations – say, substituting one side of an equation *taken as a string of symbols* for a variable in another equation – but it is essential that it is always possible to switch back into the material mode of speaking.[11]

Next, let us consider how a formal system operates on its formulas: any understanding of "formally provable(-in-T)" would be incomplete without an explicit statement of T's logical rules in terms of a recursive rule system, where each logical rule is taken to be sensitive only to the syntactic form of the expressions to which it is applied; much the same holds for the system's logical axioms. But exactly the opposite is true of proof and provability according to mathematical practice: there is no explicit statement of logical rules to be found at all, with the odd exception of introductory textbooks in analysis or discrete mathematics mentioning that $\neg\forall$ amounts to $\exists\neg$ and the like; to be sure, there might be a section on propositional logic (especially, in discrete mathematics), but that is a lesson in how to fill in truth tables systematically rather than a lesson on the meaning of "proof". In particular, no infinite recursive set of logical rules is ever laid down. Instead, the mathematical community's sense of *proving a statement from other statements* involves connecting the latter statements to the former by intermediate steps (i) which preserve truth and (ii) which make it evident why truth is preserved from one step to the next.[12] The former aspect entails that proofs are successful in getting to the truth, while the latter justifies why this is so; both conjuncts are contained in the meaning of "proof". Note that it might still be true empirically that on a "deep" unconscious or neural level, mathematical proving is nothing but recursive rule-based formula-crunching, but that is not the issue here: we wonder whether mathematicians *mean* something like that when they speak of proving, and our only claim at this point is that the answer to that question is "no". Of course, usually the logical rules of a formal system are set up in the way that one can "see" why and how their application preserves truth, but as far as formal systems are concerned that is but an accidental feature; it is not part of the definition of "formal system". These differences between formal and informal provability become even more transparent in the case of logical truths: mathematicians hardly mention logical truths in proofs at all, and indeed they hardly know how to distinguish logical truths from mathematical ones. For example, ask a mathematician whether the principle of complete induction over natural numbers is logical or mathematical. If anything, logical truths are only *implicit* in the mathematical reasoning that is part and parcel of mathematical proving. In fact, the whole division of "pure logic" on the one hand – as formalized, say, by first-order logic – and "mathematical content" on the other – as being formalized in terms of mathematical axioms – does not seem to reflect the way in which everyday mathematicians understand the expressions "proof" and "provable". The demarcation between logic and mathematics is simply not part of the informal concept of mathematical proof.

Finally, take the mathematical eigenaxioms that give each recursively axiomatized system its particular mathematical character and without which "formally provable(-in-T)" would lack mathematical content completely: As a closer look at mathematical practice reveals, informal proofs are not tied to any axiom systems at all, or not to axiom systems in the sense of the axioms of a formal system, or only to axioms which are introduced *post hoc* and which are not essential to the "proofhood" of the proofs in which they figure.[13] For the first case, consider number theory and analysis as examples: modern textbooks on these subjects may start with axioms for natural numbers or real numbers, but the question whether something is a proof in number theory or analysis is pretty much independent of whether any particular system of axioms has been put forward, let alone one in recursively axiomatized form. Indeed, when number theorists or analysts prove their theorems, they usually do not qualify or defend any steps of their proofs as "axiomatic" or "following from the axioms", and no mathematician would question every proof whatsoever from eighteenth century number theory or analysis just because the underlying system of axioms – if there were such – fails to be outlined. Secondly, consider group theory or topology or probability theory: in these cases, mathematicians do speak of axioms, but what they mean by that is nothing more than clauses in the definitions of "group", "topological space", and "probability space" – but those are not axioms in any foundational sense of the word. If there were any "real" axioms of these fields *qua* mathematical theories, then these would probably be the axioms of set theory, but once again group theorists, topologists, and probability theorists usually do not even know these set-theoretic axioms – the set theory they use is most likely a version of naive set theory – and it is not conceptually essential to their proofs that they rely on an axiomatic system of set theory. Thirdly, there is the interesting phenomenon of what appear to be deductive gaps in everyday proofs. Take the standard proof of the equivalence of "f is continuous in the ϵ-δ-sense" (Cauchy continuity) and "f is continuous in the limit-of-real-valued-sequences" sense (Heine continuity), the right-to-left direction of which needs some weak instance of the axiom of choice that is used to show the existence of a real sequence of a certain kind. (There are alternative proofs, but let us focus on the one that is presented usually.) While mathematicians have heard of the axiom of choice of course, and while they are aware of some of its applications in algebra or functional analysis, hardly any textbook on real analysis mentions the application of the axiom of choice in this elementary proof of real analysis. From the viewpoint of a formal system in which this proof were to be reconstructed, the omission of the axiom of choice would have to count as a deductive gap. However, while mathematicians might well be interested to learn about this "gap", they would not react by regarding only the amended proof, in which some reference to the axiom of choice would have been added, as an actual proof; instead, they would regard the existence of a real number sequence as demanded in that proof as "obvious", "self-evident", and beyond need of justification. Such "self-evident" statements also have to be true of course: If mathematicians found them to be false, then the original sequence of statements that had presupposed them would not be regarded as a proof anymore. While making such deductive gaps explicit is certainly a fruitful endeavour, as it leads us to discover evident principles which had been presupposed tacitly by mathematical proofs, it is not the case that only once such presuppositions have been made explicit, the resulting sequence of statements becomes a proof.[14]

The upshot of these considerations is as follows: "formally provable(-in-T)" is not synonymous with "informally provable", independently of whatever name of a formal system is substituted for the letter "T"; the two expressions simply do not have the same meaning. Not that this is a strong result yet, in fact this should be quite obvious – the really interesting question is where to go from here.

An immediate reaction would be to say that "formally provable(-in-T)" has a clear and precise definition but that "informally provable" does not, so the latter ought to be banned from our philosophical investigations, with only the former to stay. But this would be the wrong response: If *"A"* is a placeholder for mathematical statements, then *"A* is informally provable" is at least as clear and precise as *"A* is knowable", and just as modern epistemology has not decided to throw away the concepts of knowledge and knowability in view of the problems of analyzing them, we should not abandon proof and provability for analogous reasons. In fact, the latter seem at least extensionally more clear cut than the former: overall the mathematical community is pretty quick and efficient in determining whether something is a proof or not, while the philosophical community might discuss endlessly whether some particular instance of belief in the real world or in a toy story ought to count as an instance of knowledge. There have been struggles, sure – from the notorious troubles with proofs that would invoke infinitesimal quantities, over intuitionistic worries about the axiom of choice, to the more recent question whether computer-aided "proofs" are actual proofs – but all of this has been resolved or, in the last case, will be resolved. In that extensional sense, at least, informal provability does not seem to be particularly vague at all; it rather seems to be a case of *informal rigour* (Kreisel, 1967).[15]

So, if abandoning informal provability is not the right way of proceeding, then perhaps "improving" informal provability is: why not think of "formally provable(-in-T)" (for some instantiation of "T") as a Carnapian explication of "informally provable"? The answer is simple: because it is not. According to Carnap, whatever explicates an explicandum must be as similar as possible to the latter, but as our comparison from above has shown, formal provability and informal provability are just too dissimilar to satisfy this criterion. There is no reason to believe that if one could explicate informal provability at all, then this could not be done while preserving more of its essential features than any explication in terms of "formally provable(-in-T)" would ever achieve.

Perhaps informal proofs stand to formal proofs then as high-level algorithms stand to Turing machines (as suggested by Beklemishev and Visser 2005, section 2)? Or informal proofs are nothing but "indicators" of formal derivations in algorithmic systems (as claimed by Azzouni (2006), chapter 7)? (See Rav (2007) for a criticism of this view.) Or is there no conceptual link – nothing on the level of meaning – between formal and informal provability at all? Even if the latter were the case, this would still leave open the possibility that there might be an extensional link: perhaps formal provability and informal provability are extensionally equivalent? In order to make sense of this, one would first have to assign a corresponding reference to the "T" in "formally provable(-in-T)". Perhaps, then, "formally provable-in-ZFC" or the like would end up having the same extension as "informally provable"? Even if not – as one might worry about Gödelian reasons for believing that no single recursively axiomatized system could ever capture the totality of informally provable mathematical statements (more about this in Section 5) – then maybe a transfinite progression of formal provability predicates in the sense of Feferman (1962) might approximate informal provability extensionally in the limit (see Horsten 2005b, section IV)?

We will not suggest any answers to these questions here.[16] For our current purposes, let it be sufficient to point out that the "correspondence" between formal and informal provability is an intricate and difficult one, if there is a correspondence of any interesting sort at all. So we pose this as an open research question:

• In which ways does formal provability approximate informal provability?

Why should we be interested in this question? Obviously, philosophy of mathematics should be concerned with all topics that are central to mathematics, including informal provability. In order to make the wonderful tools from proof theory, model theory, recursion theory, set theory, modal logic, and so forth, applicable to the analysis of informal provability, we need bridge principles which relate informal provability to formal provability.[17] This is exactly what we ask for in the question raised above. For example, sometimes philosophers of mathematics are rather quick in drawing conclusions on mathematical, that is, informal provability, from theorems about provability in certain formal systems, that is, formal provability. But such inferences are utterly unwarranted as long as the question above remains unanswered. Even for giving partial answers to questions such as

• What do the Incompleteness Theorems tell us about mathematical provability?

we need a theory of informal provability, and indeed one that has something interesting to say about how informal provability relates to formal provability.[18] So the challenge stands. We shall return to some of its aspects in Section 5 of this chapter when we deal with truth versus informal provability.

3 A Gödelian perspective on informal provability

One way to summarize the right-hand side of the table in the last section is to say that informal provability, informal proof, and informal proving differ from formal provability and formal proof in having *semantic* and *intuitive* components or aspects. We are going to dwell on them in this section while giving our discussion a somewhat Gödelian twist.

When we say "semantic" here, we do not have in mind model theory, which is in itself a mathematical discipline, but just that the meaning of "informally provable" is tied to the meaning and meaningfulness of mathematical predicates and sentences – to concepts and propositions – rather than to predicates and sentences by themselves:[19] The terms and formulas which show up in informal proofs must be interpreted; proof steps have to preserve truth; the "foundational axioms" that are made explicit must be true. "Preserving truth" might sound like model theory again, but we think of logical consequence more along the lines of Etchemendy (1999) (and, in turn, Kreisel (1967)), whom we take to have shown that Tarski's model theoretic definition of logical consequence is but an extensional approximation of the informal notion of logical consequence which in itself is not model theoretic at all. Indeed, much of what we have to say can be viewed as an extension of Etchemendy's work on informal logical consequence to the topic of informal mathematical provability. Accordingly, when we speak of truth, this might appear to be a reference to model theory again, but it is truth *simpliciter* that we have in mind here, not truth-in-a-model. (Both were analyzed by Tarski, but only the latter is model theoretic.) As Rav (1999), p. 11, puts it: "Let us fix our terminology to understand by a *proof* a conceptual proof of customary mathematical discourse, having an irreducible semantic content, and distinguish it from *derivation*, which is a syntactic object of some formal system."

When we talk about intuition, we mean – as a first approximation – "it is intuitive that *A*" both in the traditional sense of "it is self-evident that *A*" – or, if one prefers, "it is intrinsically plausible that *A*" – as well as in Zermelo's sense of "it is tacitly presupposed that *A*" (see Shapiro (unpublished draft) for more details on this two-fold conception of intuition): As we have seen, proofs need to be divided up into intuitive steps in the first sense of "intuitive"; axioms which figure tacitly in typical mathematical proofs, such as the axiom of choice, are intuitive in both senses of the term (notwithstanding intuitionistic quandaries).[20]

What we now want to call *Gödel's insight* is the following: The semantic and intuitive components of mathematical proofs are epistemically *interdependent*; when proving mathematical statements, it is possible – and sometimes even necessary – to change from a semantic access to mathematical structures to an intuitive one and the other way round.

Rather than calling it Gödel's insight, this is more usually considered to be "Gödel's blunder"; in fact, Gödel is ridiculed for statements such as the following ones, even though, as we are going to suggest, they might have a perfectly fine interpretation:

the law of complete induction, which I perceive to be true on the basis of my understanding (that is, perception) of the concept of integer (Gödel 1951, p. 320)

[axiom/propositions] can directly be perceived to be true (owing to the meaning of the terms or by an intuition of the objects falling under them)... e.g., the modus ponens and complete induction (Gödel 1953, p. 346)

[the content of mathematics] consists in relations between concepts or other abstract objects which subsist independently of our sensations, although they are perceived in a special kind of experience (Gödel 1953, p. 351)

It is easy to see why people think that Gödel was quite confused when he made these statements. It is one thing to say that the law of complete induction over natural numbers is perceived to be true – as has been claimed before by Poincaré and others – but it sounds almost heretical to add that this perception is based on the understanding of the concept of integer, which is semantic in nature, and even more heretical to explain this understanding in terms of a perception again.

The same sway back and forth between intuition and semantics in the second quotation: the intuitive perception of the truth of a proposition is claimed to derive from semantics, that is, from the meaning of the terms that are used to express this proposition, or from an intuition of the objects that fall under the concepts that get expressed by these terms. While mentioning "complete induction" as an example is hardly a surprise, mentioning modus ponens as such is utterly controversial again – how can we *perceive* something to be true in the case of *modus ponens*? Similarly in the last quotation, in which Gödel maintains that concepts – that is, something semantic again – can be perceived in a special kind of experience.

Is it possible to make sense of this alleged correspondence between the semantic and the intuitive, between propositions and concepts on the one hand and intuitions or perceptions on the other? Here is a suggestion of how to do so; we hope this suggestion will in turn throw some interesting light on how informal proofs might actually be capable of doing their job, that is, proving mathematical statements.

Let us start with some primitive logical concepts which might be available to a hypothetical mathematical agent and which that agent might express linguistically in the familiar way:

$$\neg, \vee, \exists, \forall, =, finite, function, set\text{-}of, \ldots$$

One might think of logical concepts as being those which are invariant under bijective mappings in the sense of Tarski (1986), but that's not crucial at this point. (It is also likely to be a "merely" extensional approximation of the informal notion of *logical concept* again.) From the given primitive logical concepts, our agent is able to build complex ones by the usual compositional means; the complex concept that is expressed by the open formula

$$x_1Rx_2 \land x_2Rx_1 \land x_2Rx_3 \land x_3Rx_2 \land$$

$$\neg x_1Rx_1 \land \neg x_2Rx_2 \land \neg x_3Rx_3 \land \neg x_1Rx_3 \land \neg x_3Rx_1 \land$$

$$x_1 \neq x_2 \land x_1 \neq x_3 \land x_2 \neq x_3 \land$$

$$\forall x_4(x_4 = x_1 \lor x_4 = x_2 \lor x_4 = x_3)$$

with free individual variables x_1, x_2, x_3 and one free predicate variable R is an example. In fact, this is even a categorical concept, that is, the universe of discourse with a distinguished binary relation that falls under the concept is determined uniquely up to isomorphism. Indeed, using the terminology of so-called *ante rem* structuralism in the philosophy of mathematics (cf. Shapiro, 1997 and Resnik, 1997), we might say that this concept pins down one and only one mathematical structure: the unlabelled linear graph

with three nodes and two edges. *Ante rem* structures like these can be distinguished from standard set-theoretic systems by their invariance properties again: label the three nodes in the structure above in any way; relabel the nodes of the corresponding labelled graph along an automorphism of the underlying structure; then the resulting labelled graph is *identical*, not just isomorphic, to the original labelled graph. A graph understood as a set-theoretic system would not have this property.

In a similar way, infinite structures may be determined by complex categorical higher order concepts. As an example, take the second-order Dedekind–Peano axioms for arithmetic: replace the individual constant for 0 by the individual variable "x_0", the predicate constant for the successor relation by the predicate variable "S", and assume the scope of all quantifiers to be bounded by the predicate variable "N" (for *natural number*); by Dedekind's famous theorem, the resulting complex concept $PA_2(x_0, S, N)$ determines set-theoretic systems uniquely up to isomorphism, and hence the natural number structure – another unlabelled graph – uniquely simpliciter:[21]

This is the semantic manner in which our mathematical agent might get epistemic access to mathematical structures. Now let us compare it to another one.

As a starting point, consider a *non-mathematical object*, say, Gödel's *face*, which we assume our hypothetical agent knows: how would he or she represent Gödel's face mentally? One way of doing so would be by means of a description: there are two eyes in it; both black and grey hair hangs into it from above (it's a photo of the old Gödel); and so on. This is like the description of mathematical structures above, only involving empirical concepts. But the much more efficient and likely representation will be in terms of a mental image or something *like* an image. Say, our agent even knew Gödel personally: then his or her representation will probably be something that gets triggered not just by seeing his face from one particular angle, but it will rather encode properties of his face that are invariant under different perspectives, distances, light sources, and so forth. Indeed, it is likely that this is the primary form of representing material objects, as far as human agents are concerned. Now, here follows the thought: why not assume that our mathematical agent is capable of representing mathematical structures in a similar manner, that is, by something which does not determine structures in the way concepts do, but which stands in some sort of abstract structural similarity relation to what it represents?

For example, reconsider the unlabelled graph with three nodes from above:

It is easy to imagine that our agent might represent this little mathematical structure by means of a mental copy of it: something that consists of three node-like mental entities, two edge-like mental entities, built together by simple mental operations on representations – call this the "intuition" that our agent has of the unlabelled graph in question. This intuition ought not to be regarded as the mental counterpart of the graphical figure or drawing above but rather as something more abstract; something that stands to that figure as our mental representation of a human face stands to a photo. After all, the geometrical figure above contains way too much irrelevant information, such as the geometrical position of the nodes or the trajectories and lengths of the edges – information that is potentially misleading since it does not represent any properties of the graph. There are much more efficient representations of graphs available which also get much closer to what they ought to represent, and it would be beneficial to our mathematical agent to use one of them in order to represent graphs.

Obviously, things are not quite so straightforward in the case of the natural number structure: presumably, if our mathematical agent is subject to finitary human constraints, then his or her intuitions will be finite creatures of some kind, so that he or she will not be able to represent an infinite structure in exactly the same way as he or she may have represented the small finite graph before. However, something which is significantly like that might still be possible; for example, take the infinite structure of the natural numbers again,

●—●—●—●—●—   ⋯

and imagine the initial "0" node to be removed; the resulting *ante rem* structure would be the following (up to graphical repositioning):

●—●—●—●—●—   ⋯

In other words, the structure of the natural numbers is a fixed point of this operation. Now assume that our agent has some sort of mental representation available which represents the natural number structure as being a fixed point under this mental *remove-the-initial-node* operation; this will not pin down the structure uniquely yet, but it would already go some way to such an effect, and further invariance and minimality properties with respect to related "intuitive" operations might complete the job.[22] Intuitions in this sense would be representations which are constructed by, or invariant under, particular mental operations, and which are structurally similar to the mathematical entities they represent. Mathematical agents might get epistemic access to mathematical structures by altering and consulting these representations, while at the same time also having access to these mathematical structures by means of categorical concepts. If human agents are among such mathematical agents, then the following "naturalized" Gödelian theses seem quite plausible:

• We can informally prove propositions about mathematical structures

  – either *semantically*, by extracting propositional information from the complex categorical concepts that determine these structures,
  – or *intuitively*, by manipulating and inspecting the mental constructions that are similar to these structures.

• There is nothing "mystic" about intuition: whenever mathematicians speak of intuition, they refer to *non-conceptual/non-propositional* – in short: non-semantic – representations of mathematical structures.

• There is no reason to believe that intuition plays a role only in the context of discovery of mathematical propositions: though non-propositional representations cannot be true, they can still be *veridical*; e.g., the first non-conceptual mental representation referred to above is not similar to the natural number structure, but the second one would be. But if these representations can be veridical, then it must also be possible to use them for justificatory purposes.

- Nor is there any reason to believe that intuition is predominantly tied to Euclidean geometry, as the Kantian tradition would perhaps have it: in fact, not even geometrical intuitions should be confused with (real or imagined) concrete drawings, since not every aspect of a drawing in a Euclidean proof is actually relevant to the proof (see the seminal work of Manders (unpublished draft), Mumma (2006), and Avigad et al. (unpublished draft) on this topic). But if even geometrical representations are more abstract than concrete drawings on pieces of paper, it is not so hard to conjecture anymore that many types of mathematical structures and many instances of such types – Boolean algebras, lattices, groups, fields, ordinals, etc. – and many sorts of individuals or relations within such structures might have non-semantic representations that are in principle accessible to us.[23]

- Maintaining that there are semantic and intuitive components of mathematical proofs does not mean that these components are exchangeable arbitrarily:

Intuitions might give us direct or indirect evidence for the satisfiability of concepts and thus support existence axioms (cf. Gödel, 1953): e.g., while it would be hard to "see" that $PA_2(x_0, S, N)$ is satisfiable just by consulting the conceptual structure of that concept, this is no longer so once we gain intuitive access to the natural number structure as sketched above. This does not mean that intuitions can play this role infallibly: as with any other scientific tool or instrument, we might get things wrong when we consult our non-semantic representations, they might be misleading us in some way,[24] and they might clash with each other due to their restricted ranges of applicability.[25] But at the same time this does not mean that intuitions are "intrinsically erroneous" in any sense.

Concepts, on the other hand, are compositional and can be communicated through language in a way that goes far beyond communicating intuitions by drawing pictures or the like. Thus, even though it might be impossible to completely eliminate the role of intuitions in mathematical theorizing – and there is no reason why one would *need* to do so either – intuitions nevertheless *ought* to be "conceptualized" to some extent, as this adds to the systematicity, generalizability, and intersubjective accessibility of mathematics. For the same reason, such conceptualizations often go along with scientific progress. The arithmetization of analysis in the 19th century, Hilbert's axiomatization of geometry at the turn from the 19th to the 20th century, and finally the reconstruction of mathematics on the basis of axiomatic set theory at the beginning of the 20th century may be viewed as instances of such conceptualization processes, and each of them was crucial to enable subsequent mathematical developments.

It is an open question whether for *every* mental act of proving a mathematical statement there is a non-semantic representation of some mathematical entity that is involved in it. So far it seems like an inference to the best explanation to conclude that this is so at least in many relevant cases of proof processes, if not in all. For example, what do we refer to when we say that proofs are divided up into evident steps?

One plausible explanation would be that proofs come with a double-layered structure: the obvious and transparent layer of propositions which are said to follow one by one, and a second intuitive level on which every descriptive statement about mathematical objects corresponds to an instruction of how to generate, manipulate, and inspect non-semantic representations of these objects, where the set of possible mental operations on intuitions is restricted in some way, and where each application of a mental operation on intuitions presupposes the previous step to be taken beforehand. Or, why are there typically no references to axioms to be found in informal proofs? Because what mathematicians do when they prove – amongst other activities – is to inspect their non-semantic representations of mathematical entities; if ultimately this leaves a statement as being evident, then this is a sufficient basis for moving on, whether or not the statement in question carries the methodological weight of being called an axiom. Indeed, one might even explain why one should not necessarily expect any "first principles" to emerge from which one might naturally derive all informally provable statements in some mathematical area by mere information-extraction from propositions and concepts. Consider our non-mathematical example of Gödel's face again: What are the "axioms" on which our knowledge of his face are based? There does not seem to be any reasonable answer to this question; instead the set of statements which we are able to make about Gödel's face justifiedly seems to be bottomless, without a foundational level to stop at except for the time being. The most likely explanation for this fact is that the way in which we represent Gödel's face is not in terms of a description, that is, semantic, at all. Now, the same might be true in the case of mathematics: maybe our intuitive access to mathematical structures is just as bottomless, and no natural and ultimate set of foundational axiomatic descriptions of all mathematical structures emerges.[26] Dedekind's categorical second-order axiomatizations of the natural and the real number structure, Zermelo's quasi-categorical second-order axiomatization of the set theoretic hierarchy, and finally the successful axiomatization of large chunks of mathematics in first-order ZFC set theory seem to run counter to this view. But our claim was not that such axiomatic conceptualizations of our intuitions were impossible, only that they were by no means necessary to happen; secondly, apart from the last one, the mentioned results come with the price of another bottomless set of statements, that is, second-order logic; thirdly, the set theoretic reconstruction of mathematics in first-order set theory is still a reconstruction, the success of which might be due to the simple fact that the set theoretic hierarchy is so diverse structurally that all sorts of mathematical structures can be simulated in terms of some of its members; and finally, there is no reason to believe that ZFC set theory is the ultimate starting point of the mathematics yet to be developed nor that any future sequence of stronger and stronger foundational theories is bound to reach its ultimate limit at any finite point of time.

Although we have seen that the semantic and the intuitive components of mathematical proofs have their own specific roles to play, we nevertheless seem to be able to go back and forth between them. We can construct a non-semantic representation of a mathematical structure and conceptualize it; or we build up a complex categorical concept and "make it intuitive". In particular, the former seems to involve a transition from *intuition of* via *intuition that* (cf. Parsons, 1993) to semantic components, and it is not clear at all how this is done, and what human mathematical agents are like such that they are able to do so. So the really interesting question at this point is

- How is this *Anschauung der Begriffe* achieved?

And also, how do we know that what a particular concept determines and what a particular intuition is similar to are actually one and the same thing? Once again, we will not be able to give any answers to these questions. Instead, we merely advertise a cognitive theory of mathematical provability.

In cognitive science, it is common procedure to analyze reasoning not just in terms of symbolic representations but also by means of *mental models* (Johnson–Laird), *sub-symbolic neural structures* (Smolensky), *quasi-pictures* (Kosslyn), *perceptual symbol systems* (Barsalou), and so on. Why should we not do the same when we study informal proofs and provability? For example, if concepts turn out to have perceptual properties from the start, as Barsalou (1999) argues in terms of his "perceptual symbols" and the mental simulators that generate them, then the same should apply to logical and mathematical concepts too. Indeed the latter are amongst the example Barsalou gives; hence, perceptual components might simply be essential to mathematical proving since mathematical concepts might simply be (partially) perceptual from the start, and every analysis of what an informal proof is, of what statements are informally provable, and of how informal proving proceeds as a mental and social process would have to remain incomplete as long as this perceptual character of concepts is disregarded. The semantic and the intuitive components of informal proofs would turn out to be facets of one and the same class of entities, that is, concepts with perceptual properties or components. No wonder Gödel would jump from the one to the other without much ado, as exemplified by the quotations stated above.[27] This is a clear sense in which further progress on the empirical side of research on conceptual versus non-conceptual representations of mathematical structures might facilitate further progress in the philosophical understanding of mathematical proof.

Let me conclude this section by addressing what is likely to be the most obvious and pressing worry about such a cognitive shift of attention in philosophy of mathematics. Whatever the outcome of such a change of interest might be, wouldn't it be able to support only a *subjective*, if not psychologistic, notion of informal provability? But if such a theory would leave us merely with a better understanding of idioms such as

for proof agent x it is informally provable that A

would this not clash with the obvious objectivity of mathematics? And if so, would this not mean that a cognitive turn in the philosophy of mathematics could only be a non-starter?

Not necessarily. Perhaps future philosophy of mathematics will manage to discover a "natural" notion of *ideal proof agent* that is (i) sufficiently like a human agent in its potential semantic and intuitive capabilities and resources, but which is at the same time (ii) sufficiently idealized in order not to make any mistakes and not to be affected by any accidental boundaries and shortcomings. Much the same has happened when the informal concept of *computability* got analyzed extensively in terms of mathematically precise notions such as *Turing machine*, *register machine*, or *lambda calculus expression*. In other words, maybe there is a way of solving the equation

effectively computable : Turing machine = informally provable : x

for x, and to use the solution to study informal provability on the basis of the equivalence

- It is informally provable that A if and only if there is an ideal proof agent x such that for x it is informally provable that A.

The ideal proof agents in question might rely on non-conceptual and non-propositional representations of mathematical entities as sketched above.[28]

In order to get a better handle on such ideal proof agents, recent work on automated proof assistants might turn out to be relevant (see e.g. Avigad, 2006), and it might be useful to remind oneself that the best "real world" approximation of ideal proof agents might be the mathematical community in its totality rather than any single mathematician. But ending with such vague and more or less obvious hints is just another sign of groping in the dark.

4 The logic of informal provability

There is a now well-developed area called "Provability Logic" (cf. Boolos, 1993) in which the logical laws of provability are analyzed in terms of laws of modal logic, with the corresponding sentential modal operator □ being interpreted as "it is provable that". Since the underlying notion of provability is *formal provability in first-order Peano arithmetic*, this area is mostly about the logic of formal provability. In this section, we wonder what the logic of informal provability might look like, and how it compares with standard provability logic.

But before we do so, here is yet another worry to be dismantled: Isn't it contradictory to study the logic of informal provability in terms of a formal system of modal logic? Not if the aim is just to outline the *set* of logical truths for informal provability and to contrast it with the corresponding set for formal provability. A more general worry: Isn't it contradictory to study informal provability in terms of a formal system? No, just as it is not contradictory to study irrationality by rational means (by which we certainly do not want to imply that there would be anything irrational about informal provability).[29]

Let us start with formal provability again: Formal provability in PA satisfies the modal system GL, and GL can even be shown to be weakly complete with respect to formal provability in PA as was proven by Solovay (1976):

GL: Axioms and rules of classical logic

$$K : \Box(A \to B) \to (\Box A \to \Box B)$$

$$\text{Löb: } \Box(\Box A \to A) \to \Box A$$

$$\text{Nec: } \frac{\vdash A}{\vdash \Box A}$$

It is easy to see that these axiom schemes and rules entail all instances of the modal axiom scheme

$$4 : \Box A \to \Box\Box A$$

which is thus part of the system.

Now for informal provability: In his modal interpretation of the intuitionistic propositional calculus, Gödel (1933) suggested the modal system S4 as a logic of provability, and as Gödel made very clear, the notion of provability in question could not be *formally-provable-in-PA*, in view of his own Second Incompleteness Theorem. Especially, the T axiom scheme of S4 is bound to fail for formal provability, for instantiating $\Box A \to A$ by a logical contradiction $\bot$ would yield $\neg\Box\bot$, which would mean PA were PA-provably consistent by Necessitation, which is ruled out by the Second Incompleteness Theorem. However, it is part of the meaning of "informally provable" that whatever is provable in this sense is also true, as we have seen in Section 2, so the modal T scheme is mandatory. Moreover, the provability of $\neg\Box\bot$ does not seem to be self-refuting at all, if $\Box$ expresses informal provability. So S4 seems to be an excellent starting point for a logic of informal provability:

S4: Axioms and rules of classical logic

$$K : \Box(A \to B) \to (\Box A \to \Box B)$$

$$T : \Box A \to A$$

$$4 : \Box A \to \Box\Box A$$

$$Nec: \frac{\vdash A}{\vdash \Box A}$$

K is certainly unproblematic. As we can see, 4 and Necessitation are quite plausible closure principles for informal provability, too: If it is informally provable that A, then there must be an informal proof of it; but it is plausible that we can then also ultimately recognize and establish the latter to be a proof. Obviously, this would need some introspection *on proofs*, which might go beyond standard mathematical activities, but perhaps it is still permitted to subsume this under the label "informal provability" in a slightly broader sense. Similar thoughts apply to the case of Necessitation.[30] Therefore, S4 is sound with respect to informal provability.

Note that GL above arises from considering which general theorems on formal provability in PA (up to coding) are *derivable* in PA. If one looked instead for general theorems on formal provability in PA which are *true* in the standard model of arithmetic, then T would become a logical truth again; however, at the same time necessitation would drop out, so the difference between the logics of formal and informal provability is actually quite robust since in either case, we find that the conceptual differences between formal and informal provability show up in *logical* terms too.[31]

Before we turn to the question whether S4 is also complete with respect to informal provability, let us consider briefly a second kind of formalization of the logic of informal provability: a formalization in terms of an informal provability *predicate*. Since outside of logical contexts we tend to express both formal and informal provability by means of predicates, such a representation would be more faithful to our philosophical understanding of provability, and in the case of formal provability, it is arguably the case that formal provability predicates are conceptually prior to any formal provability operator. But this might also amount to yet another discrepancy between formal and informal provability: While formal provability can be expressed both by a sentential operator and, after arithmetization, by a type-free predicate of (codes of) formulas, Myhill (1960) – in his article on absolute provability – and Montague (1963) seem to have shown that informally provability does not allow for a similar type of predicate representation. For let $\mathcal{L}_{Prov}$ be the first-order language of arithmetic extended by an informal provability predicate *Prov*, then the system

$$\text{T: } Prov(\ulcorner A \urcorner) \to A, \qquad \text{Nec: } \frac{\vdash A}{\vdash Prov(\ulcorner A \urcorner)}$$

(for unrestricted $A \in \mathcal{L}_{Prov}$) turns out to be *inconsistent* given first-order Peano arithmetic. From arithmetic one gets diagonalization, from which one derives the existence of a "Provability Liar" sentence λ; instantiating T and necessitation above by λ leads to contradiction.[32]

This does not speak against informal provability in any way; after all, the informal provability operator is just fine. Even more importantly, this result does not even point to any substantial problem of formalizing informal provability in terms of a predicate: As Skyrms (1978) has shown, one can mirror the axioms and rules of modal operators by means of restrictions to the axioms and rules of modal predicates, without any danger of inconsistency. Indeed, when we regard T and necessitation as valid for informal provability, we think primarily of *mathematical* instances of A and maybe sentences or propositions which say something about the informal provability of such mathematical statements and maybe sentences or propositions that talk about the latter and the like, but not of self-referential or ungrounded statements such as "Informal Provability Liars". It is even possible to do better with modal predicates than one could ever do with a modal operator (in an otherwise standard modal setting) while still being free from contradiction: One can prove that there is a natural Π_1^1 set of "grounded" instances of the axioms and rules of the type-free predicate version of S4 that is consistent and which has a nice possible worlds semantics (see Leitgeb (2008), which in turn builds on Leitgeb, 2005). The set of formulas A that are used to instantiate $Prov(\ulcorner A \urcorner) \to A$, and so on is closed under propositional operations, substitutional quantification, provable equivalence in first-order Peano Arithmetic, and applications of *Prov*. Hence, a nice and perfectly reasonable S4-type logic for the informal provability *predicate* is available without worries of inconsistency. The only remaining logical differences between the formal and the informal provability predicate are the differences between the logical axioms that are valid for them – just as in the operator case above – and that the formal provability predicate for PA satisfies the predicate analogues of the axioms of the modal system GL unrestrictedly, while the predicate versions of the axioms of S4 have to be restricted in order to be valid for the informal provability predicate.[33]

Finally, let us return to the following open question that we hinted at above:

• Is S4 the complete propositional logic of informal provability? If not: What is it?

In particular, can we also justify principles of *negative* introspection – concerning the informal provability of informal *un*provability – to be logical axioms of informal provability? It is clear that this leads to a whole array of new questions. In order to justify an instance of positive introspection, such as an instance of the 4 axiom scheme, it is sufficient to deal with one informal proof at the time. But in order to prove something general about unprovability, one needs to be able to say under what conditions all potential informal proofs whatsoever can be ruled out. Is this possible at all?

If so, then only by closer study of the components of informal proofs that we have isolated in the last section, that is, their semantic and intuitive components. That is, one needs to search for general insights into informal proofs which one might turn into logical insights into negative introspection; these general insights would need to be based on what we know about the meaning of "provable" as applying to mathematical expressions in general, as well as on our general intuitions of informal proofs. Once again, we do not have anything concrete to offer, except for a vague hope that this might not be as impossible as it sounds. For example, speaking of intuitions of proofs is certainly not void of content. Proof theorists seem to rely on quite clear non-semantic representations of proofs in terms of trees or other graphs; while the proofs in question are of course formal ones, something like this might also be true of informal proofs. If so, then some formal results on the geometry or topology of formal proofs might become applicable in the realm of informal provability, too. Here follows an example: Statman (1974) extends the usual tree representations of derivations in natural deduction by edges which connect those places in a deduction in which an assumption was made with the places in which that assumption gets discharged. The resulting graphs are usually not trees anymore, therefore, it makes sense to study their topological genus, that is, the minimal number of handles one needs to put on a plane (or a sphere) in order to draw such a graph on the plane with intersections of edges only occuring in vertices. Statman then proves some very nice results about the genus of derivations thus explained. In particular, he shows that the genus of a derivation can be reduced arbitrarily, even to 0, by means of introducing explicit definitions. This is a result about natural deduction and hence about a formal system, but then again natural deduction was meant to come closer to informal reasoning than any of the axiomatic treatments of logic, so perhaps certain theorems on natural deduction might carry over to informal proofs. If so, it might be possible to analyze our intuitions of informal proofs in terms of graphs such as Statman's, and one might be able to prove general properties of informal proofs on the basis of this analysis. Following up Statman's theorem, one might, for example, think as follows: Why are mathematicians so heavily involved with introducing explicit definitions of mathematical functions and concepts at all? After all, such definitions do not add to the deductive power of mathematics (apart from re-labelling). A possible answer is as follows: Because explicit definitions allow mathematicians to reduce the genus of an informal proof to some level – perhaps the Euclidean plane? – on which the resulting informal proof can be surveyed easily as a geometrical object. This is mere speculation, of course, but this is where we stand concerning human intuitions of informal proofs.

Apart from the *propositional* logic of informal provability, one might just as well be interested in its quantified counterpart:

- What are valid principles of the predicate logic of informal provability? Even more: What is the sound and complete system of predicate logic for informal provability?

General issues concerning *de dicto* and *de re* modalities are bound to show up in new light here. For example, while $\Box\exists xP(x)$ says that the proposition expressed by $\exists xP(x)$ is informally provable, $\exists x\Box P(x)$ says *of* a mathematical entity that *it* can be proven to be P. One possibility would be to reinterpret the latter type of statement in terms of substitutional quantification into modal contexts and thus to reduce cases of *de re* modality to instances of *de dicto* modality (as suggested by Horsten, 2005a). But in light of our considerations in Section 3, the more salient interpretation of informal provability *de re* is by reference to intuition again: there is a mathematical entity, such that *through our intuitive access to it*, it is possible to prove that it has the property P. Further investigation into those back-and-forth procedures between semantic and intuitive components of informal proofs that were sketched in Section 3 might lead to plausible candidates of logical laws which connect informal provability *de dicto* and *de re*. But that is even more speculative.

5 Are there true but informally unprovable statements?

In this final section, we are not so much interested in how informal provability compares to formal provability extensionally, but rather how informal provability relates to *truth* extensionally. However, in order to say something non-trivial about the latter, it might be useful – and perhaps necessary – to invoke non-trivial insights into the former. For example, it follows from Gödel's First Incompleteness Theorem that if the set of informally provable statements is recursively enumerable, then truth exceeds informal provability extensionally. Or, by the Second Incompleteness Theorem, even if the set of informally provable statements is recursively enumerable, then it is not informally provable of a particular Turing machine that it enumerates all and only informally provable statements. (See Gödel (1951); Shapiro (1998) gives a modern reconstruction of these two corollaries to the Incompleteness Theorems.) Note that the latter result still does not rule out that it is informally provable that *there exists a Turing machine which enumerates all and only informally provable statements*. Benacerraf (1967) is aware of this, and Carlson (1984), (2000) proves the consistency of the informal provability of this existence claim with a formal system of epistemic arithmetic.[34] It is arguments and conclusions like these that we have in mind here.

So what we are after is

- In which ways does informal provability approximate truth extensionally?

Or, equivalently, are there absolutely undecidable statements (cf. Gödel, 1951)?

According to Hilbert's famous *non ignorabimus* claim, the answer to the latter question is an emphatic "no!"; according to Cohen (2005, p. 2414) it is (a somewhat less emphatic) "yes!". Let us go with Cohen for the moment: how could we then argue in favour of the existence claim

$$\text{HG } \exists p (p \wedge \neg \Box p)$$

which we express in this case in operator terms, with a sentential operator $\Box$ for informal provability?

As for every other statement, we can either support HG by inductive evidence or by informal proof. Let us focus only on the latter option here: We certainly cannot prove HG by proving one instance of HG, that is, a statement of the form

$$A \wedge \neg \Box A$$

since

$$\Box(A \wedge \neg \Box A)$$

is clearly inconsistent in S4. But it might well be possible to prove weaker claims which still entail HG, such as, for example,

- $(A \wedge \neg \Box A) \vee (B \wedge \neg \Box B)$
- $(A \wedge \neg \Box A) \vee (\neg A \wedge \neg \Box \neg A)$
- $\neg \Box A \wedge \neg \Box \neg A$

which are indeed consistent in S4.

Here is an idea of how to do so (this is joint work with Leon Horsten; cf. Horsten and Leitgeb, unpublished draft): (i) Formalize a version of the

Church–Turing Thesis (CT) by means of the informal provability operator (as has been suggested by Shapiro, Reinhardt, and others in the Epistemic Arithmetic camp); (ii) add this formalization (ECT) to S4 and background mathematics; and (iii) derive some statement such as the above in the resulting system. This might seem viable in view of the fact that CT is itself a statement that relates an informal notion – effective computability – to a formal one – Turing-computability.

In more formal terms, we presuppose:

- Language:

  - $S(x)$: x is a purely mathematical sentence
  - $Proof(x, y)$: x is an informal proof of y (for y being a purely mathematical sentence)
  - $\Box A$: it is informally provable that A (for arbitrary A)
  + "Sufficient" syntax to express:
    $TM(x)$: x is a Turing machine, etc.

- Principles:

$$T \quad \Box A \rightarrow A$$

$$K \quad \Box(A \rightarrow B) \rightarrow (\Box A \rightarrow \Box B)$$

$$Proof\text{-}\Box \quad S(\ulcorner A \urcorner) \rightarrow (\Box A \leftrightarrow \exists y Proof(y, \ulcorner A \urcorner))$$

$$Nec \quad \frac{\vdash A}{\vdash \Box A}$$

$$ECT \quad \Box \forall x \exists y \Box \varphi(x, y) \rightarrow \exists e[TM(e) \wedge \forall x \varphi(x, e(x))]$$

$+$ "Sufficient" mathematics to prove the undecidability of first-order arithmetical truth (Gödel, Tarski)

ECT is a version of the formalized Church–Turing thesis that is very much in the same ballpark as the formalizations that one can actually find in the literature on epistemic arithmetic (see Flagg (1985, p. 166) for his slightly stronger "Epistemic Church's Thesis" which he proves to be consistent with Epistemic Arithmetic, and the closely related principle BPT in Reinhard (1986, p. 44)). We do not want to go into any details here, but its rationale is to assume that a $\forall x \exists y \Box \varphi(x, y)$ statement can only be provable informally if there is an effective way of assigning to each x some y such that $\varphi(x, y)$.[35]

In the resulting formal system, the following formula can now be shown to be a theorem (the proof is contained in Horsten and Leitgeb (unpublished draft)):

Theorem 5.1. $\exists y[S(y) \wedge \forall x(\neg Proof(x, y) \wedge \neg Proof(x, \neg y))] \vee$
$(\neg \Box A_1 \wedge \neg \Box \neg A_1) \vee (\neg \Box A_2 \wedge \neg \Box \neg A_2) \vee \ldots \vee (\neg \Box A_n \wedge \neg \Box \neg A_n)$

where $A_1, \ldots, A_n$ is a finite sequence of formulas which contain the informal provability operator.[36] The undecidability of first-order arithmetic is needed as one of the premises from which the theorem can be derived. So does this show that HG from above is vindicated? Not really. Since the other principles used in the formal system are unproblematic, all it shows is that *either ECT is false or HG is the case*. Obviously, this reminds one of Gödel's (1951) dichotomy again; in fact, it might even be regarded as one possible formalization of this dichotomy. So this is *not* a Lucas–Penrose style argument, by which only the second disjunct (HG) would be claimed to be supported. Unfortunately, in this case, the obvious reaction will be to argue that ECT is false, and that ECT therefore cannot be an adequate representation of the Church–Turing thesis (which we do consider to be true).

So let us try something else. In view of our difficulties of proving HG, can we perhaps prove that

$$\text{HG } \exists p(p \wedge \neg \Box p)$$

is absolutely *un*provable?

Here is a proposal of how to do that. For that purpose, we assume that our $\Box$-language includes Hilbert-style epsilon terms (or rather epsilon formulas) for *propositions*, that is, for every formula $A[p]$ with a propositional variable p, there is a formula $\epsilon p A[p]$, such that

$$\exists p A[p] \leftrightarrow A[\epsilon p A[p]]$$

is a logical truth (compare the standard epsilon calculus with epsilon terms for individuals. See Zach, 2003). In natural language terms, "$\epsilon p A[p]$" would have to be read as follows: "a proposition p, such that $A[p]$ holds, is the case". But note that even if "$\epsilon p A[p]$" did not have a plausible rendering in natural language, there would be no obvious reason why we could not *introduce* such propositional epsilon terms on the basis of the logical axiom expressed above, with a semantics of choice functions which choose for every satisfiable formula $A[p]$ a proposition $\epsilon p A[p]$ that witnesses the satisfiability of $A[p]$.[37]

Now we reason as follows:

- As pointed out,

$$\exists p A[p] \leftrightarrow A[\epsilon p A[p]]$$

  is a logical truth.
- From this, together with S4, we get

$$\Box \exists p A[p] \leftrightarrow \Box A[\epsilon p A[p]]$$

- Now let $A[p]$ be $(p \wedge \neg \Box p)$: so it follows that

$$\Box HG \leftrightarrow \Box(\epsilon pA[p] \wedge \neg \Box(\epsilon pA[p]))$$

and hence $\Box HG$ turns out to be contradictory by the same argument as used above!

Does this show that HG cannot be proven? Not quite so: only if epsilon terms for propositions make sense at all, especially in cases in which the propositional epsilon operator is applied to an intensional $\Box$-context, which is far from being clear. In fact, given the well-known problems affecting definite descriptions for individuals in modal contexts, there is reason to believe the contrary (see, in particular, Heylen 2009). Indeed, it turns out to be possible to derive a much stronger conclusion:[38]

Abbreviate the formula $(p \wedge \neg \Box p)$ by $A[p]$ again.
By the axiom for the epsilon symbol, we derive

(1) $\exists p(p \wedge \neg \Box p) \leftrightarrow \epsilon pA[p] \wedge \neg \Box \epsilon pA[p]$
    Applying necessitation and distributing $\Box$ over $\leftrightarrow$ by K we get
(2) $\Box \exists p(p \wedge \neg \Box p) \leftrightarrow \Box(\epsilon pA[p] \wedge \neg \Box \epsilon pA[p])$
    But from KT we can also derive
(3) $\neg \Box(\epsilon pA[p] \wedge \neg \Box \epsilon pA[p])$
    which with (2) implies
(4) $\neg \Box \exists p(p \wedge \neg \Box p)$
    (This is how far we got in the derivation above.)
    We can rewrite (4) more transparently as
(5) $\Diamond \forall p(p \rightarrow \Box p)$
    Now abbreviate $\forall p(p \rightarrow \Box p)$ by B, i.e., (5) is nothing but $\Diamond B$.
    Existential introduction and K give us both
(6) $B \rightarrow \exists p[(\Box(p \rightarrow B) \vee \Box(p \rightarrow \neg B)) \wedge p]$
    and
(7) $\neg B \rightarrow \exists p[(\Box(p \rightarrow B) \vee \Box(p \rightarrow \neg B)) \wedge p]$
    by taking B and $\neg B$, respectively, as the witnesses for the existential claim.
    So proof by cases gives us
(8) $\exists p[(\Box(p \rightarrow B) \vee \Box(p \rightarrow \neg B)) \wedge p]$
    Abbreviate $[(\Box(p \rightarrow B) \vee \Box(p \rightarrow \neg B)) \wedge p]$ by $C[p]$.
    Line (8) together with the axiom for the epsilon symbol thus leads to
(9) $(\Box(\epsilon pC[p] \rightarrow B) \vee \Box(\epsilon pC[p] \rightarrow \neg B)) \wedge \epsilon pC[p]$
    Now simplify to the first conjunct and, by K, distribute $\Box$ on both implications[39] to get
(10) $(\Box \epsilon pC[p] \rightarrow \Box B) \vee (\Box \epsilon pC[p] \rightarrow \Box \neg B)$
    But (9) with propositional logic and necessitation also yields

(11) $\Box \epsilon pC[p]$
which together with (10) and propositional logic leads to
(12) $\Box B \vee \Box \neg B$
From this, with (5) and the modal system K, we can finally conclude
(13) $\Box B$
and so by T
(14) $\forall p(p \rightarrow \Box p)$

Taken together with T again this means: Informal provability and truth coincide extensionally. Thus, instead of proving "merely" that HG is unprovable, we can even derive it is *false* (as (14) is logically equivalent to ¬HG). In light of the derivation, this is no longer so surprising, but maybe it is surprising at least at first glance since the underlying assumptions might have seemed to be pretty weak. Once again: Does this show that the status of HG has been settled now? The answer is still the same: No, as it much more plausible to believe that the propositional epsilon calculus with a modal operator $\Box$ has been shown logically deficient, than thinking that the truth value of HG has been determined; instantiations of the epsilon axiom by expressions in which the propositional epsilon symbol is applied to a modal formula are simply not to be counted as logical truths and sometimes maybe not even as truths at all. So the Holy Grail in philosophy of mathematics – for this is what "HG" stands for – is still waiting to be found, as is further insight into the elusive but fundamental concept of informal provability.

Acknowledgements

We want to thank Alexei Angelides, Jeremy Avigad, Johan van Benthem, Hans Czermak, Sol Feferman, Leon Horsten, Peter Koellner, Øystein Linnebo, Pen Maddy, Grisha Mints, Richard Pettigrew, Pat Suppes, and Kai Wehmeier for their advice on this paper.

Notes

1. This strategy of separating informal provability in the core areas of pure mathematics from informal provability in foundational areas, in particular, in set theory, is of course not unproblematic: the natural number sequence and the continuum of real numbers can be reconstructed as set-theoretic systems and in this sense become objects of study in pure set theory; arithmetic and analysis use concepts such as *function* or *set* which receive their modern extensional treatment only through set theory; and in areas such as descriptive set theory, it is the case that analytic and set-theoretic methods and questions are continuous with each other (see Maddy 1997, 66–69). But since "normal mathematics" is still mainly done without using set theory as much more than just a language – and since *divide et impera* is still a very successful methodological maxim even in philosophy – we will take the liberty of disregarding the foundational aspects of mathematics in this article for the most part.

2. Myhill adds the qualification "provably", and rightly so, in a subsequent parenthetical remark; without it the statement would not follow in any obvious way from Gödel's Incompleteness Theorems.
3. Williamson (2000) is the paradigm case proponent of this view.
4. See Horsten (2000) for some of the logical properties that this kind of modal analysis of informal provability would have.
5. In Section 3, we will argue that informal proofs *do* have such non-propositional and non-conceptual components.
6. The process of *finding* or *discovering* a proof is a different type of process which we will *not* deal with.
7. Just not *too* broadly or idealized – see the "God" example below.
8. Accordingly, when Suppes (2005) starts to inquire into "What is the basis for saying that an informal proof is valid? It cannot be that it has been checked by some familiar algorithm of formal verification or computation" in the third paragraph of his paper, he immediately turns in the very first section to the question of what mental representations we form of what is, in Hilbert's phrase (cited by Suppes), "given to us in our faculty of representation".
9. Note that there seem to be obvious formal counterparts to "informally provable" and "informal proof" but only less obvious ones to "informal proving". Jeremy Avigad suggested to us in informal communication that interactions with proof assistants might provide the lacking formal counterpart.
10. If anything, the best way of making logical sense of induction is maybe in terms of open-ended schemata which can be instantiated by any "possible" mathematical expression whatsoever (as long as the expression expresses a "determinate" mathematical property); see McGee (1997) and Feferman (2006b).
11. Of course, it is possible to set up a recursively axiomatized system *to which we additionally assign some interpretation*, as in the case of formalized first-order Peano Arithmetic, but even in such a case it is not *presupposed* by the definition of "formally provable-in-PA" that any interpretation has been assigned to the terms and formulas of the system. In contrast, "informally provable" cannot be sensibly applied to uninterpreted sequences of signs.
12. We are simplifying matters here; proof by indirect assumption, for example, has a more complicated structure.
13. A similar thought can be found in Väänänen (2001), pp 509f: "Informal reasoning will remain the guiding line in mathematics ... Surely informal reasoning about infinite objects feels more like speculation than formal reasoning, where everything is spelled out in axioms. But in the latter case we can only speculate whether our formalization captures what we intend." We want to add to this that even though informal theorems and proofs are not *necessarily* based on axioms, they still *can* be if mathematicians decide to present their theories in such a way. While it is necessary for a formal system to be axiomatic, it is not necessary for an axiomatic system to be a formal system in the technical sense of the word.
14. We emphasize again that we are concerned with the core areas of mathematics here. In the foundations of mathematics, most notably in set theory, there are controversial axioms – think of "really large" cardinal axioms – which are made explicit in proofs, which are playing much the same role as axioms in formal systems, which are typically neither self-evident nor presupposed by everyday mathematical proofs, and which are argued for on non-demonstrable grounds by considering their role within mathematical theorizing as a whole. This is of

course an important and fascinating topic – see Maddy (1997) – but it is not the one that we are concerned with in this chapter.

15. See Robinson (1997) for a defense of the informal rigour of unformalized mathematical proofs.

16. More on the formalization and formalizability of informal mathematical proofs by derivations in formal systems can be found in Feferman (1979) and (1993), and Rav (2007).

17. As Mendelson (1990) argues, bridge principles between "intuitive" and "precise" mathematical notions (his terminology) need not necessarily be considered unprovable theses themselves but sometimes they can in fact be derived by rigorous argument.

18. In the foundations of mathematics, the corresponding question would be

• What do independence results in set theory tell us about mathematical provability?

Although we do not deal with set theoretic issues here, as mentioned at the beginning, we think that also questions of this sort are sometimes answered too quickly. For example, Gödel's and Cohen's celebrated result that the Continuum Hypothesis is independent of ZFC does *not* straightforwardly show that both the Continuum Hypothesis and its negation are informally unprovable. At best, it might give evidence for the thesis that they cannot be settled with the methods that we currently have at our disposal. (This is also how I suggest to interpret, and perhaps straighten out, Burgess' (1992, p. 17) appraisal of independence theorems.) A detailed discussion of the possibility of absolute undecidability by set theoretic "bifurcation" can be found in Koellner's "Truth in Mathematics: The Question of Pluralism", this volume.

19. Myhill (1960), p. 462, maintains that "the use of the word 'proof' in ordinary non-philosophical mathematical discussion is rather clearly neither a syntactical nor a semantical term". As far as we can see, this statement does not contradict our claim that informal proofs have semantic components, as we do not claim that informal provability is conceptually *reducible* to any semantic concept; but, presumably, that would be the thesis which Myhill disputed. Ironically, Myhill himself uses a truth predicate, that is, a semantic predicate, when he argues that Gödel's undecidable statement for elementary arithmetic is absolutely provable (see p. 463).

20. Once again, we want to stress that in the foundations of mathematics the defense of axioms on the basis of their direct or indirect intrinsic plausibility can only go so far in their overall justification; see Maddy (1997) for a detailed analysis.

21. We could, and maybe should, have used directed edges in this case, but never mind.

22. By the way, we are still on Gödelian grounds here, "Nor is it self-contradictory that a proper part should be identical (not merely equal) to the whole, as is seen in the case of structures in the abstract sense. The structure of the series of integers, e.g., contains itself as a proper part" (Gödel 1944, p. 130). An alternative route, which would support the potential infinity of the natural number structure rather than its actual infinity which we have focused on above, would be to draw on the intuitive possibility of adding a stroke to any stroke sequence representation of a natural number; see Parsons (1979–80).

23. The example of ordinals is not a fictional one. See Löwe (2007).

24. The standard examples are Weierstrass' function which is everywhere continuous but nowhere differentiable, or Peano's function which is space-filling despite being defined on a real interval; neither of them ought to exist according to our geometrical intuitions. See Feferman (2000) for a discussion of these "mathematical monsters" and an analysis of what they do – or do not – tell us about the "unreliability" of mathematical intuition.
25. For example, the paradoxical character of the famous Banach–Tarski paradox might be due to a clash between set-theoretic and geometrical intuitions, which in turn might be caused by our limited intuitive access to geometrical regions, that is, by facts which constrain our intuition "merely" to geometrically "very well-behaved" regions of Euclidean space.
26. This is similar to Lakatos' view of the "muddy foundations" on which mathematics rests.
27. Maddy (1990), section 2.3, and Suppes (2005) can be read as advertisements of a cognitive account of mathematical intuition, too – an account that is ultimately based on some sort of neural "pattern cognition".
28. Robinson (1997, p. 64), must have something like this in mind when he writes, "The conclusion I draw from this and other similar introspective cognitive experiments with actual proof scenarios is that such examples of rigorous but unformalized mathematical thinking provide a rich source of material for the development of a more authentic model of mathematical reasoning and communicating results than the more limited model offered by the rigid formal systems of traditional mathematical logic. It seems clear that in any such realistic model there will have to be a central role for the thinking subject. That is to say, we must explicitly include features in the model which capture the way the intuition and imagination deal directly (without the mediation of formal definitions) with spatial and temporal relationships, with order relations, with comparison and combination of quantities, and with elementary manipulations of symbols."
29. We owe the irrationality–rationality example to Michael Sheard.
30. We are glossing over some potential problems here which are similar to the problems that affect applications of the introspective 4 or KK principle in epistemology; see Williamson (2000) for further details.
31. Artemov (2001) has developed an elegant logical system of explicit formal proofs into which 34 can be embedded in a way such that every instantiation of □A is rendered as an expression of the form "proof term t proves A". However, this result does not show that S4 is a logic for formal provability after all – □A *does* have a genuine interpretation in terms of "it is *provable* that A", but the type of provability in question is informal provability.
32. We use an informal provability predicate for *sentences* here, where we should actually use one for *propositions*, but never mind.
33. The results contained in Halbach et al. (2003) "almost" amount to a formal proof that formal provability predicates are the only type-free modal predicates of sentences for which the predicate versions of modal axioms can be maintained unrestrictedly.
34. At the end of our chapter, we will outline an argument which – at least *prima facie* – casts some doubt on the informal provability of this existence claim since maybe the informal provability of existential theses may be turned into the informal provability of corresponding instantiated claims with the help of additional logical resources.

35. It is easy to read this wrongly and to argue: clearly, there are provable $\forall x \exists y \varphi(x, y)$ statements for which no Turing machine could determine for every x a y such that $\varphi(x, y)$; so ECT is obviously false. However, what is really at issue here is to assume that for all x there is a y *of which* it is provable that $\varphi(x, y)$, and furthermore to assume that this latter assumption is provable. It does not seem clear at all that given this assumption the consequent of ECT can still be false.

36. $\neg$ denotes the arithmetically definable function which takes the code of a sentence B as an argument and which maps it to the code of $\neg B$.

37. Horsten (2009) combines an absolute provability operator with definite descriptions for individuals – and a Carnapian understanding of definite descriptions – in order to draw conclusions on absolute provability (though not the conclusion at which we are aiming). We are very much indebted to him for presenting his argument to us which in turn inspired the argument to be given. Accordingly, our collapse argument to be stated further down below was inspired by collapse arguments for definite descriptions in modal contexts which we had the pleasure to discuss with Leon Horsten and Jan Heylen.

38. The axiomatic system on which this result rests is (i) the modal system KT (including classical propositional logic); (ii) an existential generalization or introduction axiom scheme for the propositional existential quantifier: $A[B] \rightarrow \exists p A[p]$; and (iii) the axiom scheme for the propositional epsilon symbol: $\exists p A[p] \leftrightarrow A[\epsilon p A[p]]$. Note that the instances of the latter two axiom schemes are considered to be proper logical axioms, so that necessitation may be applied to them. Whenever $A[p]$ is a given formula, and we write for example, $A[B]$ or $A[\epsilon p A[p]]$, then this is meant to be the formulas that result from $A[p]$ by replacing all occurrences of p by B or by $\epsilon p A[p]$, respectively.

39. We want to thank Grisha Mints for pointing out to us a simplification at this step of the proof.

References

Artemov, S., "Explicit provability and constructive semantics", *Bulletin of Symbolic Logic* 7 (2001), 1–36.

Auslander, J., "On the roles of proof in mathematics", in: B. Gold, and R. Simons (eds.), *Proof and Other Dilemmas: Mathematics and Philosophy*, Spectrum, The Mathematical Association of America, 2008, 61–77.

Avigad, J., "Mathematical method and proof", *Synthese* 153 (2006), 105–159.

Avigad, J., E. Dean, and J. Mumma, "A formal system for Euclid's *Elements*", unpublished draft.

Azzouni, J., *Tracking Reason. Proof, Consequence, and Truth*, Oxford: Oxford University Press, 2006.

Barsalou, L.W., "Perceptual symbol systems", *Behavioral and Brain Sciences* 22 (1999), 577–660.

Beklemishev, L.D. and A. Visser, "Problems in the logic of provability", in: D. Gabbay et al. (eds.), *Mathematical Problems from Applied Logics. New Logics for the XXIst Century*, International Mathematical Series Vol. 4, New York: Springer, 2005.

Benacerraf, P., "God, the devil, and Gödel", *The Monist* 51 (1967), 9–32.

Boolos, G., *The Logic of Provability*, Cambridge: Cambridge University Press, 1993.

Bundy, A., M. Jamnik, and A. Fugard, "What is a proof?" *Philosophical Transactions of Royal Society A* 363 (2005), 2377–2391.

Burgess, J., "Proofs about proofs", in: Detlefsen, M. (ed.), *Proof, Logic and Formalization*, London: Routledge, 1992, 8–23.

Carlson, T.J., "Epistemic arithmetic and a conjecture of Reinhardt", *Abstracts of papers presented to the American Mathematical Society* 5 (1984), 200.

Carlson, T.J., "Knowledge, machines, and the consistency of Reinhardt's strong mechanistic thesis", *Annals of Pure and Applied Logic* 105 (2000), 51–82.

Cohen, P.J., "Skolem and pessimism about proof in mathematics", *Philosophical Transactions of Royal Society A* 363 (2005), 2407–2418.

Detlefsen, M. (ed.), *Proof, Logic and Formalization*, London: Routledge, 1992.

Detlefsen, M., "Proof: Its nature and significance", in: Gold, B. and R. Simons (eds.), *Proof and Other Dilemmas: Mathematics and Philosophy*, Spectrum, The Mathematical Association of America, 2008, 3–32.

Etchemendy, J., *The Concept of Logical Consequence*, Stanford: CSLI Publications, 1999.

Feferman, S., "Transfinite recursive progressions of axiomatic theories", *Journal of Symbolic Logic* 27 (1962), 259–316.

Feferman, S., "What does logic have to tell us about mathematical proofs?", *The Mathematical Intelligencer* 2 (1979), 20–24. Reprinted with minor changes in: S. Feferman, *In the Light of Logic*, Oxford: Oxford University Press, 1998, 177–186.

Feferman, S., "What rests on what? The proof-theoretic analysis of mathematics", in: J. Czermak (ed.), *Philosophie der Mathematik*, Volume I, Akten des 15. internationalen Wittgenstein Symposiums, Vienna: Hölder-Pichler-Tempsky, 1993, 147–171. Reprinted with minor changes in: S. Feferman, *In the Light of Logic*, Oxford: Oxford University Press, 1998, 187–208.

Feferman, S., "Mathematical intuition vs. mathematical monsters", *Synthese* 125 (2000), 317–332.

Feferman, S., "Are there absolutely unsolvable problems? Gödel's dichotomy", *Philosophia Mathematica* 14 (2006a), 134–152.

Feferman, S., "Open-ended schematic axiom systems", lecture at the ASL Annual Meeting 2005, Stanford. Abstract in *Bulletin of Symbolic Logic* 12 (2006b), 145.

Flagg, R., "Church's thesis is consistent with Epistemic Arithmetic", in: Shapiro, S. (ed.), *Intensional Mathematics*: Elsevier, 1985b, 121–172.

Giaquinto, M., *Visual Thinking in Mathematics*, Oxford: Oxford University Press, 2007.

Gödel, K., "Eine Interpretation des intuitionistischen Aussagenkalüls", *Ergebnisse eines Mathematischen Kolloquiums*, 4 (1933), 39–40. Translated as "An interpretation of the intuitionistic propositional calculus", in: S. Feferman et al. (eds.), *Kurt Gödel Collected Works*, Vol. I, Oxford: Oxford University Press, 1986, 300–302.

Gödel, K., "Russell's mathematical logic", 1944, in: S. Feferman et al. (eds.), *Kurt Gödel Collected Works*, Vol. II, Oxford: Oxford University Press, 1990, 119–141.

Gödel, K., "Some basic theorems on the foundations of mathematics and their implications", 1951, in: S. Feferman et al. (eds.), *Kurt Gödel Collected Works*, Vol. III, Oxford: Oxford University Press, 1995, 304–332.

Gödel, K., "Is mathematics syntax of language?", 1953, in: S. Feferman et al. (eds.), *Kurt Gödel Collected Works*, Vol. III, Oxford: Oxford University Press, 1995, 334–364.

Gold, B. and R. Simons (eds.), *Proof and Other Dilemmas: Mathematics and Philosophy*, Spectrum, The Mathematical Association of America, 2008.

Halbach, V., H. Leitgeb, and P. Welch, "Possible worlds semantics for modal notions conceived as predicates", *Journal of Philosophical Logic* 32 (2003), 179–223.

Heylen, J., "Carnapian Modal and Epistemic Arithmetic", in: M. Carrara and V. Morato (eds.), *Language, Knowledge, and Metaphysics. Selected Papers from the First SIFA Graduate Conference*, London: College Publications, 2009, 97–121.

Horsten, L., "In defense of epistemic arithmetic", *Synthese* 116 (1998), 1–25.

Horsten, L., "Models for the logic of possible proofs", *Pacific Philosophical Quarterly* 81 (2000), 49–66.

Horsten, L., "Canonical naming systems", *Minds and Machines* 15 (2005a), 229–257.

Horsten, L., "Remarks about the content and extension of the notion of provability", *Logique et Analyse* 189–192 (2005b), 15–32.

Horsten, L., "An argument concerning the unknowable", forthcoming in *Analysis* (2009).

Horsten, L. and H. Leitgeb, "Church's thesis and absolute undecidability", unpublished draft.

Koellner, P., "On the question of absolute undecidability", *Philosophia Mathematica* 14 (2006), 153–188.

Koellner, P., "Truth in mathematics: The question of pluralism", this volume.

Kreisel, G., "Informal rigour and completeness proofs", in: I. Lakatos (ed.), *Problems in the Philosophy of Mathematics*, Amsterdam: North-Holland, 1967, 138–171.

Lakatos, I., *Proofs and Refutations*, New York: Cambridge University Press, 1976.

Leitgeb, H., "What truth depends on", *Journal of Philosophical Logic* 34 (2005), 155–192.

Leitgeb, H., "Towards a logic of type-free modality and truth", in: C. Dimitracopoulos et al. (eds.), *Logic Colloquium 05*, Lecture Notes in Logic, Cambridge University Press, 2008, 68–84.

Löwe, B., "Visualization of ordinals", in: T. Müller and A. Newen (eds.), *Logik, Begriffe, Prinzipien des Handelns*, Paderborn: Mentis Verlag, 2007, 64–80.

Maddy, P., *Realism in Mathematics*, Oxford: Clarendon Press, 1990.

Maddy, P., *Naturalism in Mathematics*, Oxford: Clarendon Press, 1997.

Manders, K., "The Euclidean diagram", unpublished draft, 1995.

Mayo-Wilson, C., "Formalization and justification", unpublished draft.

McGee, V., "How we learn mathematical language", *Philosophical Review* 106 (1997), 35–68.

Mendelson, E., "Second thoughts about Church's thesis and mathematical proofs", *Journal of Philosophy* 87 (1990), 225–233.

Montague, R., "Syntactical treatments of modality, with corollaries on reflexion principles and finite axiomatizability", *Acta Philosophica Fennica* 16 (1963), 153–167.

Mumma, J., *Intuition Formalized: Ancient and Modern Methods of Proof in Elementary Geometry*, PhD Thesis, Carnegie Mellon University, 2006.

Myhill, J., "Some remarks on the notion of proof", *Journal of Philosophy* 57 (1960), 461–471.

Parsons, C., "Mathematical intuition", *Proceedings of the Aristotelian Society* 80 (1979–80), 145–168.

Parsons, C., "On some difficulties concerning intuition and intuitive knowledge", *Mind* 102 (1993), 233–246.

Rav, Y., "Why do we prove theorems?", *Philosophia Mathematica* 7 (1999), 5–41.

Rav, Y., "A critique of a formalist-mechanist version of the justification of arguments in mathematicians' proof practices", *Philosophia Mathematica* 15 (2007), 291–320.

Reinhardt, W.N., "Epistemic theories and the interpretation of Gödel's incompleteness theorems", *Journal of Philosophical Logic* 15 (1986), 427–474.

Resnik, M., *Mathematics as a Science of Patterns*, Oxford: Oxford University Press, 1997.

Robinson, J.A., "Informal rigor and mathematical understanding", in: G. Gottlob et al. (eds.), *Computational Logic and Proof Theory*, Lecture Notes in Computer Science 1289, Berlin: Springer, 1997, 54–64.

Shapiro, S., "Epistemic and Intuitionistic Arithmetic", in: S. Shapiro (ed.), 1985, 11–46.

Shapiro, S. (ed.), *Intensional Mathematics*, Amsterdam: Elsevier, 1985.

Shapiro, S., *Philosophy of Mathematics: Structure and Ontology*, New York: Oxford University Press, 1997.

Shapiro, S., "Incompleteness, mechanism, and optimism", *Bulletin of Symbolic Logic* 4 (1998), 273–302.

Shapiro, S., "We hold these truths to be self-evident: But what do we mean by that?", unpublished draft.

Skyrms, B., "An immaculate conception of modality or how to confuse use and mention", *Journal of Philosophy* 75 (1978), 368–387.

Solovay, R.M., "Provability interpretations of modal logic", *Israel Journal of Mathematics* 25 (1976), 87–304.

Statman, R., *Structural Complexity of Proofs*, PhD Thesis, Stanford University, 1974.

Suppes, P., "Psychological nature of verification of informal mathematical proofs", in: S. Artemov et al. (eds.), *We Will Show Them: Essays in Honour of Dov Gabbay*, Vol. 2, College Publications, 2005, 693–712.

Tarski, A., "What are logical notions?", *History and Philosophy of Logic* 7 (1986), 143–154.

Tieszen, R., "What is a Proof?", in: M. Detlefsen (ed.), *Proof, Logic and Forumalization*, London: Routledge, 1992, 57–76.

Väänänen, J., "Second-order logic and foundation of mathematics", *The Bulletin of Symbolic Logic* 7 (2001), 504–520.

Williamson, T., *Knowledge and Its Limits*, Oxford: Oxford University Press, 2000.

Zach, R., "The practice of finitism. Epsilon calculus and consistency proofs in Hilbert's Program", *Synthese* 137 (2003), 211–259.

13
Quantification Without a Domain

Gabriel Uzquiano

1 Introduction

Much recent work in the philosophy of mathematics has been concerned with indefinite extensibility and the problem of absolute generality. It is not uncommon to take the set-theoretic paradoxes to illustrate a phenomenon of indefinite extensibility whereby certain concepts, for example, *set* and *ordinal*, are indefinitely extensible. What is less clear is how to articulate this response to paradox or what it means for the prospects of absolute generality. While some philosophers take the indefinite extensibility of certain concepts to imply the unavailability of unrestricted quantification over their instances, others seem inclined to conclude that there is no comprehensive domain.[1] These conclusions are importantly different: One seems to place a serious limitation on thought and language while the other might seem a subtle ontological discovery.

The purpose of this chapter is twofold. One aim is to clarify what is at stake in the debate over the existence of a comprehensive domain. We distinguish the thesis that there is a comprehensive domain from more metaphysically laden theses from which one may be tempted to derive more substantive ontological conclusions such as metaphysical realism. The other aim is to clarify what it would take to deny the existence of a comprehensive domain as we propose to understand it. We will articulate a general schematic route from the Russellian argument on which indefinite extensibility considerations are generally based to the thesis that there is no comprehensive domain. The purpose of the journey is to identify what exactly would be required for a reasonable defense of the thesis and to suggest that there are reasons of principle why the prospects for such a defense do not look too bright.

2 The problem of absolute generality

The problem of absolute generality is the question of whether there are—or could be—instances of absolutely general quantification. Call an occurrence of a quantifier "absolutely general" if and only if the quantifier is completely unrestricted and ranges over a comprehensive domain, by which we mean, as usual, a domain of all objects. We have two questions. One is the linguistic question of whether there is—or could be—completely unrestricted quantification. The other is the ontological question of whether there is a comprehensive domain for unrestricted quantifiers to range over.

The first question has received much attention in the literature. Most parties in the debate agree that even if available, unrestricted quantification is relatively uncommon. To rehearse a familiar point, in ordinary contexts, our quantifiers are often both restricted explicitly by a noun phrase and tacitly by context. When one utters the following:

(1) Every passenger has boarded the plane.

in an ordinary context, the quantifier phrase "every passenger" combines the determiner "Every" and the noun "passenger" and, as result, is restricted to passengers. In most contexts in which the sentence is uttered by a speaker, there is, in addition, a tacit restriction to a domain of contextually salient passengers. Otherwise, we would have to interpret a speaker's utterance of the sentence to claim that every passenger in the entire world has boarded a given plane, which would be uncharitable to say the least. It is better to interpret ordinary utterances of (1) to be tacitly restricted to a certain domain of contextually salient passengers. The question of how exactly this restriction is supposed to take place is highly controversial, but what is important for present purposes is just the observation that context often restricts the range of our quantifiers. We generally gather from context what restrictions are in force, guided by our presumption that what a speaker says is meant to make sense.

The linguistic question is whether completely unrestricted quantification is—or could be—still available in other circumstances. The context of ontological investigation provides a case in point. When Quine answers "Everything" to the fundamental ontological question "What is there?" in Quine (1948), he presumably intends the quantifier to carry no restriction whatever.[2] But even if we concede this point, the question remains whether the occurrence of the quantifier is absolutely general. This is the ontological question of whether the quantifier may be taken to range over a perfectly comprehensive domain of all objects.

In what follows, we are primarily concerned with the ontological question. This is not to say that the questions can be separated and discussed in isolation. Much research on absolute generality has presupposed the existence of a comprehensive domain and has focused on the availability of completely unrestricted quantification. Instead, we will proceed in the opposite direction. We will assume that completely unrestricted quantification is available in certain contexts and proceed to focus on the ontological question of whether there is a perfectly comprehensive domain.

But we should first delimit the question by providing a gloss on the phrase "a comprehensive domain". As we use the term, to speak of a domain of certain objects is just to speak of the objects themselves; and to speak of a comprehensive domain is just to speak of zero or more objects such that *every* object is one of them. The thesis that there is a comprehensive domain becomes

(2) There are some objects such that every object is one of them,

where the plural quantifier "there are some objects" and the singular quantifier "every object" are completely unrestricted.[3] This assumption is permissible in the present context in which we have taken for granted the availability of completely unrestricted quantification in order to focus on the question of whether there is a comprehensive domain.[4] But not only does (2) as a statement of the absolutist thesis presuppose the availability of completely unrestricted quantification, unless there is a comprehensive domain in the first place, statement (2) will simply fail to articulate the position.

It will be helpful to distinguish two different grounds on which the existence of a comprehensive domain has commonly been questioned in the literature.

One is the Carnapian tradition on which ontology is relative to a certain perspective and one should not expect to make sense of a perspective-independent domain of all objects. In Carnap (1950), he took ontological questions to be relative to a linguistic framework. He tied his view to a distinction between internal and external questions by which he meant a distinction between theoretical questions asked within a linguistic framework and practical questions as to what framework to adopt. One may ask, for example, whether numbers exist against the background of a certain framework in which case the answer is rather trivial and not what the metaphysician probably expected. However, it would be a mistake to raise the practical question of whether to adopt a framework from within which there are numbers as a theoretical question on what there is. More to the point that concerns us, it would be nonsensical to speak of a linguistic-framework-independent domain of all objects. The most one should hope for is a domain that is comprehensive relative to some linguistic framework or another.

A variety of views in the spirit of Carnap (1950) remain alive and well in the literature.

One strand is broadly Quinean in spirit and locates the source of relativity in the theoretical enterprise. John Burgess (2004a, 2005b) and Robert Stalnaker (2003) have recently described an ontological relativist outlook that does without external questions as Carnap conceived of them. Instead, the ontological relativist acknowledges that the choice of a framework or perspective is not a matter of convention but rather takes place against the background of certain theoretical presuppositions. Such choices are never merely practical decisions, but rather fall in line with other theoretical choices within a framework. The point remains, however, that there is no comprehensive domain. At most, we should acknowledge domains that are comprehensive relative to one framework or another.

Another strand locates the source of relativity in language. Thus Hilary Putnam (1994) describes his "thesis of conceptual relativity" as the thesis that "there isn't one privileged use of the word 'object'". And Eli Hirsch (2002) formulates his thesis of "quantifier variance" as the denial that there is a "metaphysically privileged sense of the quantifier". At the heart of their respective views lies the thesis that there is a multiplicity of uses for the quantifiers on which different sentences have different truth values. Thus a sentence "there are composite objects" is true with respect to a certain use of the quantifiers and false with respect to another. In Putnam's words

> What logicians call "the existential quantifier", the symbol "(∃x)", and its ordinary counterparts, the expressions, "there are", "there exist" and "there is a", "some", etc., do not have a single absolutely precise use but a whole family of uses.[5]

There is, in particular, no privileged language with which to assess ontological questions. And because there is no absolute quantificational structure to accompany such a privileged language, it is non-sensical to speak of a language-independent domain of all objects. The most we should hope for is to speak of a domain that is comprehensive relative to some use of the language or quantificational structure or another.

What the views under consideration have in common is the inchoate thought whether a linguistic framework, a conceptual scheme or a quantificational structure. This thought has been implemented differently by different philosophers, but we will use the broad label "ontological relativism" to apply indiscriminately to all such views.

What is the import of ontological relativism for the question of absolute generality? The ontological relativist need not disagree with the absolutist over the availability of completely unrestricted quantification; he or she may insist that what lies in the domain of discourse will nevertheless be relative to one perspective or another, but this is no reason to think that every occurrence of a quantifier must be restricted either explicitly or by context.[6]

The question of a comprehensive domain is more delicate. Parsons (2006) and Hellman (2006), for example, have suggested that there is a tension between ontological relativism and the existence of a comprehensive domain. But maybe there is no need for the relativist to object to (2) and indeed some pressure to accept it. For the following is a theorem of the logic of identity:

(3) Everything is self-identical,

even when we take the occurrence of "everything" to be completely unrestricted. However, statement (2) is a trivial consequence of (3) together with the following consequence of plural comprehension:

(4) There are some objects such that an object is one of them if and only if it is self-identical.

But nothing in the ontological relativist position requires a revision of the logic of identity of the logic of plural quantification. Even when constrained by one perspective or another, domain relativists remain free to acknowledge the existence of a comprehensive domain consisting of whatever objects are recognizable as potential values of variables from within the perspective. When the quantifiers that occur in statement (2) are interpreted to have unrestricted range over a domain that is comprehensive relative to a certain perspective, statement (2) should still strike many relativists as a truism.

What the ontological relativist will object to is not quite the existence of a comprehensive domain as characterized by (2) but rather the thought that what lies in the domain of all objects is perspective-independent. Part of the problem with this thought is that it is difficult to articulate in a way that is acceptable to all parties in dispute. For the ontological relativist, the problem arises from the observation that no matter what thesis they state, it will be, by their own lights, stated within a certain perspective and will thereby be constrained to quantify only over what is available relative to that perspective. The problem for his or her opponent is that he or she will claim not to understand what it is for what there is to be relative to a certain perspective or another.

One option for the ontological relativists is to treat "absolutely" as an irreducible operator somewhat like "in every perspective". Perhaps they could then characterize their foe as the thesis:

(5) There are some objects such that *absolutely* every object is one of them.

You may perhaps question the intelligibility of the operator, but once you grant its use, (5) takes us far beyond a truth of the logic of plural quantification. What (5) tells us is that some objects are all the objects there are no matter what perspective one adopts. The thesis that there is an *absolutely* comprehensive domain doesn't seem to be what is at stake in the debate over absolute generality. It may be argued of course that (5) has ramifications for broader metaphysical debates such as the question of realism. But to identify the existence of a comprehensive domain with (5) is to change the subject to a debate that is orthogonal to the question of absolute generality.

The other broad reason to doubt the existence of a comprehensive domain stems from considerations of indefinite extensibility and is our main topic in the remainder of the chapter. Having identified (2) as what is at stake in the debate over a comprehensive domain, the question arises: Who would ever deny a consequence of the logic of plural quantification? Indefinite extensibility may still lead one to deny (2), which we tendentiously characterized as a truth of the logic of plural quantification. If one thinks this, then one will deny that every instance of the general schema of plural comprehension is a logical truth. The general line of objection to be explored in the remainder of the chapter takes the set-theoretic paradoxes to show that no matter what items are identified as a putative comprehensive domain, there must be at least another item not among them. The picture that emerges instead is that of an open-ended series of ever more comprehensive domains, each of which is extended by its successor in the series.[7] Unfortunately, it is not entirely clear how best to articulate the route from indefinite extensibility to the thesis that there is no comprehensive domain, particularly in view of the fact that the phenomenon of indefinite extensibility is sometimes taken to speak not against the existence of a comprehensive domain but rather against the availability of completely unrestricted quantification.[8]

3 Indefinite extensibility without a domain

It is time to look at the question of whether considerations of indefinite extensibility may be used to question the existence of a comprehensive domain. Indefinite extensibility is often motivated with the help of Russell's paradox, which consists in the observation that the principle of naïve comprehension, namely, (6), entails (9):

(6) $\exists y \forall x (x \in y \leftrightarrow \phi(x))$,   where y is not free in $\phi(x)$

(7) $\exists y \forall x (x \in y \leftrightarrow x \notin x)$

(8) $\forall x (x \in r \leftrightarrow x \notin x)$

(9) $r \in r \leftrightarrow r \notin r$

But what should we conclude from this argument? All parties agree that the Russellian argument is a reductio, but there is no agreement on the object of the reductio. While some take the argument to provide a reductio of (6), others take the argument to call into question the unrestricted nature of the occurrence of the universal quantifier in (6).

In what follows, we would like to look at two main questions. One is the question of whether the Russellian argument can be taken to refute the presupposition that there is a comprehensive domain. Once we identify a route from Russell's paradox to the negation of (2), we will ask what to make of the linguistic question of whether completely unrestricted quantification is—or could—ever be available.

You may have thought that Russell's paradox is a reductio of (6), which is the familiar principle of naïve set comprehension:

$$(\text{N-Comp}) \qquad \exists y \forall x (x \in y \leftrightarrow \phi(x)),$$

where, as usual, $\phi(x)$ does not contain the variable "y" free. But how could this help anyone question the existence of a comprehensive domain? Part of the problem is that the existence of a comprehensive domain is a consequence of a principle of plural comprehension which appears to play no role in the derivation of the contradiction. The key to force plural comprehension into the picture is to factor naïve set comprehension into the following two principles:

$$(\text{Pl-Comp}) \qquad \exists xx \forall x (x \prec xx \leftrightarrow \phi(x)),[9]$$

where $\phi(x)$ does not contain "xx" free, which is a plural comprehension schema, and:

$$(\text{Collapse}) \qquad \forall xx \exists x \forall y (y \in x \leftrightarrow y \prec xx),$$

which is itself a principle of set comprehension.[10]

Once we make this distinction, Pl-Comp becomes a potential target for blame. The orthodox response to the Russellian argument would be to place the blame on Collapse and to keep Pl-Comp as a logical truth. But there is room now to take the opposite tack as well. One could deny Pl-Comp and accept Collapse instead. Since the existence of a comprehensive domain is a consequence of Pl-Comp, the move would be a first step from the Russellian argument to the rejection of a comprehensive domain.[11]

More precisely, once we break (6) down into Pl-Comp and Collapse, we need to distinguish an intermediate step in the transition from (6) to (7):

(10) $\exists yy \forall x (x \prec yy \leftrightarrow x \notin x),$

But maybe this is the fatal step in the argument. The suggestion would be that what we learn from the Russellian argument is that it is not the case that there are some sets such that they are all and only those sets that are not members of themselves. In general, given a condition, ϕ, there need not be any objects that are all and only those objects that satisfy the condition.

What motivates the move is the intuitive thought that all it takes for some objects to form a set is for them to exist. Once the members have been characterized, their set has been specified. A set is, after all, completely characterized by its elements. Once we have them, there is nothing to bar them to form a set. No further fact is required. So COLLAPSE has great initial appeal, and one may prefer to sacrifice plural comprehension in order to restore consistency. In particular, if we allow some sets to be all and only the non-self-membered sets, then nothing else stands in the way of their set, which, we know, will yield a contradiction.

The connection with indefinite extensibility should be apparent. The concept set is indefinitely extensible because no matter what sets we specify as a candidate extension, we will find that they are not all and only sets. By COLLAPSE, they themselves constitute a set. But, by the Russellian argument, this set cannot be one of the sets we started with and we have a more comprehensive candidate extension for the concept set. By COLLAPSE, those sets do themselves constitute a set and give rise to an even more comprehensive candidate extension for the concept set, and so on. The difference with other accounts of indefinite extensibility is that it explicitly does away with a comprehensive domain for set theory.[12]

Thus far we have a schematic response to the Russellian argument. In order for an answer to be complete, it must provide some characterization of the conditions for which PL-COMP fails. In other words, given a certain condition expressed by a formula ϕ of the language, we would like to know whether some objects are all and only the objects satisfying the condition. In what follows, we look at two broad options. One takes some objects to become available only when they occur as members of a some set. The other is based on a requirement of predicativity inspired by Michael Dummett's (1991) diagnosis of Russell's paradox. I will suggest that none of them is completely satisfactory for the task at hand. The reasons, however, will be importantly different in each case.

3.1 Two candidate restrictions of plural comprehension

One candidate restriction gives voice to the thought that a set is very intimately connected to its members. The thrust of COLLAPSE is that a set is to become available as soon as their members do. But when do the members become available? One hypothesis has to do with the iterative conception of set. Sets and their members are almost simultaneously formed in stages of a cumulative hierarchy. We start with urelements, and at each new stage, we form all sets that become available from members that occur at earlier stages.

But one may think further that a condition successfully generates some objects as *the* items that satisfy the condition only when it applies to items bounded by some level of the cumulative hierarchy. Or, in other words, only when it applies to members of a set. In particular, one may consider the following restriction of PL-COMP:

$$\text{(PL-SEP)} \qquad \forall t \exists yy \forall x (x \prec yy \leftrightarrow (x \in t \land \phi(x))).$$

If t is a set and ϕ is a condition, then there are some objects such that they are all and only the members of t that satisfy ϕ. This principle allows us to *separate* some objects from a set as *the* members of the set that satisfy a certain condition.

PL-SEP provides a direct route from the Russellian argument to the non-existence of a comprehensive domain. For (10) is a consequence of the following thesis in combination with PL-SEP and COLLAPSE:

$$\exists yy \forall x (x \prec yy \leftrightarrow x = x),$$

If some items are *the* self-identical objects, then, by COLLAPSE, they must form a set. But, by PL-SEP, some objects are *the* non-self-membered sets, and, given (10), a contradiction follows.

There are at least two difficulties with this restriction of PL-COMP. One is that it tells us nothing of the case in which there is no set from which to separate the objects that satisfy a certain condition. But we may need some guidance in such situations. For example, when Zermelo (1930) characterized the succession of models of second-order Zermelo–Fraenkel set theory with urelements (ZFCU), he made no assumption that the urelements should form a set. Some of his models make vivid the possibility that they do not form a set in which case PL-SEP will not tell us whether some objects are all and only the urelements.

Maybe the answer to this concern is that set theory is ultimately responsible for providing such guidance and we should remain silent when set theory does.[13] But do we really need the detour through set theory to determine whether some objects are all and only the inhabitants of London?

A more serious problem is that, presumably, an account of what instances of plural comprehension are safe is conceptually prior to an account of what sets there are. The iterative conception of set, for example, presupposes answers to the first question that cannot be delivered by PL-SEP. Start with urelements again. One would like to say that at each stage of the cumulative hierarchy *any* sets formed at earlier levels become available for the formation of a set. But how should we understand this claim without plural comprehension? In order to know what sets of urelements exist, we must determine what conditions defined on urelements give rise to some urelments as *the* urelements that satisy the condition. But to know what conditions do this, we need to know what sets of urelements exist? The combination of COLLAPSE and PL-SEP cannot even make sure that *every* condition defined on urelements gives rise to a plurality or a set.[14] We could perhaps make do with less by claiming only that whenever we have a condition ϕ and a level of the cumulative hierarchy, there is a set of sets of level V that satisfy the condition.[15] But if conditions are merely given by first-order formulas, this would clearly fall short of what we want.

The second candidate restriction of PL-COMP I would like to consider is closely related to Michael Dummett's (1991) diagnosis of Russell's paradox. Russell's paradox originally emerged as a fatal problem for the combination of Frege's axioms for second-order logic augmented and Basic Law V:

$$(\text{BASIC LAW V}) \qquad \forall F \forall G(ext(F) = ext(G) \leftrightarrow \forall x(Fx \leftrightarrow Gx))$$

In the context of *Grundgesetze*, the paradox arises when we ask whether the extension of the concept $R = [x : \exists F(x = ext(F) \wedge \neg Fx)]$, $ext(R)$, falls under R. If $r = ext(R)$ and $\neg Rr$, then $\forall F(r = ext(F) \rightarrow Fr)$, whence Rr. But if Rr, then there is some F such that $r = ext(F)$ and $\neg Fr$. By Basic Law V again, R and F must be coextensive and thus $\neg Rr$, which leads to a contradiction.

The source of the problem for Dummett lies in the formation of the concept R as an instance of second-order comprehension. When we fix a domain D and obtain the Russellian concept R over D as an instance of comprehension, we are thereby forced to find an extension for R outside D. But if D is a universal domain, then we will find ourselves in Frege's predicament. Thus far Dummett's response to the paradox is perfectly analogous to the rejection of PL-COMP above.

The difference is that in Frege's system, the concept R is generated by an instance of the second-order schema of impredicative comprehension:

$$(\text{I-COMP}) \qquad \exists X \forall x(Xx \leftrightarrow \phi(x)),$$

where, as usual, ϕ does not contain the variable "X" free. In particular, R is generated by an impredicative instance of comprehension which contains a bound second-order variable "Y":

$$\exists X \forall x(Xx \leftrightarrow \exists Y(x = ext(Y) \wedge \neg Yx))$$

This suggests a strategy for avoiding the paradox by insisting on a restriction of I-COMP from which the Russellian concept is not obtainable. The simplest restriction is to instances of comprehension in which no bound second-order variables occur:

$$(\text{PRED-COMP}) \qquad \exists X \forall x(Xx \leftrightarrow \phi(x)),$$

provided ϕ contains no bound second-order variables. Absent impredicative instances of comprehension, Basic Law V may no longer force us to extend the initial domain, and paradox is averted. As Dummett writes

Without second-order quantification, Frege's formal system would be paralysed, but the set-theoretic paradoxes would not be derivable.[16]

Dummett's diagnosis eventually gained much momentum from Terence Parsons' (1987) observation that the first-order fragment of Frege's system in *Grundgesetze* is consistent.[17] More recently, there has been much interesting work on various second-order predicative subsystems of Frege's system: Richard Heck (1996) and Fernando Ferreira and Kai Wehmeier (2002) have extended Parsons's results to such systems. For a summary and refinement of many such results, see Burgess (2005b).

One suggestion at this point is to transpose Dummett's suggestion to the case at hand and replace P-COMP with some predicative restriction the simplest of which would be as follows:

$$(\text{Pred-Pl-Comp}) \qquad \exists xx \forall x (x \prec xx \leftrightarrow \phi(x)),$$

where ϕ contains no bound plural variables. This will not be very helpful if we insist on keeping $\in$ as a primitive predicate of the language since "$\neg x \in x$" would give rise to a legitimate instance of comprehension, which in combination with COLLAPSE, would yield a contradiction. So, for this response to even have a chance, we would have to define membership as involving covert plural quantification:

$$x \in y \equiv \exists yy \ (y = ext(yy) \land x \prec yy),$$

in which case "$\neg x \in x$" would become: "$\neg \exists xx (x = ext(xx) \land x \prec xx)$".[18] But the sensitivity of the solution to the stock of primitive terms in the language should probably give us pause.

One initial difficulty is to provide a justification of the predicative restriction of plural comprehension. A condition of application is, after all, much more tied to a concept than it is to the objects to which it applies. In the context of plural comprehension, a condition is just supposed to give us a means to single out some objects as the objects that satisfy the condition. The existence of the objects is independent from the nature of the condition by means of which we manage to single them out. But on this picture, the impredicativity of a given condition can at most tell us that it is not suited to single out some objects as the objects that satisfy the condition. It is less clear that it could tell us of some objects that they could not exist as those and only those items that satisfy one condition or another.[19]

Unfortunately, a more serious problem for the present strategy arises from the observation that little mathematics can be interpreted in such predicative theories. Basic Law V with predicative monadic second-order logic interprets Robinson arithmetic. But as Burgess (2005b) makes clear, there is a limitative result: functional arithmetic with what has been called "superexponentiation" proves the consistency of Basic Law V in simple, double, triple, and so on monadic predicative second-order logic. Furthermore, functional arithmetic with "superduperexponentiation" proves the consistency of Basic Law V in full ramified predicative second-order logic.[20] This sets a stringent upper bound on the amount of mathematics that is interpretable in predicative versions of Frege's system in *Grundgesetze* which suggest that we may have averted contradiction by unduly impoverishing the system by brute force.

There is a further problem. Even if one finds predicative restrictions of P-COMP attractive, they seem to imply the existence of a comprehensive domain of discourse. For the following has a predicative appearance after all:

$$\exists yy \forall x(x \prec yy \leftrightarrow x = x),$$

provided we treat "$=$" as a primitive predicate of our language. But maybe this is not compulsory and one could take identity to be defined along the following lines:

$$x = y \equiv \forall xx(x \prec xx \leftrightarrow y \prec yy).$$

While this would render the preceding instance of comprehension impredicative and would bar the formation of a comprehensive domain, the choice of a primitive stock of predicates remains ad hoc and unmotivated.

3.2 Quantification without a domain

Whatever the replacement for PL-COMP, you may wonder whether it forces a negative answer to the linguistic question as well. If there is no domain of sets to begin with, how can we ever be in a position to quantify unrestrictedly over sets? We mentioned in the introduction that the linguistic question is independent from the question of whether there is a domain of *absolutely* all objects. An ontological relativist can deny the existence of such a domain and nevertheless agree that there is no linguistic obstacle for some of our quantifiers to be be unconstrained by any linguistic or contextual restrictions. So, unrestricted quantification may be available even if what exists is relative to a certain perspective.

At the same time, we made some effort towards distinguishing the question of a comprehensive domain:

(2) There are some objects such that every object is one of them.

from the ontological relativist foe:

(5) There are some objects such that *absolutely* every object is one of them.

We know that (5) is orthogonal to the linguistic question. But is the linguistic question similarly independent from the overly modest (2)? The answer may appear obvious. If there is no domain of all objects, how can we ever be in a position to quantify unrestrictedly over *them*. You may think that in order to quantify unrestrictedly over a domain, we need to assume that such a domain exists in the first place in which case the linguistic question does presuppose an answer to the question of a comprehensive domain.

I would like to suggest otherwise. If one is tempted to reject PL-COMP, one should simultaneously reject the formulation of the linguistic question as the question of whether we can quantify unrestrictedly over a domain of all objects. There is an alternative formulation of the question that avoids this presupposition. Take unrestricted quantification over sets. What is in question for someone who denies PL-COMP is not quite whether we can quantify unrestrictedly over some items which are all and only sets because there are no such items. Instead, what is in question is whether we can quantify unrestrictedly over *items* that satisfy a certain condition. But then whether unrestricted quantification over sets is precluded by the failure of PL-COMP will turn on what one thinks of what we could call the All-in-Many Principle:

> To quantify over objects satisfying a certain condition is to presuppose that there are some objects which are all and only those objects that satisfy the condition.

Compare with what Richard Cartwright (1994) calls the All-in-One Principle:

> To quantify over certain objects is to presuppose that those objects constitute a 'collection' or a 'complete collection'—some one thing of which those objects are members.

On the latter principle, quantification over sets requires the existence of some set-like item containing all and only sets as members; on the former, quantification over sets requires the existence of some items which are all and only sets. Cartwright (1994) and Boolos (1993) have forcefully argued against the latter principle but they appear to presuppose the former principle in the course of their arguments. However, if one is to take seriously the rejection of PL-COMP, one can in fact reject the former principle as well. If, given a condition, there is a further fact as to whether there exist some objects which are all and only those objects satisfying the condition, then one may wonder whether this further fact is required in order for us to be able to quantify unrestrictedly over objects that satisfy the condition.

One option is to say that we can quantify unrestrictedly over all sets even if there is no domain of sets to begin with. In fact, set theory appears to presuppose the availability of such quantification. When we say

(11) No set is a member of the empty set

or

(12) No set is self-membered

we want to speak of all sets whatever whether or not there are some items which are all and only them. The failure of PL-COMP need not threaten the availability of unrestricted quantification over sets. Nor will it presumably stand in the way of a determinate truth value for unrestricted quantified statements. At this point, the disagreement between our theorist and the absolutist may appear merely terminological. While the absolutist characterizes the problem of absolute generality in terms of the existence of a comprehensive domain, which we cashed out in terms of plural quantification, our opponent here may refuse to accept the terms of the problem and propose instead to characterize the question of a comprehensive domain as the question of whether there is some universal condition of which everything is an instance. If self-identity is such a comprehensive condition, then this may be enough by the lights of our theorist to think that there is a comprehensive domain and no further obstacle, in fact, to the occurrence of absolutely general quantifiers in some of our utterances.

Should we take seriously the possibility of such a reformulation? Not if the rejection of plural comprehension is too costly to be feasible. In what follows, we suggest that this is indeed the case. The combination of COLLAPSE with a restriction of PL-COMP and the denial of the All-in-Many principle is not a very stable position. Before we do, however, let me note for the record that there are other ways in which one might have attempted to rescue the spirit of COLLAPSE if not the letter. I have in mind Linnebo (2008), which replaces COLLAPSE with a modalized version that is in fact compatible with PL-COMP. None of what follows applies to Linnebo's position, which, however, falls outside the scope of this chapter for a different reason. To the extent that it endorses PL-COMP, it poses no threat to the existence of a comprehensive domain.

4 The costs of the restriction

We have looked at a schematic response to the Russellian argument and we have suggested that whatever instances one considers, they should be compatible with the availability of unrestricted quantification. The purpose of this section is to ask how revisionist this general outlook might turn out to be. Different instances of this schematic response provide different replacements for PL-COMP. They all, however, allow for unrestricted quantification over sets. In what follows, I would like to point to at least three serious costs associated with the position. They are each instances of a single problem which has to do with the loss of expressive power of plural quantification caused by the rejection of PL-COMP.

4.1   Second-order ZFC(U)

The first problem that arises is that second-order ZFC(U) with the first-order quantifiers taken to range unrestrictedly over sets may be thought to be inadequate as a formalization of the axioms of second-order set theory.[21] Take a plural interpretation the second-order quantifiers of ZFC(U) as plural quantifiers over sets. Since the rejection of PL-COMP deprives plural quantification from its expressive power, it renders a plural interpretation of the second-order axiom of replacement unsatisfactory as a formalization of the principle behind all instances of the first-order schema of replacement:

$$\forall x \exists ! y \Phi(x, y) \to \forall u \exists v \forall r (r \in v \leftrightarrow \exists s (s \in u \land \Phi(s, r)))$$

But you may have thought that we should be able to express the single thought that underlies every instance of the axiom schema, one which is generally expressed with the help of second-order quantification as follows:

$$\forall X (Func(X) \to \forall u \exists v \forall r (r \in v \leftrightarrow \exists s (s \in u \land X \langle s, r \rangle)))),$$

where *Func(X)* abbreviates: "*X* is a functional relation". But the question remains of how best to understand such quantification over predicate position. In George Boolos (1984), he suggested we do this in terms of plural quantification:

$$\forall xx (Func(xx) \to \forall u \exists v \forall r (r \in v \leftrightarrow \exists s (s \in u \land \langle s, r \rangle \prec xx))),$$

where *Func(xx)* abbreviates: "*xx* are ordered pairs which constitute a function". Unfortunately, his interpretation presupposes PL-COMP and is incompatible with COLLAPSE.

Moreover, whatever the replacement for PL-COLLAPSE, the mere presence of COLLAPSE suffices to make sure that no plural interpretation of the second-order axiom of replacement will be up to the task. For given COLLAPSE, the plural formulation of replacement will collapse into a theorem of Zermelo set theory:

$$\forall x (Func(x) \to \forall u \exists v \forall r (r \in v \leftrightarrow \exists s (s \in u \land \langle s, r \rangle \in x))).$$

True, but trivial given all the other axioms. Otherwise put, when the axioms of second-order ZFC(U) are given a plural interpretation that is compatible with COLLAPSE, they no longer outstrip ZC(U) and are satisfiable in $\langle V_{\omega+\omega}, \in \cap (V_{\omega+\omega} \times V_{\omega+\omega}) \rangle$. What is even more dramatic, they suffer from the same inadequacies as the axioms of second-order ZC(U), and, in particular, they are not even able to guarantee the existence of V_ω or to make sure that the universe of set theory is well-founded.[22]

It is no answer to blame this problem on the plural interpretation of second-order quantification. For even if one finds a Fregean interpretation intelligible and proceeds to read second-order quantification as quantification over Fregean concepts, one should presumably be moved by the Russellian argument to reject the Fregean interpretation of

$$(\text{I-Compr}) \qquad \exists X \forall x (Xx \leftrightarrow \phi(x)),$$

where, as usual, ϕ does not contain the variable X free. On this interpretation, there is a Fregean concept whose instances are all and only those objects that satisfy a certain condition ϕ. But if there is such a concept and we accept the Fregean version of Collapse whereby the instances of every Fregean concept make up a set, we face exactly the same dilemma we had when we looked at the combination of PL-Comp and Collapse. But it is not clear what could justify opting for a different solution to the conflict in the Fregean case.

4.2 Proper classes

One way to reduce the schema of replacement to a single axiom is by appeal to first-order quantification over classes. In conventional axiomatizations, replacement is formulated as a single axiom some of whose quantifiers are intended to range over classes:

$$\forall X (Func(X) \to \forall u \exists v \forall r (r \in v \leftrightarrow \exists s (s \in u \wedge \langle s, r \rangle \in X))),$$

where "X" is now a first-order variable that ranges over classes and $Func(X)$ abbreviates: "X is a functional class of ordered pairs". The question of course is what to make of singular quantification over classes since it appears to be in tension with a certain picture of set theory as the most comprehensive theory of collections. If classes are not just a different sort of collection, then what are they and how are we to understand apparent singular quantification over them?

Theories of classes generally come in two varieties. Predicative theories of classes such as von Neumann–Bernays–Gödel class theory (NBG) are characterized by a *predicative* comprehension schema:

$$\exists X \forall x (x \in X \leftrightarrow \phi(x)),$$

where, as usual, $\phi(x)$ does not contain the variable "X" free and, in addition, *no bound class variables are allowed in* $\phi(x)$. Predicative theories of classes pose no particular problem. We know, for example, that NBG class theory is a conservative extension of first-order ZF. Moreover, there is a simple substitutional interpretation of predicative classes where quantification over classes is understood in terms of substitutional quantification over formulas of first-order ZFC. Alternatively, we could replace the substitutional quantifiers with countable conjunctions and disjunctions in a version of first-order ZFC formulated in a suitable infinitary language. Talk of classes is ultimately dispensable when it comes to predicative theories. Unfortunately, it is not clear we can make do with predicative classes.

Matters are different when we relax the predicative restriction on class comprehension. Impredicative theories of classes such as Morse–Kelley class theory (MK) are characterized by an impredicative comprehension schema:

$$\text{(I-CL COMPR)} \qquad \exists X \forall x (x \in X \leftrightarrow \phi(x)),$$

where, as usual, ϕ does not contain the variable X free. Here, in contrast to the predicative case, we allow for formulas in which class variables sometimes occur as bounded variables. The removal of the predicative restriction has profound consequences, which can be summarized by the observation that unlike NBG, first-order MK enables one to define truth for first-order ZF and, by Tarski's indefinability theorem, it is far from a conservative extension of ZF. More importantly for our purposes, there is no substitutional interpretation available for MK and we face a choice between a plural interpretation of quantification over classes and the appeal to a distinction between sets and classes as instances of a more general sort of collection.

Unfortunately, COLLAPSE renders the plural interpretation of first-order MK unhelpful for the purpose of understanding talk of impredicative classes. Much like before, it is unclear that an alternative interpretation of such talk is available. This problem is significant because impredicative classes are regularly appealed to in heuristic motivations for *large cardinal* axioms. In some cases, talk of proper classes is not dispensable and plural quantification becomes a helpful expedient to make sense of set-theoretic practice.[23] Without the plural interpretation, we have no choice but to take talk of impredicative classes at face value as talk of collections that are much too comprehensive to count as sets. This, however, would be at odds with the view of set theory as the most comprehensive theory of collections. Moreover, once proper classes are allowed, even more comprehensive superclasses, hyperclasses, and the like are not far behind by perfectly analogous considerations.

4.3 Reflection

The proponent of COLLAPSE faces a choice between a conventional infinite axiomatization of ZFC(U) and a finite axiomatization of MK(U) which

requires one to abandon the general picture of set theory as the most comprehensive theory of collections. This is a genuine dilemma since we know
from Montague (1961) that no consistent extension of first-order ZFC(U) is
finitely axiomatizable. Richard Montague established this by an application
of reflection. ZFC(U) proves every first-order instance of the general schema
of reflection:

$$\Phi(x_1, \ldots, x_n) \rightarrow \exists t \Phi^t(x_1, \ldots, x_n).$$

In the reflection schema, $\Phi^t(x_1, \ldots, x_n)$ stands for the *relativization* of
$\Phi(x_1, \ldots, x_n)$ to t, that is, the formula that results when the quantifiers and
parameters are restricted to t.[24] By reflection, ZFC(U) proves the existence of
a model for every finite set of theorems of ZFC(U). So if we could axiomatize
ZFC(U) by means of a finite number of axioms, we would contravene Gödel's
second-incompleteness theorem.

Reflection principles are motivated by the inchoate thought that the universe of set theory cannot be characterized from below: if a statement is true
of the universe V, then it is true of some set and, indeed, true of some proper
initial segment of the universe V_α. When formalized in terms of higher-order
languages, this thought has given rise to a variety of reflection principles
that lead to a natural hierarchy of indescribability hypotheses. For example,
in the second-order case, we have the following schema:

$$\Phi(X_1, \ldots, X_n, x_1, \ldots, x_m) \rightarrow \exists t \Phi^t(X_1, \ldots, X_n, x_1, \ldots, x_m),$$

where $\Phi(X_1, \ldots, X_n, x_1, \ldots, x_m)$ is a formula of the language of second-order
set theory and $\Phi^t(X_1, \ldots, X_n, x_1, \ldots, x_m)$ is the result of the relativization of
the quantifiers and first- and second-order parameters to t.[25]

In general, second-order reflection takes us well beyond second-order ZFC.
In the language of second order ZFC, we can express the statement that the
universe of pure sets, V, forms a strongly inaccessible class, and, by the
appropriate instance of second-order reflection, there is a set t of strongly
inaccessible size; indeed, there is a strongly inaccessible κ such that V_κ
is a model of ZFC. Successive applications of second-order reflection soon
yield strongly inaccessible, Mahlo cardinals of all orders, weakly compact
cardinals and, more generally, Π_n^1-indescribable cardinals for each $n \in \omega$.[26]
However, second-order reflection is still very weak by the lights of large
cardinal axioms and lies at the bottom of an entire hierarchy of stronger
and stronger indescribable cardinals which are known to be compatible with
$V = L$.[27]

Even if second-order reflection is in effect a modest indescribability
hypothesis, we should proceed with caution when we formulate reflection
in terms of third- and higher-order languages. In a third-order language, the
question immediately arises whether we should allow third-order parameters

in the formulas we want to reflect. But as noted by Reinhardt (1974), we face an inconsistency if we allow third-order parameters unrestrictedly.[28] There is, however, no apparent threat of inconsistency for the entire hierarchy of higher-order reflection principles that arises when we restrict attention to second-order parameters and explicitly exclude third- and higher-order parameters from the formulas we reflect.

To summarize, second-order reflection is a very modest large cardinal hypothesis motivated by an attractive thought as to the nature of the cumulative hierarchy. It seems to me it would be unfortunate if we turned out to be forced to make a choice between such a modest principle and COLLAPSE. But this appears to be the situation we find ourselves in once we restrict PL-COMP to make room for COLLAPSE. In this context, we had better not describe reflection as the hypothesis that no formula characterizes the domain of all objects—urelements and sets—because there is no such thing. Nonetheless, we can rephrase the inchoate thought behind reflection as the hypothesis that no universal condition, that is, a condition that is satisfied when its quantifiers are completely unrestricted, is satisfied when restricted to some set or another. Could this thought serve as a heuristic guide as to how high the cumulative hierarchy is supposed to reach? In connection to Burgess (2004a), one may likewise wonder whether reflection is still able to generate much of set theory in the absence of PL-COMP.

The prospects don't look too bright. Linnebo (2007) has recently noted a tension between the acceptance of COLLAPSE and second-order reflection. For consider what happens when we reflect COLLAPSE in the appropriate instance of second-order reflection schema:

$$\forall xx \exists x \forall y (y \in x \leftrightarrow y \prec xx) \rightarrow \exists t \forall xx \in\in t \exists x \in t \forall y \in t (y \in x \leftrightarrow y \prec xx),$$

where: "$\forall xx \in\in t(\ldots)$" abbreviates: "$\forall xx(\forall y(y \prec xx \rightarrow y \in t) \rightarrow \ldots)$". Since COLLAPSE is just the antecedent of the conditional, we have the following:

$$\exists t \forall xx \in\in t \exists x \in t \forall y \in t (y \in x \leftrightarrow y \prec xx).$$

But this tells us that there is a set t such that no matter what some members of t are, t contains a set whose members are exactly them.

You may be tempted to immediately derive a contradiction from the existence of a set r which consists of all and only those non-self-membered members of t, that is, $\{x \in t : x \notin x\}$. For if r is such a set, then we have the following: $r \in r \leftrightarrow r \notin r$, which leads to a contradiction. But notice that the existence of r does not automatically follow from COLLAPSE. This principle tells us that whenever we have some items, we have a set of them. But if a set t is a witness to this instance of reflection, we still need some guarantee that some sets exist that are all and only the non-self-membered members of *t*. Absent PL-COMP, we need to ask whether we have enough comprehension to derive the existence of such sets.

We isolated two different weakenings of PL-COMP. One is PL-SEP, which will immediately yield the existence of *r* as a consequence. Because every non-self-membered member of *t* is a member of a set, for example, *t*, there are some sets which are all and only those non-self-membered members of *t* and a contradiction follows.

The answer is less obvious when one chooses a predicative restriction of PL-COMP which somehow accommodates all the necessary provisos we noticed above. Since such a restriction will presumably come accompanied by a different choice of primitive predicates for the language, the condition *being a non-self-membered member of t* will presumably count as impredicative and the predicative restriction of comprehension will not itself generate the existence of the members of *t*. This, however, is little consolation in view of the existence of another route to *r*. In particular, *r* seems to immediately follow from a certain instance of the axiom schema of separation:

$$\forall x \exists y \forall z (z \in y \leftrightarrow (\phi(z) \wedge z \in x)).$$

In particular

$$\exists y \forall z (z \in y \leftrightarrow (z \notin z \wedge z \in t)).$$

And once we have the existence of *r*, a contradiction will immediately follow. So, given a predicative restriction of PL-COMP, one may keep second-order reflection at the cost of rejecting instances of the axiom schema of separation. The situation is completely analogous to the one that arises when one allows for the existence of a universal set on the grounds that it does not by itself require an impredicative instance of naïve comprehension but rejects separation in order to avoid the existence of the Russell set, which would be generated by an impredicative instance of naïve comprehension. To summarize, we have the following three principles that are inconsistent: (i) COLLAPSE (ii) second-order reflection, and (iii) the axiom of separation. The upshot is that if you want to preserve COLLAPSE, then you must make a choice between an axiom that is probably not negotiable and an attractive hypothesis as to the extent of the cumulative hierarchy. Unless one is prepared to entertain a blanket rejection of separation motivated perhaps by the existence of a universal set, it seems to me that there is little choice but to reject reflection. Thus, second-order reflection cannot hope to play the heuristic role it is generally thought to play in set theory, which seems a cost.

5 Conclusion

We have clarified what is at stake in the debate over a comprehensive domain and the prospects of unrestricted quantification without a domain as motivated by considerations of indefinite extensibility. This has taken us to articulate a less traveled route from Russell's paradox to the rejection of a comprehensive domain. Although tortuous, the journey has been fruitful in at least two ways. We have seen, first, that there is another way in which the linguistic question of unrestricted quantification is independent from the question of a comprehensive domain. And we have a clearer understanding of what choices have to be made at various junctures for the combination of unrestricted quantification and the rejection of a comprehensive domain to serve as a background to set theory. They involved a combination of COLLAPSE with a restriction of PL-COMP, which we have suggested is both highly revisionary and in tension with a certain perspective on set theory as the most comprehensive theory of collections.[29]

Notes

1. I have in mind Glanzberg (2004), Glanzberg (2006), and Parsons (2006) as examples of the former theorists and Fine (2006) and Hellman (2006) as examples of the latter.
2. You may wonder whether the determiner is not explicitly restricted by the noun "thing" in Quine's answer. However, it is better to think of "thing" as a universally applicable noun and thus treat the determiner as carrying no restriction whatever.
3. In English, it is natural to take "some objects" to mean "two or more objects" or perhaps "one or more objects". A more radical approach due to Burgess and Rosen (1997) is to read "some objects" as "zero or more objects", which is the choice we make here. On the first two readings, the thesis that there is a comprehensive domain would entail the existence of at least one object. But as Gonzalo Rodriguez-Pereyra reminded me, this is still a substantive ontological claim. Moreover, it is not clear that we could not make sense of quantification over an empty domain and it might therefore be unduly restrictive to exclude this possibility at the outset. As for the question of how to understand the locution "zero or more objects", there is a familiar expedient due to Boolos (1984) that allows us to define it in terms of "one or more" or "two or more". In addition, Burgess and Rosen (1997) show how any locution can be taken as a primitive from which to define the other two.
4. This will not be appropriate as a characterization of absolutism by the lights of a domain absolutist, who, moved perhaps by considerations of indefinite extensibility takes all quantification to be restricted quantification to a less-than-comprehensive domain. I have in mind a position in the vicinity of Glanzberg (2004).
5. Putnam (2004).
6. This point was first emphasized by Kit Fine. For further discussion, see Fine, Hellman and Parsons' contributions in Rayo and Uzquiano (2006).
7. I have in mind the views expressed in Fine (2006) and Hellman (2006).
8. For example, Glanzberg (2004), Glanzberg (2006), and Parsons (2006).
9. Recall our decision to read "$\exists xx$" as follows: "there are zero or more objects". Had we chosen one of the alternative interpretations of plural quantification, we would have required an antecedent to make sure the condition is not vacuous.

10. I borrow the label from Øystein Linnebo (2008) who has recently defended a modalized version of the principle.
11. Both Øystein Linnebo (2008) and Stephen Yablo (2006) have recently suggested a similar move for different reasons.
12. The difference is more subtle in fact. The account of indefinite extensibility given in Glanzberg (2004), Glanzberg (2006), and Parsons (2006) would assent to utterances of (2) while denying that the quantifiers are in fact completely unrestricted for indefinite extensibility reasons. This makes the accounts difficult to compare.
13. Thanks to Øystein Linnebo.
14. In fact, as we noted before, it is even possible that PL-SEP prevents certain conditions on urelments to give rise to a plurality.
15. This, incidentally, would quickly give us a set of urelements.
16. See Dummett (1991), p. 219.
17. This had been conjectured by Schroeder–Heister (1987).
18. Read "$y = ext(xx)$" as "the xx form an extension y" or "y is the extension of xx".
19. Thanks to Øystein Linnebo for suggesting the concern.
20. "Simple", "double", "triple", and so on allude to the number of rounds whereby new second-order variables are being introduced in the systems with ramified predicative theory being their union. See Burgess (2005b) for details and definitions of "superexponentiation" and "superduperexponentiation".
21. We write "ZFC(U)" to mean Zermelo–Fraenkel set theory with choice and with or without urelements.
22. Uzquiano (1999) looks at these and other similar problems.
23. See Uzquiano (2003) and Burgess (2004a).
24. Formally, we relativize Φ to a set t by replacing each quantification of the sort shown below by the quantification shown beside it:

$$\exists x(\ldots) \quad \text{with} \quad \exists x(x \in t \,\wedge\, \ldots)$$
$$\forall x(\ldots) \quad \text{with} \quad \forall x(x \in t \rightarrow \ldots)$$

In the relativized formula, we implicitly assume that the the first-order parameters to range over t.
25. The relativization is obtained by the method sketched in note 24 supplemented with two more clauses designed to deal with second order quantification

$$\exists X(\ldots) \quad \text{with} \quad \exists X(\forall x(Xx \rightarrow x \in t) \,\wedge\, \ldots)$$
$$\forall X(\ldots) \quad \text{with} \quad \forall X(\forall x(Xx \rightarrow x \in t) \rightarrow \ldots).$$

And just like in the relativized formula, we take the first-order parameters to range over t, we implicitly assume the second-order parameters to range over subsets of t.
26. In general, second-order reflection amounts to Π_n^1-indescribability for every $n \in \omega$. By a result due to Hanf and Scott, weakly compact cardinals are exactly the Π_1^1-cardinals. For details, see for example Kanamori (2003), Theorem 6.4, or Jech (2006), Theorem 17.18.
27. That Π_n^m-indescribability is compatible with $V = L$ is noted, for example, by Kanamori (2003), ch. 1, Theorem 6.6. Much stronger indescribability principles—some of which reflect formulas of transfinite order (with no third- or higher-order parameters)—fall under the consistency result proved in Koellner (To appear) that certain generalized reflection principles are consistent relative to

the Erdös cardinal $\kappa(\omega)$. This result sets an upper bound on reflection principles that is compatible with $V = L$.

28. Perhaps all is not lost. William Tait has motivated a restriction of reflection with third- and higher-order parameters in Tait (2005) to a special class of formulas, that is, positive formulas, and he has proposed a corresponding hierarchy of reflection principles. These principles turn out to take us well beyond Π^n_m-indescribability—or even beyond reflection on formulas of transfinite order without third- or higher-order parameters—as they lead to ineffable cardinals. However, Peter Koellner has recently proved a dichotomy theorem to the effect that generalized reflection principles, which include Tait's reflection principles, are either weak (i.e. consistent relative to $\kappa(\omega)$ and hence with $V = L$) or inconsistent. For details of these recent developments, see Tait (2005) and Koellner (2009).

29. Earlier versions of this material were presented at the third Paris-Oxford workshop on language and ontology and at the New Waves conference at the University of Miami. I thank participants in those events for valuable comments. I am grateful to Thomas Hofweber, Peter Koellner, and Øystein Linnebo for valuable comments and suggestions on an earlier draft of this paper.

References

Benacerraf, P., and H. Putnam, eds. (1983) *Philosophy of Mathematics*, Cambridge University Press, Cambridge, second edition.

Boolos, G. (1984) "To Be Is to Be a Value of a Variable (or to be Some Values of Some Variables)," *The Journal of Philosophy* 81, 430–449. Reprinted in Boolos (1998).

Boolos, G. (1993) "Whence the Contradiction?" *Aristotelian Society, Supplementary Volume* 67, 213–233. Reprinted in Boolos (1998), 220–236.

Boolos, G. (1998) *Logic, Logic and Logic*, Harvard, Cambridge, Massachusetts.

Burgess, J. (2004a) "E Pluribus Unum: Plural Logic and Set Theory," *Philosophia Mathematica (3)* 12, 13–221.

Burgess, J. (2004b) "Mathematics and Bleak House," *Philosophia Mathematica (3)* 12, 18–36.

Burgess, J. (2005a) "Being Explained Away," *Harvard Review of Philosophy* 13, 41–56.

Burgess, J. (2005b) *Fixing Frege*, Princeton University Press, Princeton NJ.

Burgess, J., and G. Rosen (1997) *A Subject With No Object*, Oxford University Press, New York.

Carnap, R. (1950) "Empiricism, Semantics and Ontology," *Analysis* 4, 20–40. Reprinted in P. Benacerraf and H. Putnam (1983), 241–257.

Cartwright, R. (1994) "Speaking of Everything," *Nous* 28, 1–20.

Dummett, M. (1991) *Frege: Philosophy of Mathematics*, Harvard University Press, Cambridge, MA.

Ferreira, F., and K. Wehmeier (2002) "On the Consistency of the Δ^1_1-CA Fragment of Frege's *Grundgesetze*," *Journal of Philosophical Logic* 31, 301–311.

Fine, K. (2006) "Relatively Unrestricted Quantification," in A. Rayo and G. Uzquiano, eds., *Absolute Generality*, Oxford University Press, Oxford, 20–44.

Glanzberg, M. (2004) "Quantification and Realism," *Philosophy and Phenomenological Research* 69, 541–572.

Glanzberg, M. (2006) "Context and Unrestricted Quantification," in A. Rayo and G. Uzquiano, eds., *Absolute Generality*, Oxford University Press, Oxford, 45–74.

Heck, R. (1996) "The Consistency of Predicative Fragments of Frege's *Grundgesetze der Arithmetik,*" *History and Philosophy of Logic* 17, 209–220.

Hellman, G. (2006) "Against 'Absolutely Everything'!" in A. Rayo and G. Uzquiano, eds., *Absolute Generality,* Oxford University Press, Oxford, 75–97.

Hirsch, E. (2002) "Quantifier Variance and Realism," *Philosophical Issues* 12, 51–73.

Jech, T. J., ed. (1974) "Axiomatic Set Theory, Part II," of *Proceedings of the Symposia in Pure Mathematics,* Volume 13, American Mathematical Society, Providence, R.I.

Jech, T. J. (2006) *Set Theory,* Springer, 3rd edition Springer-Verlag, Berlin.

Kanamori, A. (2003) *The Higher Infinite: Large Cardinals in Set Theory from their Beginnings,* Springer.

Koellner, P. (2009) "On Reflection Principles," *Annals of Pure and Applied Logic* 157, 206–219.

Linnebo, Ø. (2007) "Burgess on Plural Logic and Set Theory," *Philosophia Mathematica (3)* 15:1, 79–93.

Linnebo, Ø. (2008) "Pluralities and Sets," presented at a Workshop in Philosophical Logic in Oxford, March 2008 Unpublished typescript.

Montague, R. (1961) "Semantic Closure and Non-Finite Axiomatizability I," in *Infinitistic Methods: Proceedings of the Symposium on the Foundations of Mathematics* (Warsaw 1959), New York, Pergamon, 45–69.

Parsons, C. (2006) "The Problem of Absolute Universality," in A. Rayo and G. Uzquiano, eds., *Absolute Generality,* Oxford University Press, Oxford, 203–219.

Parsons, T. (1987) "On the Consistency of the First-Order Portion Frege's Logical System," *Notre Dame Journal of Formal Logic* 28, 161–168.

Putnam, H. (1994) "The Question of Realism," in J. Conant, ed., *Words and Life,* Harvard University Press, Cambridge MA.

Putnam, H. (2004) *Ethics Without Ontology,* Harvard University Press, Cambridge MA.

Quine, W. V. (1948) "On What There Is," *Review of Metaphysics* 2, 21–38. Reprinted in Willard V. Quine, *From a Logical Point of View.*

Quine, W. V. (1980) *From a Logical Point of View,* Harvard University Press, Cambridge MA.

Rayo, A., and G. Uzquiano, eds. (2006) *Absolute Generality,* Oxford University Press, Oxford.

Reinhardt, W. (1974) "Remarks on Reflection Principles, Large Cardinals, and Elementary Embeddings," in T. J. Jech (1974).

Schroeder-Heister, P. (1987) "A Model-Theoretic Reconstruction of Frege's Permutation Argument," *Notre Dame Journal of Formal Logic* 28:1, 69–79.

Shapiro, S. (1987) "Principles of Reflection and Second-Order Logic," *Journal of Philosophical Logic* 16, 309–333.

Stalnaker, R. C. (2003) *Ways a World Might Be: Metaphysical and Anti-Metaphysical Essays,* Clarendon Press, Oxford.

Tait, W. (2005) "Constructing Cardinals from Below," in W. Tait, ed., *The Provenance of Pure Reason: Essays in the Philosophy of Mathematics and Its History,* Oxford University Press, Oxford.

Uzquiano, G. (1999) "Models of Second-Order Zermelo Set Theory," *Bulletin of Symbolic Logic* 5:3, 289–302.

Uzquiano, G. (2003) "Plural Quantification and Classes," *Philosophia Mathematica* 3, 67–81.

Yablo, S. (2006) "Circularity and Paradox," in T. Bolander, ed., *Self-Reference,* CSLI Publications, Stanford CA.

Zermelo, E. (1930) "Über Grenzzahlen und Mengenbereiche: Neue Untersuchungen iiber die Grundlagen der Mengenlehre," *Fundamenta Mathematicae* 16, 29–47.

Index